Praise for *Oceans of Fate*

In *Oceans of Fate* Dan Black has produced a most welcome addition to the largely unknown story of one of Canada's grand passenger liners — part of a once great fleet that no longer exists. The story of the *Empress of Asia* is a masterly portrait of the ship, its crew, and passengers. Black's skillful writing makes thousands of people who passed through the *Empress* during its 30-year existence come alive, and the narrative is enhanced by maps and period photographs. *Oceans of Fate* is a thoroughly researched and well-written book about a little-known aspect of Canadian history.

— COLONEL JOHN BOILEAU (ret'd), author of *Samuel Cunard: Nova Scotia's Master of the North Atlantic*, *Halifax and Titanic*, and *The Lucky and the Lost: The Lives of Titanic's Children*

There's a deeply haunting quality to *Oceans of Fate*. It's a heart-wrenching yet uplifting tale that has us eavesdrop upon those who spoke for thousands of private lives, all associated with the *Empress of Asia*, a Canadian ocean liner that embraced the pride, frolic, and brutal maelstrom of the first half of the 20th century. The painstaking research that went into the book unearthed diaries, journals, and personal letters of crew and passengers. The ship itself is emblematic of the last great ocean liners that also found purpose in war. *Oceans of Fate* will reacquaint you with major historical events and have you linger on the "still, sad music of humanity." Reading it and getting to know the characters within is a humbling experience.

— PETER JOHNSON, author of *Quarantined: Life and Death at William Head Station, 1872–1959*

Dan Black's book *Oceans of Fate* is beautifully written. It transforms the steel and wood of a hull and deck into flesh and bone, weaving together the legacy of a ship that was not simply sailing the oceans carrying passengers and cargo but experiencing the tension of wars and the ultimate savagery of warfare. With deep research, the book presents a tapestry of captains and crews, of renowned passengers, an Olympic medalist, people lost at sea, and soldiers answering the call to arms — many never to return home. His ability to bring alive the life and times of the *Empress of Asia* speaks to the quality of Black's work and writing style. As an educator, *Oceans of Fate* offers exemplars of industrialization and technology, conflict and travel, all while imagining ourselves standing on the deck of the *Empress of Asia*. I can only hope that Dan Black continues to chronicle the other ships that were a part of the *Empress* class. We are fortunate to have such a gifted writer in our midst.

— A. BLAKE SEWARD, M.S.M., creator of the Lest We Forget Project, and recipient of the Prime Minister's Award for Teaching Excellence and the Governor General's History Award for Excellence in Teaching

With *Oceans of Fate*, Dan Black once again shows his immense talent for finding overlooked and under-reported stories and sharing them — and the people at the forefront of each narrative — in a way that fully engages readers and leaves them astonished by the depth of research.

— BRIAN JEFFREY STREET, author of *The Parachute Ward*, and co-author of *Champagne Navy* and *Invasions Without Tears*

Dan Black's writing, focused so often on the impact of war on individuals, is known for its compassion and clarity. In *Oceans of Fate*, a sweeping and meticulously researched study of the life of the steamship *Empress of Asia* and the men and women whose lives were bound up with it, those qualities stand out once again. From the ship's birth in a Scottish shipyard to its death from Japanese bombing in the Second World War, the reader is presented with a colourful, intricate, and absorbing panorama that relates not only what happened to the ship and the people it touched but also describes the tumultuous age in which it, and they, existed. The professional passion for detail and accuracy in Black's writing deserves ringing applause; even more so does the heartfelt care he shows for the men and women touched by events of the ship's life. This is historical writing at its finest.

— VICTOR SUTHREN, director general, Canadian War Museum (1986–1997)

Dan Black's detailed chronicle of the transpacific liner *Empress of Asia* is a captivating, insightful, and carefully researched story, not just of the ship, but of the people who knew the *Asia* well, in peace and war: officers and crew, passengers, military personnel, and many more. Their personal stories and memories bring the *Asia* — life on board, her times and adventures — vividly to life, including her tragic loss near Singapore early in the Second World War. This book will be a great addition to the maritime story of Canada and the World Wars.

— ROBERT D. TURNER, FRCGS, LL.D, curator emeritus, Royal BC Museum, and author of *The Pacific Empresses: An Illustrated History of Canadian Pacific Railway's Empress Liners on the Pacific Ocean*

OCEANS OF FATE

OCEANS OF FATE

Peace and Peril Aboard the Steamship *Empress of Asia*

DAN BLACK

FOREWORD BY JAMES P. DELGADO

Publisher: Meghan Macdonald | Acquiring editor: Kathryn Lane | Editor: Michael Carroll
Cover designer: Laura Boyle
Cover image: Vancouver Maritime Museum, PA005.010

Library and Archives Canada Cataloguing in Publication

Title: Oceans of fate : peace and peril aboard the steamship Empress of Asia / Dan Black ; foreword by James P. Delgado.
Names: Black, Dan, 1957- author | Delgado, James P., writer of foreword
Description: Includes bibliographical references and index.
Identifiers: Canadiana (print) 20240460138 | Canadiana (ebook) 20240460146 | ISBN 9781459752511 (softcover) | ISBN 9781459752528 (PDF) | ISBN 9781459752535 (EPUB)
Subjects: LCSH: Empress of Asia (Steamship) | LCSH: Ocean liners—Canada—History—20th century. | LCSH: Passenger ships—Canada—History—20th century. | LCSH: Armed merchant ships—Canada—History—20th century. | LCSH: Merchant marine—Canada—History—20th century. | LCSH: World War, 1914-1918—Naval operations, Canadian.
Classification: LCC HE566.O25 B53 2025 | DDC 623.82/4320904—dc23

Canada Council for the Arts
Conseil des arts du Canada

We acknowledge the support of the Canada Council for the Arts and the Ontario Arts Council for our publishing program. We also acknowledge the financial support of the Government of Ontario, through the Ontario Book Publishing Tax Credit and Ontario Creates, and the Government of Canada.

Printed and bound in Canada.

Dundurn Press
1382 Queen Street East
Toronto, Ontario, Canada M4L 1C9
dundurn.com, @dundurnpress

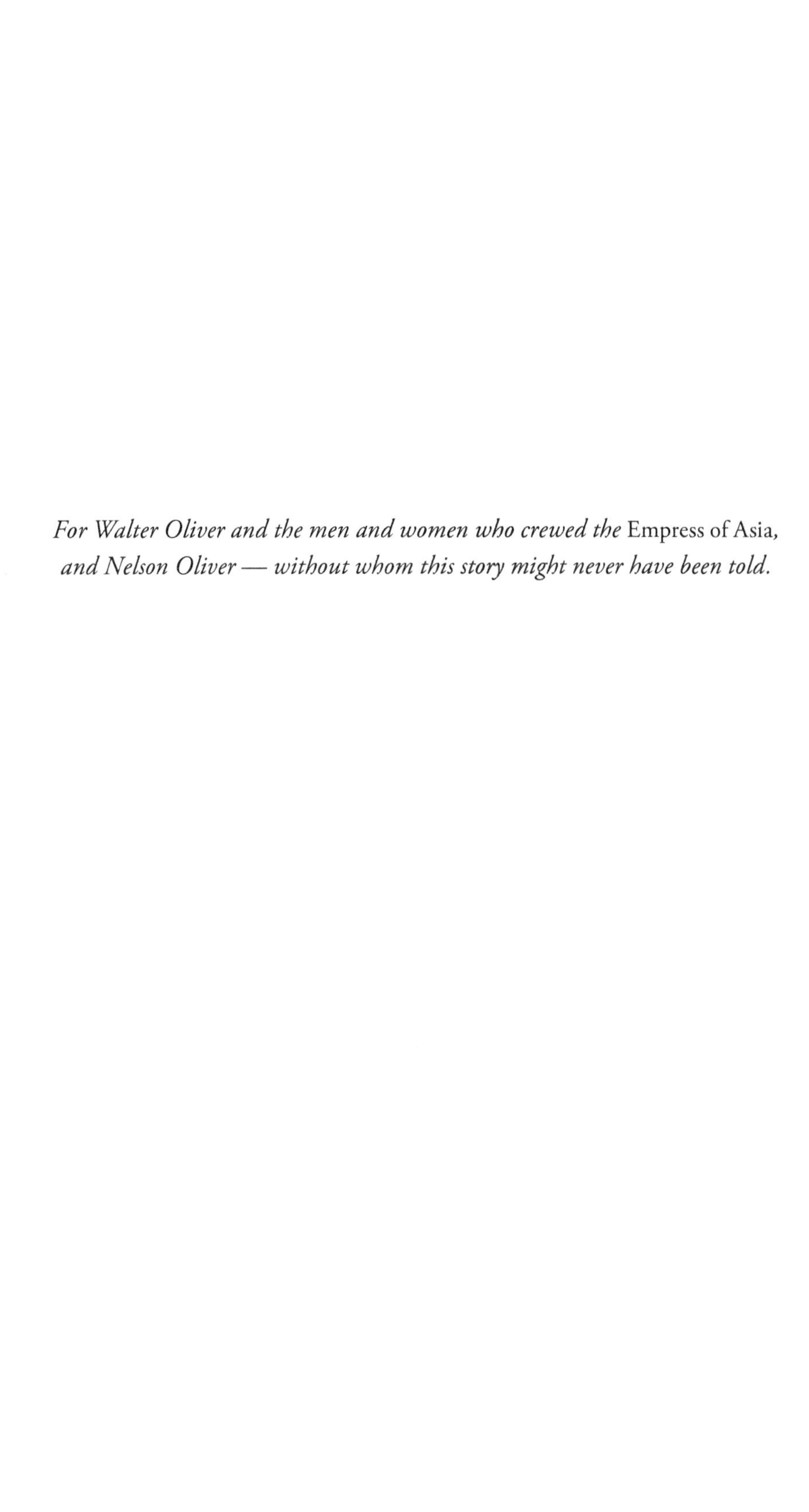

For Walter Oliver and the men and women who crewed the Empress of Asia,
and Nelson Oliver — without whom this story might never have been told.

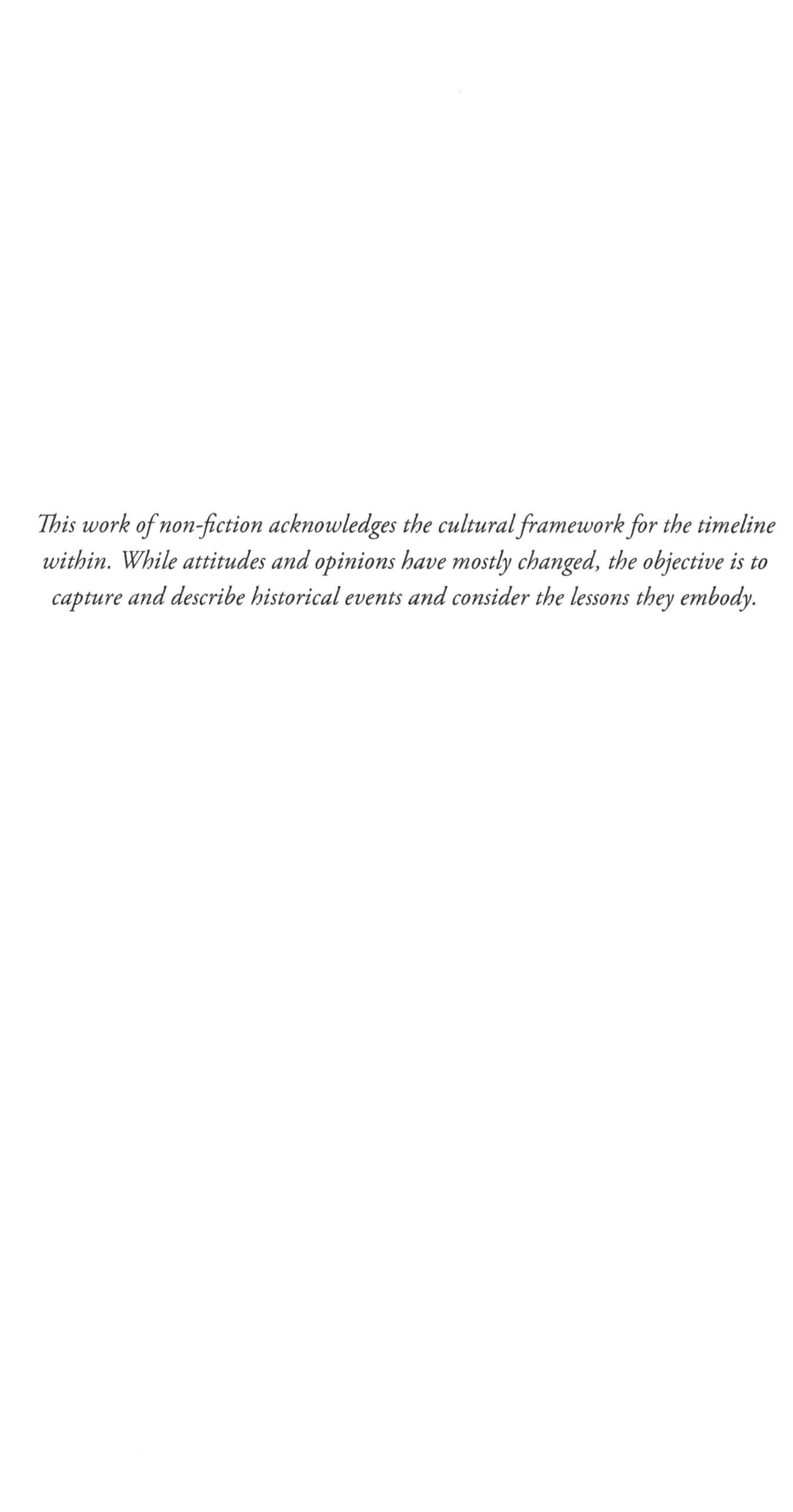

This work of non-fiction acknowledges the cultural framework for the timeline within. While attitudes and opinions have mostly changed, the objective is to capture and describe historical events and consider the lessons they embody.

CONTENTS

FOREWORD

Ultimately, the story of a ship is the story of people. With a long history, a vessel such as the *Empress of Asia* is a focal point for thousands of people's stories, from birth, or to death for some, and for the ship itself. From the beginning to the end, every ship's story encompasses so many other tales, and finally, connects the world to those tales. Dan Black has masterfully told the story of the *Empress of Asia* through the eyes, words, and experiences of many, and through their experiences, that long-dead ship lives again in these pages. What emerges, in addition to the human stories, is how the ship connected not only to various global ports but also was home to a diverse crew from around the world. Some might view the ship and its saga as quintessentially Canadian, especially with its powerful connection to Vancouver, but like most ships it was a floating "citizen" of the ocean.

For 15 years, I was privileged to be one of the stewards of Vancouver's archival legacy of the ship and its various fleet mates of the Canadian Pacific Railway. The Vancouver Maritime Museum was that repository where founding curator Leonard G. McCann had collected an incredible array of images, documents, and ephemera. At the time of my arrival as the museum's director in late spring 1991, Len, working with the museum's exhibit designer and fabricator, Gary Tucker, and with guest curator, Dr. Wallace B. Chung, were working on completing *Empress to the Orient*, a major exhibition.

That show included key artifacts, ship models, posters, and ephemera from Dr. Chung's incredible collection. The pièce de résistance was the newly restored figurehead of the first *Empress of Japan*, resurrected from dry-rotted large fragments to resplendent glory; it remains on display as a premier object of the museum these decades later. The magnificent doors of Piers B-C — the portals into Canada for immigrants brought on the Empresses — were rescued by Wally Chung when the piers were demolished to make way for Canada Place and became the entrance to the Vancouver Maritime Museum's W.B. and M.H. Chung Library, which we opened and dedicated as one of the exhibition's legacies in 1993.

At the same time, Len and Gary were working with Dr. W. Kaye Lamb to republish his pioneering scholarship as a new book, also entitled *Empress to the Orient*, which we released that same year. In that fashion, I learned far more than I had ever thought possible about the Empresses and their significance. But I confess that I have learned far more from reading Dan Black's incredible manuscript.

As I read the manuscript, my mind travelled back to sitting with Len, Kaye Lamb, Gary, and others as we went through Samuel Robinson's 18 photograph albums. I never met the man, for he died the year I was born, but I was able to see, through the eye of his lens, the ships, the places he visited, and the people. Earthquake-ravaged Japan, deck scenes on the Empresses, the wrecked and burnt-out hulk of the German cruiser *Emden*, the people of Japan, China, the waterfront of Hong Kong, and so much more were all saved and preserved in those precious volumes of images.

The adage of pictures being worth many words aside, what you are about to read merges the best of both words and images to bring you, the reader, on a voyage of your own to a ship, and its times, but especially, of its people.

James P. Delgado, Ph.D., FRGS, FRCGS

PREFACE

The conversation began this way.

"For your next book, you should write about the *Empress of Asia*, as she is quite a story," urged Nelson Oliver, who had been researching the ship for 20 years.

"I don't know," I said. "I'm familiar with the ship's history, but there might not be enough human interest to sustain me. I like to write about people — not things."

The conversation nearly ended there until I was reminded that the "thing" — while operating as a troopship — was destroyed by Japanese dive-bombers off Singapore in the Second World War, had broken trans-pacific speed records, transported Chinese labourers bound for the Western Front, faced terrific storms and lost people overboard, was used for opium smuggling, and yes, "accidentally" bombed more than a year before Pearl Harbor, prompting an international incident. Oh, and she also transported vast numbers of immigrants and refugees to North America at a time when South Asians in particular were not welcomed.

Still, with all that, I did not want to write an operational history. It had been done and such histories, although vital, are not what I do. Then one day I was staring at two photographs of the ship. In one, she was a gleaming white steel cliff ready to sail with stylishly dressed passengers waving from behind a web of farewell streamers. In the other, she was scarred — covered in dreary wartime grey — passing through the Panama Canal with thousands of depleted and wounded soldiers returning home.

That was the moment I decided a book could reflect how one ship — in her 30-year life — not only intersected the lives of many but also transcended a period of history that passed from peacetime through one world war, the Roaring Twenties, the Great Depression, and into another larger world war. So the challenge was to capture enough human-interest stories while paying close attention to the working lives of crew on a ship built for peace but used in war — twice.

Oceans of Fate is a story that could apply to many ships, but I chose the *Empress of Asia* because one of her officers had written a contextual account of his time at sea. I liked, too, that the ship was Canadian-registered and had that peacetime-wartime split personality. I was also drawn to her destruction — in a violent confrontation far removed from the Battle of the Atlantic and European battlefields. Finally, there is the dramatic story of how her crew narrowly escaped besieged Singapore or languished — and in some cases died — in prisoner-of-war camps.

As it turns out, the *Empress of Asia* — and more importantly her human side — is a remarkable story.

PART ONE

OCEAN BOUND

1

HEATER BOYS AND HOLDER-ONS

In that time he learned the powerful engines of the craft to a point where the slightest cause of variation of any of the parts was to him an open book.

— ***Vancouver Sun***

The *Empress of Asia* is dead, half sunk, all 16,909 tons — except for bits blown away or turned to ash. Fourth Officer Walter Oliver takes one final look as he and the rest of the crew struggle to escape besieged Singapore on a small coastal vessel. It is February 11, 1942, barely a week after their former passenger liner turned troopship came under attack — for the second time in two days — by Japanese dive-bombers.

Oliver, who has spent 17 years at sea, stares at the burned-out hulk — her hull fatally breached and taking on water. There is zero time to waste — Japanese planes are thundering overhead, trying to kill him and his crew with bombs and machine gun bullets. Just seconds earlier, the concussion from one explosive nearly succeeded, spraying a column of seawater over the men and violently rocking their boat side to side. Despite the havoc and destruction, Oliver takes a few seconds to say farewell to the wreck of the *Asia*

Throughout his many years at sea, merchant sailor Walter Oliver could be counted on for his courage and ship-borne knowledge.

as she slips astern. He knows that within the scorched and mangled steel, there are stories to tell, and perhaps one day — when he is old and ashore — he will write about her and his time at sea.[1]

○ ○ ○

The *Asia*'s First and Second World War experiences bookend the history of a peacetime passenger liner, one that through the 1920s and 1930s touched thousands of lives, including those who stoked her fires, fought through storms from the bridge, or vacationed to distant lands. Her history also embraces the people who built her, from designers and engineers to platers, heater boys, riveters, electricians, plumbers, and upholsterers. So for Walter

Oliver she was more than the dark seawater sloshing over worn rivets, gutted cabins, and coal-fired furnaces; she embodied the passage of people and time, not only through war and peace but across social class, race, business, and politics as well.

In 1911 young men and older lads spilled from Glasgow and surrounding countryside, looking to join the bustling yards on Scotland's Upper Clyde. Ship construction had long since evolved from wood to iron to steel, and soon at least 100,000 skilled and general labourers were working in 38 shipyards and ancillary shops spanning the river west to the Firth of Clyde.[2] Many found themselves on Govan Road in the smoky outskirts west of Glasgow, where they gazed at a three-storey, red-sandstone building, home of Fairfield Shipbuilding and Engineering Company. The impressive entrance was flanked with life-size statues of an engineer and a shipwright but added a bit of nautical whimsy with carved mermaids and a bust of Neptune over the doorway.

That year, on December 4, the keel for a massive and modern passenger liner ordered by Canadian Pacific Railway (CPR) was laid less than a month after the keel of a nearly duplicate ship ordered by the same company.[3] In offices, in workshops, and on scaffolding, hundreds of workers busily applied the skills and labour required to complete the project. It was an exciting time and while basic pay hardly kept some men and families afloat, entry-level work often led to something better, perhaps as an apprentice in the mechanics' department or cabinetmakers' shop, or as a marker-off or plater's helper. At day's end, "when the stopping whistle" blared across the yard, the pockets of the former bulged with shards of French chalk used to mark steel plates for cutting.

As the *Asia* gradually took shape, the slipways and fitting-out basins remained dangerous amid all weathers, as toiling without head protection or harness on the slippery scaffolding above the slips could result in injuries, or worse. The identities and life stories of those fellows — risking life and limb — are mostly lost to time, but their work remains embedded, anonymously, in the company's history as a major passenger liner and warship builder. In the beginning, Randolph, Elliot & Company began by manufacturing engines and machinery in 1834. When marine engineer John Elder joined 18 years later, it became the Randolph, Elder & Company with a focus on marine engines. The mathematically gifted Elder was the son of a successful marine engineer, but also owed much to Robert Napier, who had become an iconic figure in the rising marine engineering and shipbuilding industry. Napier was both an innovator and tutor. One of his earlier projects, which lost money meeting Admiralty specifications, was completion of Her Majesty's Ship *Black Prince*, which along with HMS *Warrior* became the earliest armour-plated, iron-hulled warships.

As a young apprentice, Elder benefited from Napier's knowledge, and through sheer genius achieved stunning mechanical breakthroughs. Among them development of the celebrated compound steam engine and redevelopment of a cylinder casing or jacket filled with hot steam. This engineering work proved a boon for maritime trade and passenger service, allowing ships to not only reach higher speeds but also make long voyages using less coal. Elder remained with the company until his early death in 1869 at age 45. Succeeding him was Sir William Pearce — a future Conservative Party Member of Parliament — who nearly made it to 56 before his death in 1888. It was under Pearce's management that Fairfield Shipbuilding and Engineering Company, acquiring its name from an old farmstead, became a limited company and grew into one of the world's leading shipbuilders.

Gripping heavy tools, from mechanical riveting hammers to large spanners, the men building the *Empress of Asia* sweated, smoked, laughed uproariously at one another's coarse jokes, and cursed fluently. The incessant noise had robbed men of hearing, and many here and in other shipyards along the Clyde and in Belfast nursed cuts, bruises, sprains, scars, and old fractures, earning epithets like "Nail in the Boot" for a fellow with a limp or who worked too slowly, or "Pointless" for one who had lost a finger.[4] Focusing on the job and surroundings was essential as girders and massive steel plates dangled and swung from derricks sometimes just feet above workers' heads.

The plates forming the ship's hull and the holes punched for rivets had to be clean — free of burrs and rust — or they would not fit. Furthermore, the size and length of a rivet depended on the thickness of the adjoining steel plates that had to line up before being driven through — an orchestrated process that when it came to hand-riveting began with the heater boy, who everybody hoped was on his game or had not been out too late the night before. Standing on the partially completed deck next to an open, coal-fired forge, the heater boy — with or without a protective leather apron — reached into the red-hot pit with tongs and extracted a glowing rivet from beneath the coals. This he flung through the air into the gloved hands of the catcher, who then passed it down to the riveter. Working quickly in tight spaces, the riveter placed the still-glowing bolt into the punched hole, and while hammering it home, the holder-on — the lad crouched on the other side of the adjoining plate — gripped a piece of iron with a heavy flat head held fast against the hole to absorb the blows. Then, as the rivet cooled, contraction pulled the plates further together, and slowly, methodically, the ship's hull, like so many before, rose above the yard. It was tough, unforgiving work engineered to reduce margin of error while building the hull from ground up according to specifications.

○ ○ ○

The CPR's dream for two transpacific sister ships to replace three original Empress liners — the *Empress of China*, *Empress of India*, and *Empress of Japan* — became evident early in 1910. By then, it was clear the order could not come fast enough, and within a year, on July 27, 1911, pressure grew when the *China* became a total loss after running aground in heavy fog near Tokyo Bay. She, along with the *India* and *Japan*, had been steadfast and reliable since entering service in 1891, but they were no longer the fastest, a crucial advantage when it came to competing for Royal Mail and lucrative raw silk contracts. Nor could they compete in size and comfort against newer ocean liners operating from the United States and Japan.

Using the mysterious code name UBAB, the CPR approached Fairfield to prepare preliminary, secret designs for both liners that together would cost a whopping $5 million. Only a few top officials on either side of the Atlantic knew the plan had originated with the CPR, and as designs evolved, overall dimensions shifted from 700 feet in length to 650 feet, then 592 feet with a 68-foot beam.[5]

But even then one major goal seemed certain, and when the contract was signed that summer between the CPR and Fairfield, it opened the way for a higher, faster standard of ocean travel, supported by a requirement that called for top speeds of 21.5 knots over the measured mile and 20 knots over a 600-mile sea trial.

Combining beauty and luxury, the graceful ships would feature three massive funnels lending a sporty swept-back look above a deck that capped an uninterrupted flow of steel deckhouses. Resplendent in white, the "palatial" Empresses would boast modern, practical cruiser-style sterns for more stability while running down swells.

Each vessel would possess eight decks and a lower hold comprising a hull depth of 46 feet, the same roof-to-ground distance of a typical four-storey building. From the top, the decks were Fiddley, Promenade, Bridge, Shelter, Upper, Main, Lower, and Orlop, the last being above the hold. Gross tonnage wise, the *Asia* would be 99 tons more than her sister, the *Empress of Russia*.

In the years ahead, anyone viewing the sisters from afar would have to look closely to distinguish them. As it turned out, people purchased illustrated postcards of the *Asia* thinking it was the *Russia*; however, the forward-facing portholes on the centre of the bridge or wheelhouse offered a clue — the *Asia*'s being round and the *Russia*'s square.

Interior architecture and sumptuous furnishings would also generate separate identities: the *Asia* with its Jacobean and Georgian decor, while the *Russia* would adopt a Louis XIV and XV style. The mastermind was 47-year-old British interior designer George Abraham Crawley, who had collaborated with American architect Grosvenor Atterbury on the stunning Westbury House, a Charles II–style mansion on Long Island, New York. Into his Westbury designs, Crawley had incorporated the works of British artists, including wood carvings by sculptor Francis Derwent Wood, who later designed metal masks for men with ghastly facial wounds suffered in the First World War.[6]

Surprisingly, the sisters would burn coal at a time when newer ships relied less on stokers by using oil with its cleaner, more consistent burn rate. One explanation was the proximity and abundance of quality coal from mines on Vancouver Island not far from the ships' home port. Not all coal, however, was the same quality, and like most ships, the *Asia* would have to "bunker" or load coal at various ports around the world. Some coal, including that from South Africa, produced a thick, flour-like ash and consequently "smoked" more than normal and did not burn well. Too much ash cut off the airflow to the fire, which meant the furnaces had to be "cleaned" more often. Other coal, including that from Sydney, Nova Scotia, had a heavy metal content — good for metallurgical purposes — but when burned formed "clinkers" or slabs that came out like "clumps of taffy." Better coal produced higher heat because it burned faster and somewhat cleaner.[7] Therefore, the farther a ship could travel after loading quality coal the better.

Those considerations aside, trust in Fairfield was high. Six CPR ships came from there: the *Princess Charlotte* and *Princess Adelaide* serving the West Coast, the *Assiniboia* and *Keewatin* on the Great Lakes, and the *Empress of Britain* and *Empress of Ireland* on the transatlantic route.[8] Spearheading the project at Fairfield were the company's naval architect, Dr. Percy Hillhouse; W.D. McLaren, head of Fairfield turbine research and design; and Major Henry Maitland Kersey, a decorated South African War veteran overseeing planning and construction for the CPR.

In mid-1911, the press speculated the ships' names might honour CPR magnates William Cornelius Van Horne and George Stephen. However, it was soon revealed that the names of the new Empresses would continue to reflect and promote the geography of the CPR's transpacific trade routes.

Fairfield had at least a dozen ships under construction, and as the sounds of industry rose above the yard, there came shocking news of the White Star Line's *Titanic* disaster on April 15, 1912. Although built at Belfast by Harland & Wolff, the *Titanic* and her loss sent chills around the world — felt deeply in shipyards. For the *Asia* and *Russia*, designs called for a double hull that extended the length of the ship. Watertight bulkheads would rise 18 feet above the ship's maximum load line while two others reached higher to the Bridge deck. The builders were confident four of the ship's watertight compartments could be open to the sea without the vessel sinking.[9]

To achieve the desired speed and endurance, the steamships would feature six double-ended and four single-ended Scotch boilers, cylindrical in shape with a combined heating surface of 54,251.2 square feet. With three furnaces at either end, each double-ended boiler measured 21.5 feet with a diameter of 17 feet. The plates forming the shell were 1.6 inches thick. Each single-ended boiler measured over 11 feet with the same diameter and shell thickness.

Located within watertight compartments deep in the hold, the boilers were the heart of the ship and had to withstand 190 pounds of pressure per square inch. As thick black smoke and hot gas escaped up the funnels, steam would expand through four compounded Parson Turbines that drove four, four-bladed propellers made of solid manganese bronze — each weighing three and a half tons and measuring nine feet in diameter.

Anticipating each ship would burn approximately 311 tons of coal per day to average a service speed of 18.5 knots, the bunkers would be enormous and positioned for access and stability. Coal capacity for each ship was 4,625 tons, and McLaren calculated a single round trip to the Far East from Vancouver would consume more than 10,000 tons. This meant the ships would require coaling in Nagasaki, a port city in southwestern Japan.

Arriving at Fairfield that year was the dark-bewhiskered James Adamson, former chief engineer on the clipper-bowed *Empress of India*. Adamson, who had served on the *India* as second engineer during her maiden voyage, had embarked on a working holiday to his hometown of Glasgow. Although he had not taken a full vacation in 21 years, he was looking forward to his time at Govan assisting with the supervision of engine installation. Even more exciting was that his experience had put him in line to become chief engineer on one of the sisters.

Up to then, Adamson had covered close to a million miles of ocean to and from the Far East. "In that time he learned the powerful engines of the craft to a point where the slightest cause of variation of any of the parts was to him an open book," noted the *Vancouver Sun*.[10] The "open-book" reference was apropos for a man whose deep knowledge leaped from his mind to the printed page. With stunning precision, Adamson created an owner's manual of specifications on the *Asia* and *Russia*, and older *India*. Using slide rule, fountain pen, and red pencil, he painstakingly recorded every detail, from the smallest piston on a bulkhead's self-closing emergency stop valve to the coal volume of the largest bunker. He would know the ship better than anyone.

Adamson's work rubbed off on another young engineer who arrived before the *Asia* was completed. Robert Henry Shaw had apprenticed as a millwright at Crewe, Cheshire, England, before joining James Chambers & Co., a Liverpool-based shipping company. The 27-year-old would sail with the *Asia* on her maiden voyage and years later fill the remaining pages of Adamson's manual with specifications on the 26,032-ton *Empress of Japan II*, later renamed *Empress of Scotland*. Shaw's dedication and courage would be clearly on display decades later after a German bomber attacked the *Japan* off Ireland. He and his crew remained below, working in the dark to keep the crippled ship moving. It is hard to underestimate the value of Adamson and Shaw, and the way in which their accumulated experience kept the ships running and led to such an important and comprehensive owner's manual.

Progress on the *Asia* turned to celebration on November 23, 1912, when 29-year-old Augusta Alleyne Bosworth, wife of CPR vice-president George Morris Bosworth, took a deep breath in the cool, misty air and swung the champagne bottle to christen and launch the ship. Some three months earlier, Alice Beauclerk of Montreal, daughter of CPR president Sir Thomas Shaughnessy and Elizabeth Shaughnessy, did likewise for the *Russia*.[11]

With the *Asia* gliding down the greased slipway toward the Clyde, dragging a knot of rumbling steel chains to slow her, there was immense pride on the faces of those who stood or clambered down from scaffolding. Front and centre were the elegantly dressed guests of honour like the Lord Provost of Glasgow, prominent CPR officials, and British naval officers. From the start, the Admiralty had taken great interest in the ships, and the braided presence of high-ranking officers did not dispel rumours the *Asia* and *Russia* would be useful for wartime naval deployment. "The plans of the steamers are understood to be of such a nature that they could easily convert the steamers into fast and serviceable cruisers … it is claimed the Canadian Pacific railway [*sic*] company has seen an advantageous proposition in taking up this question with a view of having its big fleet of ships subsidized by government for the purpose, if vessels should be required."[12]

Expectations aside, the ships required weeks of "fitting out" with attention to detail, followed by demanding sea trials. It was also time for the *Asia*'s first appointed master to call and offer his appraisals based on years at sea.

Clean-shaven Samuel Robinson, a captain with the CPR, stepped through the yard's ornate entrance on the morning of Monday, April 14, 1913. A billowing dark greatcoat hid his five-foot-six frame, although the 42-year-old looked taller, more powerful. Robinson was anxious to see the ship, but he was not in the best of moods. He had been suffering from an itchy bacterial skin infection on his back, which he and his ever-patient wife, Jessie (usually called Jess), were treating with copious amounts of lime and sulphur.[13]

Born north of the Humber Estuary in the port city of Hull, England, in 1870, Robinson was 14 when he went to sea on the sailing ship *Imbros*. He

The *Empress of Asia*'s first master, Captain Samuel Robinson.

reached Canada in 1894, and after a bout of British Columbia gold fever, signed on with the CPR in 1895. While serving on the *Empress of Japan*, he fell for Jess, daughter of a tough New England seafaring family.

They were married in 1901 at Yokohama, although their time together nearly ended in 1906 while he was master of the CPR's 6,163-ton *Monteagle*. Robinson had gone ashore at Hong Kong, and while returning in a launch, misjudged the arrival of a typhoon. Intentionally or not, the launch's anchor snagged a submerged cable, and while he rode the storm, 13 large ships broke free and piled up on shore.

Robinson's path to the sea owed much to his father, whose seafaring yarns coloured his youth. The prosperous old port and its connection to wool and woollen exports, whaling, and the importing of timber and wine was also a predominant influence. Along the waterfront, Robinson had witnessed the diaspora of arrivals from Northern Europe — people who had crossed the treacherous North Sea seeking settlement in Great Britain or beyond — so revisiting Hull had been on his itinerary well before he and Jess boarded a transcontinental train at Vancouver in March 1913.[14]

En route to Boston with a connection in Montreal, Robinson survived the journey by staring through the carriage window at the shifting Canadian landscape while taking breaks from Mark Twain's *Pudd'nhead Wilson*. At Montreal, the couple had "tiffin" with the Bosworths, learning more about the *Asia*'s November christening before heading south to Boston where in grey cloudy weather Jess caught a nasty cold and Samuel's skin erupted painfully once more.

From New England, the Robinsons travelled northeast to Saint John, New Brunswick, for passage to Liverpool aboard the *Empress of Britain*. Heavy fog delayed their departure on March 22, but overall it was a pleasant crossing and afforded Robinson time to play bridge in the saloon with long-time friend Lionel Dale Douglas, who had served on the *Empress of India* and would become the *Asia*'s chief officer. Douglas was in a cheery mood. He had just married his sweetheart, Christine Pierce, in New York with the charming ceremony dutifully signalled at Vancouver by the hoisting of blue ensigns on the masts of his former ship, the *Empress of Japan*, which coincidentally was also Robinson's last ship.[15]

Ashore in Liverpool, the two cast their eyes on the *Empress of Russia* that was alongside awaiting her maiden voyage under Captain Edward Beetham, with reliable Adamson in place as chief engineer. Robinson and Douglas heard again how the ship surpassed expectations during her seven-day sea trials. On the mile-long course, *Russia* hit 21.178 knots and later met the required average speed of 20 knots on the 600-mile test. This was good news, because the *Asia* was only weeks away from her critical trials.

Cold rain and "the itch" continued to plague the Robinsons on their cross-country train journey to Hull, and then lingered relentlessly as they travelled north through Berwick, Edinburgh, and west to Glasgow and Govan. Within days, the captain's pain would be intolerable. "Rain all day — itching very bad & seeming to spread. In agony all day, trying new stuff. Went back to lime wash in evening."[16]

Despite this, Robinson visited the yard daily where he toured and carefully inspected the *Asia* and spoke with executives, including Major Kersey, who was 11 years his senior, and like the itch, would one day get under Robinson's skin. Originally from Suffolk, England, Kersey had seen much prior to joining the CPR as general manager of steamers in 1890. Greatly credited with fostering plans to build the *China*, *India*, and *Japan*, he nevertheless saw opportunity in New York where he joined the White Star Line.[17] Kersey continued to cultivate powerful friends, and on the evening of February 1, 1898, he was sitting near U.S. President William McKinley in the Waldorf-Astoria during the National Association of Manufacturers' banquet.

For Kersey, it was a rosy affair until a deputy sheriff approached and tapped him on the shoulder, stating, "You are my prisoner!" The *Los Angeles Herald* noted that Kersey spun around and replied, "How dare you lay your hand on me!"[18] He initially perceived it as a practical joke until the officer produced an arrest warrant signed by a Supreme Court justice. The newspaper stated the order was "granted at the instance of a Miss Julia Gleason in a suit she has instituted against Mr. Kersey for $5,000 damages for an alleged assault" while she was employed as his cook. Kersey did not have the cash to cover bail, so after writing a cheque to the hotel's cashier, he handed $500 to the sheriff, who released him to the banquet hall.

The next day, while involved in business associated with the Klondike Gold Rush, Kersey packed a trunk and returned to British Columbia as a representative of European financiers. As for the alleged assault, research for this book uncovered no news of a follow-up court appearance and it appears the charge disappeared somehow. Two years later, Kersey was fighting in South Africa with the 15th Battalion Imperial Yeomanry, rising to major and earning a mention in Dispatches, the King's Medal with two clasps, and Companion of the Distinguished Service Order.[19]

At Govan, the decorated major supervised construction of the high-end passenger quarters and prestigious interior amenities that would proudly distinguish the two sisters from other transoceanic liners. The plans called for 284 first-class, 100 second-class, and "Oriental Steerage" for 808.[20] The best cabins, including six *suites de luxe*, were to be located on the Bridge and Promenade decks with hot and cold running water

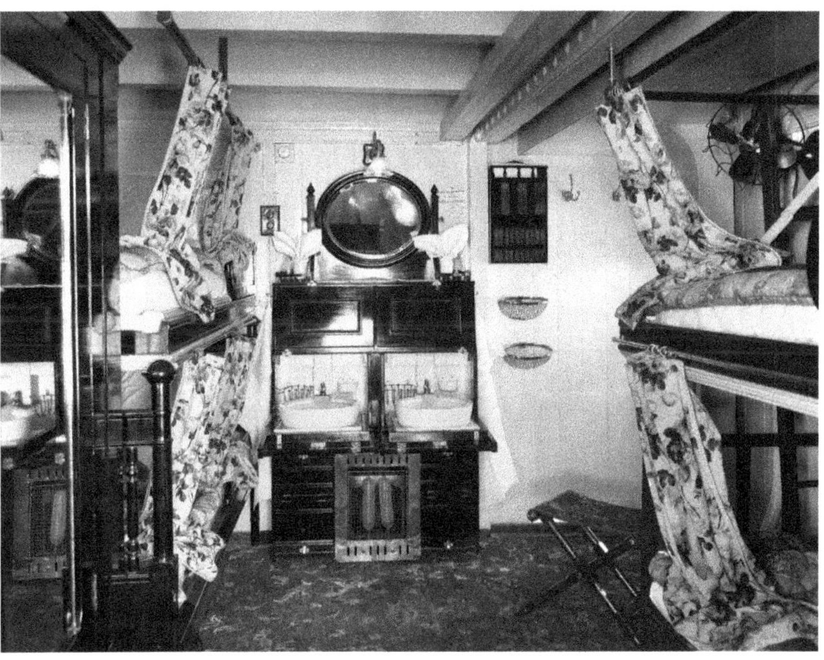

The well-appointed and functional interior of a four-berth stateroom. Economy of space was crucial.

and enough space for two or three passengers, plus an option to connect adjoining suites. Second-class cabins, also designed with comfort in mind, were to be midship on the starboard side of the Upper deck. In steerage, plans called for a comfortable space with limited access to the upper decks as well as access to smaller open spaces at the bow and stern of the Shelter deck.

As a gifted designer, George Crawley had ambitious plans for the spacious first-class lounge on the Promenade deck. Measuring 50 by 36 feet, it would incorporate quiet alcoves, tasteful decorative chairs, plush sofas, and an expanse of luxurious broadloom. Elegant replicas of ancient Roman columns with their distinctive volutes, along with rich wall panelling, would showcase English art, gleaming brass fixtures, and lavish drapery — all lit by a spectacular overhead circular skylight.

Entering the lounge for the first time could take one's breath away, although experiencing such luxury was considered the norm for many first-class passengers.

On the starboard side of the Promenade deck near the lounge, the ship's first-class, wood-panelled, Georgian writing room measured a generous 43 by 43 feet. It would possess the same attention to detail and include handsome Chippendale furniture and sturdy desks. Meanwhile, plans for the spacious Jacobean-style smoking room, near the stern, called for finely woven broadloom surrounded by oak panelling with intricately carved details. In addition to soothing colour and warmth, the thick broadloom would help reduce engine noise and vibration from below. Even while under construction, it was hard not to imagine guests blowing curls of blue smoke toward a half-timbered ceiling that drew attention to the incorporated glass light bulbs.[21]

But it was the ship's dining saloon on the Shelter deck beneath a massive domed ceiling that would satisfy wealthy appetites for luxury and fine dining. The room was enormous — 72 feet long by 62 feet wide with a large

Lavish table settings and expensive decor speak volumes of the quality of service in the *Asia*'s stylish first-class dining saloon.

reception hall aft, a starboard library, and portside private dining room reserved for special occasions.

Pausing on the tiled floor next to a wrought-iron balustrade above the winding grand stairway leading to the reception and dining rooms below, Kersey was impressed. He could envision 200 stylishly dressed first-class passengers descending toward servings of Noisettes of Lamb-Washington or Roast Turkey-Farcie with Cranberry Sauce.[22] For the time being, however, his field of vision overlooked a workforce of diligent tinsmiths, cabinet-makers, and interior decorators, most sustaining themselves with oatcakes and Scottish tea bread as they worked against the clock.

As fitting-out proceeded, suppliers and installers arrived with tables and cushioned chairs, potted palms, floral arrangements, rolls of lint for the hospital, wooden vaulting horses and barbells for the gymnasium, and to Robinson's delight, a high-gloss grand piano for the lounge. By late April, the captain's focus turned to seaworthiness. He examined watertight doors and the engine room, and inspected the massive holds toward the stern for hauling raw silk. Morning yard visits switched to afternoon teas and quiet evenings with Jess. Meanwhile, the press on both sides of the Atlantic continued to praise, describing the ships as being "extremely pleasing to the eye, shaped as they are like warships without the ugly and sinister guns and torpedo nets."[23]

Toward the end of May, the weather turned ugly when Robinson stood, arms crossed on the bridge, to take his ship for a run on the Firth of Clyde. It was the start of her trials on the measured mile. At one point, the ship's turbines surpassed expectations, producing 27,280 shaft horsepower. "Fastest runs 21.11 & 21.76 as was 21.43," noted Robinson in his diary. "Returned Greenock 4 PM."[24] Squalls and mist dampened the *Asia*'s 30-hour, 600-mile trial at 11:30 p.m. "Strong head gale & high sea all afternoon," Robinson noted on May 30. "Started 30-hour trial 2 AM but had to stop for fog. Started again 9:10 AM & made 19.27 against gale to fastest at 10:02 PM."[25]

Although it was uncomfortable, the wretched weather and rough seas made for a true test of endurance, and once the weather cleared, the *Asia* continued to impress by accelerating across the water with intent. She averaged 20 knots for the remainder of the 600 miles, leading to the lowering

A rare photo shows the *Empress of Asia* — with her bridge still incomplete — pouring on the coal during trials on Scotland's River Clyde.

of the Fairfield flag and raising of CPR's red-and-white checkerboard-style house flag, designed by Van Horne. A poignant detail in that rheumatism would soon limit Van Horne's own mobility.

Incredibly, it had taken fewer than 18 months to build the *Asia*. Robinson and his chief engineer, William Auld, had their ship and charted course for Liverpool in preparation for a 19,912-nautical-mile maiden voyage to Vancouver via the Cape of Good Hope.

2

DANCING LIKE A HURRICANE

Captain Robinson is the most cheery, genial, and popular man one could wish for.... He sings charmingly — nice, soft sweet tenor ... [and] dances like a hurricane.

— **Passenger Joseph Daniel Noonan**

After kissing Jess goodbye on June 14, 1913, the ship's master, with a vastly undersold-passenger list of just 59, directed his ship through the mouth of the Mersey to the Irish Sea. They would not cross the Mediterranean or the Red Sea's blistering heat, although the ship and her captain would become very familiar with the latter during the impending war. Instead, Robinson chartered a course to Vancouver via the rocky headland known as the Cape of Good Hope, and the first leg would cover 1,429 nautical miles to Madeira, north of the Canary Islands. Jess, meanwhile, would remain in the old country, before visiting Paris and returning home to Canada in August.

Working in shifts, known as watches, the *Asia*'s crew of approximately 515 easily outnumbered passengers. In addition to officers, engineers, and a

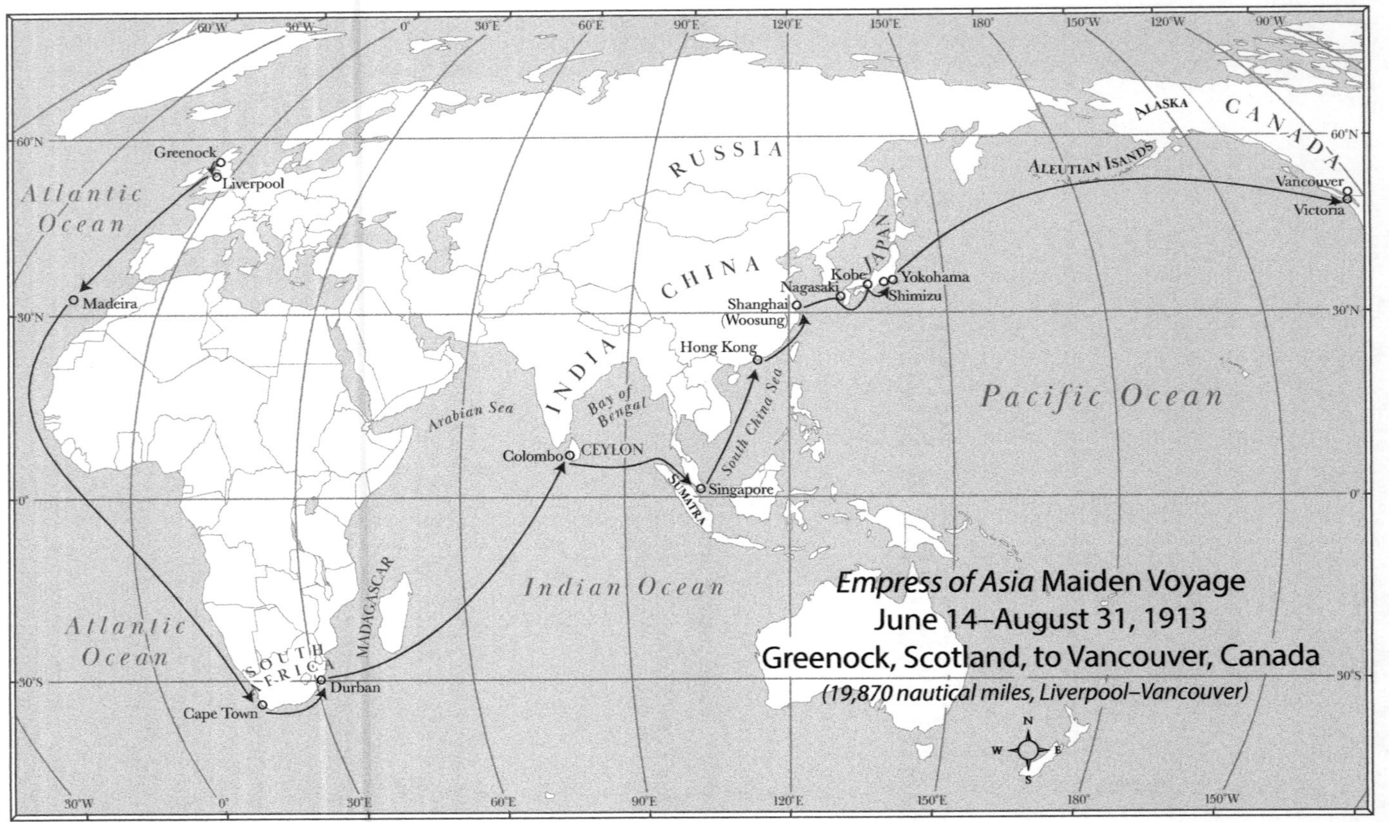
Empress of Asia Maiden Voyage
June 14–August 31, 1913
Greenock, Scotland, to Vancouver, Canada
(19,870 nautical miles, Liverpool–Vancouver)
Greenock
Liverpool
Atlantic Ocean
Madeira
RUSSIA
CHINA
INDIA
Arabian Sea
Bay of Bengal
Colombo
CEYLON
SUMATRA
Singapore
South China Sea
Hong Kong
Shanghai
(Woosung)
Nagasaki
Kobe
JAPAN
Yokohama
Shimizu
ALASKA
CANADA
ALEUTIAN ISANDS
Vancouver
Victoria
Pacific Ocean
Indian Ocean
Atlantic Ocean
SOUTH AFRICA
MADAGASCAR
Durban
Cape Town

surgeon, there were stewards, stewardesses, pursers, storekeepers, and assorted tradespeople, including a barber, a chef, cooks, an electrician, and a carpenter. Down in the stokehold were the firemen, trimmers, and greasers. The former, known as stokers on navy ships, maintained the furnaces, while the trimmers wheeled coal from the bunkers to the red-hot furnaces. Using a slice bar — the heaviest tool in their arsenal — the firemen reached into the furnace and lifted the bottom of the fire (coals) to break up the clinkers and ash. This allowed air to get into the fire for a faster burn. Cleaning a furnace meant moving or "winging over" the hot coals to one side and then using a rake to remove the exposed ash. Despite the tremendous heat, long-sleeved shirts and trousers were essential for firemen; scenes of bare-chested, soot-smeared men sweating against the glow of a furnace were mostly a latter-day Hollywood fabrication. Clothes helped protect the skin against backflash, which occurred when a shovelful of dusty coal ignited in the air above the fire and blew back out the furnace door. It was also for this reason that firemen chewed on the sweat rag around their neck, which was pulled up to protect the lower part of their faces.[1] Still, eyebrows and eyelashes could disappear in a flash.

The location and weight of the coal in the bunkers was crucial, and it was the trimmer's job to ensure good distribution so the ship could maintain an even keel or proper "trim." Meanwhile, ducking about with his oilcan amid the engine room's chattering machinery was the greaser, who squirted or slopped lubricant on a multitude of moving parts, including turbine and shaft bearings, and levers controlling steam. The greaser and engineer also kept a close eye on the water gauge at the back of the boiler because a dry boiler could explode. If the glass tube on the gauge indicated the boiler was low, adjustments were immediately made to the injection pump that forced fresh water, not salt water, from the tanks at the bottom of the ship into the boiler.[2]

When it came to adjusting the speed, the process began with an order and telegraphed message from the bridge to the engine room, setting off a series of bells. The requested speed appeared on the telegraph, but counting the bells when the telegraph rang also told an experienced crew how many revolutions per minute were required. After acknowledging the order, the engineer or greaser adjusted speed by opening or closing the valve that

The *Asia*'s bridge: Visible are the wheel, binnacle, inclinometer, and telegraph to forward orders to the engine room. Of note are the bridge's round portholes facing the bow.

controlled the amount of steam entering the turbines. The more steam, the greater the speed.[3]

By Friday, June 20, the *Asia* had skirted Tenerife, the largest of the Canary Islands, and spent an uncomfortable night on the 23rd encountering heavy squalls and wet "doldrummy weather" before crossing the magnetic equator where Robinson "swung" his ship to check and make adjustments to his compass. "Light Sly wind & fine," he wrote in his diary.[4]

When not on the bridge or relaxing in his cabin, Robinson enjoyed "yarning" or tactfully sliding up to the piano to sing with prominent guests, among them the Russian-German baroness Hildegard Schilling and John Neville Woodcock, a lecturer and later professor of classics from Toronto's Trinity College travelling with his wife and two young children.[5]

Although recorded on the manifest as 43 and "single," the baroness was actually 10 years younger and recently widowed, born July 19, 1880, to

German-born parents at Bucha, near Kyiv in present-day Ukraine. Smartly dressed with stylish brown hair, blue eyes, and a fair complexion, the baroness presented a captivating personality.

Robinson also rubbed shoulders with Passenger No. 38630, Quebec politician Narcisse Pérodeau, a liberal who despite never running for office earned an appointment to Quebec's Legislative Council and was later sworn in as minister without portfolio.[6] In the 1905 Legislative Council, the civil law graduate from Montreal's McGill College (now McGill University) introduced a bill proposing reform to the law of intestacy aimed at allowing surviving spouses to share in the estate along with blood relatives.

Pérodeau pushed for this on behalf of widows whose late husbands had not provided for them. Although the bill had not received royal assent by the time he boarded the *Asia*, Pérodeau remained optimistic, and within two years despite serious opposition, the act to amend the Civil Code passed and became known as the Pérodeau Law.[7] Robinson was all too familiar with sad cases involving wives of men lost at sea or to alcohol and was suitably impressed with Pérodeau's efforts.

Retreating to his cabin on June 29, the captain removed his cap and overcoat, loosened his tie, and closed the door. Thinking of Jess many miles away in England, he was grateful that the intense itching on his back had eased thanks to an ointment she had gone out of her way to secure in Scotland. The sea, meanwhile, remained calm with a fine southeast trade wind. His ship would soon call at Cape Town where more than a dozen first- and second-class passengers would board, including the Honourable William Ross, chairman of the Cape Province Council. As he settled into the heavy wooden chair behind his desk, Robinson contemplated his recent purchase and considered it as good a time as any to begin using the typewriter.[8]

Leaving Johannesburg on July 3 were three well-dressed gentlemen — Gordon Grimmer, Joseph Daniel Noonan, and George Wille — with steamer trunks and tickets to board the *Asia* at Durban, her next port. Their train had been last out before a riotous crowd smashed and looted gun shops and jewellery stores and set the station ablaze. Tension between trade unionists and employers had finally boiled over. In May, miners at the Kleinfontein had gone on strike after feeling ignored by owners and government, and now the men resorted to riotous tactics. A blast killed one man, a 13-year-old street vendor was shot through the back, and several horses died in the crossfire.

While the Johannesburg railway station burned, the *Asia*'s first-class passengers at Cape Town assembled in the ship's dining saloon for a five-course, Fourth of July meal honouring all things American. The menu featured an illustration of the ship with a watercolour of Mount Fuji and opened with hors d'oeuvres followed by tantalizing selections such as Potage Americaine, Halibut ala Rockaway, Roast Sirloin and Ribs, Ox Tongue, and Quail ala Roosevelt. For the sweet tooth, there were confections such as Long Island Pudding and Mousse Frappe completed with assorted cheeses, coffee, or tea.[9] All of it prepared in the ship's gleaming, well-equipped galley that easily rivalled the kitchens in top-rated hotels and restaurants.

On the dry, dusty harbourfront at Durban, the three gentlemen from Johannesburg fanned themselves with their hats while staring, incredulously, at people heading north, intent on visiting the city they had narrowly escaped. "Some … on the Empress who had come from England proceeded to Johannesburg from Capetown [*sic*] and were right in the thick of things," noted the 33-year-old Noonan in his diary.[10]

With 3,615 nautical miles between Durban and Colombo, the capital of Ceylon (today's Sri Lanka), the *Asia* got underway at 12:15 p.m. on July 9. While Noonan and his friends boarded with ease, the same could not be said for a less punctual fellow. "He went bathing at about eleven-thirty, the boat being timed to leave at twelve and was still in the water when the hooter went. He then deemed it advisable to finish his bath, dressed, and calmly strolled down to the Point, munching apples en route, and arrived just in time to see us steaming out. Fortunately, he got a tug to take him out and caught us just as we were about to go full steam."[11]

Born at Kimberley, South Africa, home of the famous diamond mine, Noonan was a successful attorney and notary with the Supreme Court of Cape Colony and an attorney of the Transvaal. His career launched in 1903, and after articling and opening a practice in Mafeking, he moved to Johannesburg where a succession of law firms welcomed him. Whereas his dear friend, George Wille, would prevail well into his eighties, Noonan's life was cut tragically short on the Western Front as a second lieutenant with the 6th Battalion Royal Munster Fusiliers. Also called to the bar in 1903, Wille would go on to become the first full-time professor of law at the University of Cape Town, lecturing on Roman-Dutch law. The library and its users would one day benefit from his vast collection of papers, including letters describing his world tour and that of Noonan's.

As the *Asia* left Durban, a powerful westerly spawned a "fearful" dust cloud and high breakers at the harbour's narrow entrance. Soon Noonan and fellow passenger Claude Nesser were leaning over the rail, tossing up lunch:

> He [Claude Nesser] and I began to feel funnier than ever. At least I admitted to him that there was just a possibility of my becoming seasick. The boat was pitching quite a lot for her [the ship] and far too much for JOSEPH DANIEL [Noonan], who shortly afterwards caused mirth to the said Nesser and the occupants — or some of them — of the mighty deep in the usual manner. Subsequently on comparing notes with the rest of the party I discovered I had a companion in distress in the person of Our George [Wille], who also jumped about a bit.[12]

On the evening of July 11, the *Asia*, buffeted by trade winds, had reached the south coast of Madagascar, followed 367 nautical miles later by the remote shores of Réunion Island. By then, passengers had introduced themselves or could at least nod or say hello to familiar faces while strolling the deck, visiting the hairdresser and the lounge, or escaping the heat on the veranda café with its breeze and leaded windows. "There are comparatively few people (second- and first-class passengers) on board — about one hundred and ten altogether — but they are quite a congenial lot,"[13] noted the astute Noonan, who identified the baroness and Pérodeau among notables.

The three gents loved the ship and shared a cabin on the Promenade deck until Noonan and Grimmer decided to yield it to Wille and move into larger digs with a porthole. "It is the most luxurious ship one can imagine," wrote Noonan. "The fittings, cabins, lounges, tables, bath rooms, and all conveniences being excellent. All the steward [ratings] except first [chief steward] and second [steward] are Chinese, and they make splendid servants. They are aggressively tidy and one cannot put a thing down for two minutes in one's cabin without having it put away."[14]

Wille assessed the ship a "perfect beauty," complimented the Chinese crew, and noted the marvellous food, which at times was a little too rich. "I am told there are some 400 Chinese on board as stewards, waiters, stokers, etc. and only forty white men."[15] There is no manifest listing the Chinese crew, and most of the time they blended into the background or stayed below, but among the "white men" was 39-year-old boilermaker Thomas McMurtrie, a proud, pipe-smoking Scot who would remain with the ship until 1937. If the *Asia* needed repairs to her massive boilers, the proficient, contemplative, and mustachioed McMurtrie would take care of it.

Long-time boilermaker Thomas McMurtrie enjoys a puff or two while surveying the sea.

Bachelor Noonan had quickly made note of the small number of women on board, including a beautiful Miss Sinclair, scheduled to wed in Singapore, and a Mrs. Gilder, who "helps to beguile the tedium of the voyage." Deck tournaments, including apple bobbing, biscuit whistling, and monkey slinging, which did not involve a monkey, kept passengers amused.

While the *Asia* steamed northeast, her master continued to make the rounds. "Captain Robinson is the most cheery, genial, and popular man one could wish for and all the passengers are very fond of him — particularly some of the ladies," noted Noonan. "He sings charmingly — nice soft, sweet tenor, plays his own and others' accompaniments, dances like a hurricane, joins in sports, and does everything a Commander could do to make a trip enjoyable. He sang two songs at the concert last night and afterwards installed himself at the piano and we gathered around and sang choruses."[16]

With her Upper deck crew on watch, firemen and trimmers sweating rivers in the stokehold, scullion mates clearing waterlogged food from clogged drains, and well-dressed stewards swishing in and out of the dining saloon with armfuls of dishes, the first- and second-class passengers assembled for an extravagant costume ball on July 17. Next day, clad in duck trousers and his thinnest tennis shirt, the circumspect George Wille sat in the writing room composing a letter while rivulets of perspiration trickled down his flushed cheeks. "Nearly everyone was in costume," he began, trying to make sense of it all. "The captain came in a clown's costume, Noonan and I as pierrots, Gordon Grimmer as a gondolier. The first gentleman's prize went to a Canadian as a rickshaw boy from Durban and the first ladies' to a Miss von Schlicht from the Cape as a Red Indian Squaw."[17]

In his diary, Robinson was more succinct: "Hot but still fine. Fancy dress ball in evening. Dressed up as a clown & felt like one."[18]

At Colombo on July 19, the South Africans were joined by another accomplice, 22-year-old passenger Lawrence Cosgrave, described by Wille as a "very entertaining young Canadian." Cosgrave tagged along on a railway journey to Kandy, Ceylon's capital in the central highlands where the Temple of the Tooth and an artificial lake added to its Buddhist history. Later that evening, the men piled into a rickshaw intended for two, while their silent, smiling, and no doubt labouring puller conveyed them around the moonlit lake.[19]

Born at Toronto in 1890, Cosgrave attended the Royal Military College of Canada before completing his education at McGill University in Montreal where he met Dr. John McCrae. Both joined the Canadian Expeditionary Force and, in fact, signed each other's attestation papers. In little over a year, the officers would be overseas where Lieutenant-Colonel John McCrae wrote a first draft of "In Flanders Fields" on a scrap of paper resting on Cosgrave's shoulder.[20] During the next three years, Cosgrave would twice earn the Distinguished Service Order.

The young man's future was to hold more. On September 2, 1945, he became a signatory to the Japanese Instrument of Surrender signed on the USS *Missouri* in Tokyo Bay. Incredibly, the war veteran, who had been blinded in one eye during his service, would perhaps become more known for signing the historic document on the wrong line. Although some attributed the mistake to his old eye wound, he had also signed McCrae's attestation paper on the wrong line.[21] That aside, one of Cosgrave's proudest moments came at age 73 when he returned to Belgium to unveil the Colonel John McCrae Memorial at Boezinge, Ypres, West Flanders.[22]

As the *Asia* entered Malacca Strait between the Malay Peninsula and Sumatra at the end of July 1913, the unbroken hum and whirl of electric ceiling fans did little to alleviate the scorching heat that could lead, with alarming speed, to dehydration and shortness of breath. One of the first to fall victim was the gregarious Mrs. Gilder, who grew light-headed and collapsed and was quickly carried to the captain's cabin to recuperate. At 6:00 p.m. on the 25th, the ship was alongside teeming Singapore, billed in the travel literature as the "Liverpool of the East." Noonan wrote: "The approach is very beautiful … several small islands being passed, all of them being covered with thick foliage. Boys in Sampans — little frail canoes — accompanied us to the wharf and dived most successfully for coins, scorning, however, to go for coppers."[23]

The promise of exploring exotic foreign cities was what sold ocean travel. Like the modern cruise, it took passengers beyond lavish onboard dinner parties and playing make-believe at costume balls into raw, bustling streetscapes of reality that saturated the senses with everything from the smell of ripe and not-so-ripe fruit and fish and fragrant incense to containers stuffed with odd, natural remedies. From the CPR's viewpoint, one could "see the world" and be culturally educated from the relative safety of a ship that catered to white passengers on a journey for the "thinking man or woman."[24] Such voyages were, absolutely, yet another popular extension of the upper class in the realm of white colonial power.

In stops and starts, a small electric tram transported the wide-eyed Noonan and his friends into the vibrant and sweltering heart of the island city where they shopped for curios before straying into a crowded, poorer neighbourhood that rudely expanded their experience beyond the classroom and courtroom. "In the worst parts it was a conglomeration of the concentrated essences of all the worst smells that have ever been smelt by persons with delicate olfactory organs and how on earth the Singapore people stand them beats Cockfighting."[25]

Famous establishments with glamourous entrances, like the high-end, colonial-style Raffles Hotel, were in glaring contrast to the seedier gambling houses and opium dens. "In Singapore we saw several opium dens and the four of us went into one to see what it was like," explained a perturbed Wille. "Just a number of men lying on wooden benches and every now and then inhaling smoke from long pipes. This vice has been absolutely abolished in China, and England still permits it!"[26]

It took the *Asia* four days to transit the South China Sea to Hong Kong where, despite being new, she entered dry dock at Kowloon for a 12-day overhaul. She had logged 13,553 nautical miles, but from the moment she was launched into salt water and air, rust had formed. However, the liner still looked good on July 29 when the forecast of another bad storm was "somewhere about." But memories of the 1906 typhoon that nearly killed Robinson did not deter him from singing in the lounge until it became too hot and the "Terrific Lighting & thunder squalls" arrived.[27]

Noonan, Grimmer, and Wille were intent on seeing more than what the *Asia*'s itinerary allowed, so bidding farewell to their new Canadian friend, Cosgrave, they embarked upon another vessel for Japan. Many others, including the baroness, patiently stuck to the itinerary after briefly switching to waterfront hotels from which they toured the thriving colony's shops and teeming markets hung with wares and food that went beyond their taste. There was much to take in, and while the amiable and attentive captain accompanied some of them to Hong Kong's famous landmarks, he continued to cherish conversations with the baroness.

From Hong Kong, the *Asia* plied northeast across the South and East China Seas until dropping anchor in the muddy Yangtze River at Woosung. Shanghai was just a short trip by lighter, labelled the "Paris of the East" with its boisterous, Westernized waterfront known as the Bund.

Reaching Nagasaki, Japan, on August 17, passengers lined the rail and tossed coins and fruit to women and children whose faces were smeared black with coal dust. At times, only the whiteness in their eyes shone as the women climbed wooden planks and ladders, passing wicker baskets of coal scooped up, with the help of youngsters, from flat-bottomed skiffs pressed against the towering ship. Each basket held two large shovelfuls, and it took

The tough, dirty work of coaling the *Asia* and other steamships at Nagasaki, Japan, included the assistance of women and children who loaded and passed baskets of coal from skiffs up and into the towering passenger ship.

just over eight hours to load 3,530 tons, but less time for the dust to cover workers' arms, legs, and faces, and infiltrate their lungs.[28]

After "bunkers," heavy rain accompanied the ship as she tracked north through the archipelago to Kobe, Shimidzu (Shimizu), and Yokohama. On board for her first Pacific crossing were 140 in first-class, 80 in second, and 629 in steerage. She also carried 4,223 tons of cargo, including 3,000 bales of raw silk valued at $3 million. Her smooth North Pacific crossing ended at William Head Quarantine Station the evening of August 30, 1913. After lying on her chains overnight to wait out a gale, she reached Victoria at 6:00 a.m., and by 2:30 p.m., the white Empress had completed the shortest distance between two ports — Victoria to Vancouver — and successfully navigated her all-important 19,912-nautical-mile maiden voyage.

It had taken 79 days, and although she had left Liverpool with 59 passengers, she disembarked 845 at Vancouver.

Robinson brimmed with pride. There would be many more years of life at sea where he would meet with grateful, interesting passengers and seafarers — sometimes with people and events he found disagreeable. Not all would be as gracious as the baroness, who prior to boarding a train for Niagara Falls, promised Robinson she would write often.

The captain's negative impression of Major Kersey surfaced soon after the company official boarded the *Asia* at Victoria for the celebratory homecoming. Kersey was a week from assuming the role of managing superintendent of CPR steamships, and on September 1, while Robinson and Jess watched the Labour Day parade, Kersey took centre stage with the press. "I am delighted that these two ships [*Asia* and *Russia*] should have made so good an impression, and I hope they will enable us to carry out our ideal of attracting the very best class of travellers to the C.P.R. steamships. I watched their construction from the time their keels were laid; I was present at the launching of both … and took a run down the Mediterranean on the [*Russia*], as well as being on the trial trip of the [*Asia*]."[29]

Stepping onto the pier, the Honourable William Ross of Cape Town focused on race and immigration when reporters caught up to him. Asked about Asiatic immigration to South Africa, Ross stated that in the minds of the people of South Africa there should be no more immigration. Pressed to comment further, he excused himself from the scrum, citing the critical stage of relations between imperial and South African governments. "We are determined that Hindu immigration shall stop," he offered, "but we may not have yet discovered the right way to do it."[30] In just over two months, another attorney, 44-year-old Mahatma Gandhi, would be arrested and imprisoned in South Africa for leading a march protesting a tax on Indian immigrants.

An eminent party of elated CPR directors arrived on September 10. Among them, Vice-President George J. Bury, federal politician Sir Edmund Osler, and R.B. Angus, president of the Bank of Montreal. Kersey was also there, and this time a dispute with Jess, who was boarding, unleashed Robinson's wrath. "Kersey acting like a b … f … with baggage and jess. Directors on board.… Row with Kersey over everything. Feeling pretty mad & I'm inclined to throw the whole d-d thing up."[31]

The captain's anger never made headlines, but another incident did, involving a child the *Vancouver Sun* dubbed "Maggie." The newspaper, which did not publish the girl's surname, stated there had been a "bold attempt" to kidnap Maggie, but it failed when the *Asia* pulled away from the pier earlier than expected to make time against an 11-foot tide. By then, the VIPs would have been admiring the ship that would take them to Victoria. According to the newspaper, Maggie had freed herself and at the last moment "leaped for life … and sprang across the perilous gulf [between ship and wharf]. For a week she had been kidnapped on the great liner, either lost in its cavernous internals or cuddled in a corner of the stewards' quarters."[32] Without proof, suspicion fell on the Chinese crew while longshoremen received credit for the rescue.

From September 10, the *Asia* completed five return voyages to the Far East and one outbound voyage before war intervened. Her last regular outbound voyage departed Vancouver on July 9, 1914, with Robinson in charge. Just 11 days earlier, at Sarajevo, assassin bullets claimed the Austro-Hungarian Empire's Archduke Franz Ferdinand and his wife, Sophie.

After calling at Yokohama and coaling at Nagasaki, the *Asia* steamed southeast toward Manila, and while belabouring through heavy seas off Formosa (Taiwan), Robinson learned of a fatal stabbing. However, he does not state in his diary whether the victim was male, female, passenger, or crew. From Manila, the *Asia* ploughed through another storm before reaching Hong Kong where authorities took the accused into custody. It was, however, news of the pending war in Europe that frightened most, and on August 3, Robinson wrote: "Fine & bright. Office AM. Got orders come over to Dock Buoy [Hong Kong Harbour] and hand ship to Naval Authorities."[33]

The next day, after Germany ignored Britain's ultimatum to withdraw its forces from Belgium, a state of war was declared. Endorsement and commitment followed from the British Empire dominions of Canada and Newfoundland, and in British Columbia, the premier authorized spending more than $1 million on two submarines originally built for the Chilean navy.

Below and above the waterline, the 21-month-old rivets binding the *Asia*'s steel plates had to hold because, like her sister, she was about to become a warship.

3

STRIPPED FOR WAR

The beach is strewn with wreckage and lots of corpses in various stages of decomposition.

— **Assistant Executive Officer Herbert James**

Walter Oliver was four when the world went to war in the summer of 1914. He was living in the Southwestern Ontario farming community of Tillsonburg, after moving from Durham in northeastern England with his mother, Eliza, in August 1910. His father, Walter Oliver, Sr., who had worked as a market gardener, was there already, employed by the family that gave the town its name.

He was far too young to have memories of Tillsonburg or the transatlantic voyage to Canada, but impressions likely formed on the White Star Line's *Zeeland* bound for Liverpool in February 1915. With war underway, Walter Sr. had decided to move everyone back to Durham. This was risky because by then U-boats were taking a toll and Germany had declared the waters around the British Isles a war zone, making all merchant ships and passenger liners, including those from neutral countries, vulnerable to attack without warning.

Impressed by the size of the 11,905-ton *Zeeland*, Oliver was nearly five when he first glimpsed wartime troop movement on a ship carrying more than 300 soldiers, 65 nursing sisters, Belgian reservists, and hundreds of civilians, including 19 children younger than him. While the ship, soldiers, and the ocean captured his imagination, he was unaware that within a matter of days he would be bidding farewell to his 32-year-old father, who would be shot and gassed in 1916 while with the Durham Light Infantry.

In recalling those life-changing events, Oliver recognized that by 1915 the *Empress of Asia* would have been well into her war service.

○ ○ ○

With her outer decks and superstructure slick from rain and the purple-black coal dust, the *Empress of Asia* slipped into her berth at Hong Kong on July 29, 1914. The skies would eventually clear, but when Captain Samuel Robinson went ashore the next day, his disposition matched the cheerless grey sky. Hunched beneath his damp greatcoat, the ship's master was intent on lodging a complaint over the imminent seizure of his ship by the British Admiralty.[1]

Robinson's dark mood deteriorated further when Royal Navy captain Colpoys Walcott came aboard for a tête-à-tête. Walcott, who had just turned 36, had orders to take command of the ship the second she became an armed merchant cruiser. As a peace offering, Walcott invited the simmering Robinson to remain as navigation officer, and although desperately tough for any master to relinquish command, Robinson knew the *Asia* and oceans as well as anyone, and that alone made him an asset entering the war.

Summoned to Robinson's cabin that afternoon, Chief Officer Herbert James readily agreed when asked by Walcott if he would serve as the cruiser's assistant executive officer.[2] The British-born sailor was married with three little girls at home in Canada, the youngest, Barbara, under 16 months.

Like many sailors of his generation and older, James had experienced the quantum leap in ocean travel. "I told him [Walcott] most certainly," noted James, whose life as a deep-sea mariner began at 15. History benefited from his decision because the man kept an invaluable journal chronicling the *Asia*'s time as an armed merchant cruiser, including her pursuit of the German light cruiser *Emden* and the fatal, accidental ramming of a dhow in the Gulf of Aden.

Born at Barnet, Hertfordshire, now a suburb of London, on May 16, 1875, James entered Queen Elizabeth's Grammar School at age 11 where his teacher described him as "dull and backward. Conduct and industry — moderate." It is important to note, however, that the same teacher's dark sarcasm extended to 18 of the 30 students with remarks that included "very heavy & slow"; "speech impediment"; "volatile and unsteady."[3] Perhaps the problem lay more with the teacher than the class.

After leaving school in July 1890, James became the only one from his family to go to sea. His parents — William, a builder, and Mary — cared for seven children, although two died as infants. Herbert was their second youngest and made the most of his early sea time, travelling to the far reaches of the world before arriving at Vancouver in 1907 where he signed onto the *Empress of China* as fourth officer.[4]

Life was good along the south shore of English Bay in the middle-class neighbourhood of Kitsilano; however, sitting still was hardly an option for someone eager for adventure. Within months, at age 32, he switched to the SS *Monteagle* as third officer and then chief officer on the *Empress of India*.[5] A stickler for cleanliness and detail, James was always immaculately dressed, and as a naval reserve officer, he knew what made a good ship and crew. Furthermore, he was never timid about expressing his thoughts. This meticulousness opened doors, including the one to the *Asia*'s chief officer's cabin less than a year after the ship was completed. Even before Walcott had vetted him for a second time on August 3, James had been painstakingly noting happenings of the previous day — how dry-dock workers "commenced pulling things to pieces" and "dock people very busy working all last night without ceasing. The ship a perfect pandemonium."[6]

With the ship commissioned on August 3, gun platforms fore and aft were hastily riveted in place on the 5th, followed by eight old 4.7-inch guns two days later. The next morning, most of the ship's Chinese crew quit, not wishing to experience war first-hand — it would be another three years before China entered the fight against Germany. "They [the Chinese] do not seem to like the idea of going with us," noted James.

This development posed a serious problem, and not just for the engine room, which required approximately 170 men. To cover the shortage, a wide net was cast that brought an odd mix of Chinese crew, regular naval officers, engineers, Royal Naval reserves, 20 members of the Royal Garrison Artillery, and 25 members of the 40th Pathans.[7]

It was quite a sight on August 8 when the ship's company mustered on the afterdeck for the ceremonial hoisting of the White Ensign and pendant. Resplendent in traditional peaked cap and immaculate uniform with gleaming braid and polished medals, Walcott addressed his officers and crew, reading the king's Message to the Fleet followed by a reply sent from the commander of China Station. "All hands then sang the National Anthem (very flat) and gave three cheers for the King," noted James.[8]

Although she weighed anchor for some target practice, the warship's lack of qualified stokers remained problematic. "Experienced great difficulty in obtaining firemen in spite of offering much higher wages," wrote James on the 9th. "The Chinese not caring to take the risk, however by much strategy we managed to obtain a fair complement and proceed to sea at 9:10 p.m."[9]

Like James and Robinson, Walcott was a boy when he went to sea. Born at the old Thames-side village of Cookham, northeast of Maidenhead, he entered Eastman's Royal Naval Academy at age 12. Eastman's was a top private preparatory school for boys wishing to join the Royal Navy. In addition to rudimentary seamanship, the course exposed him to Latin, Greek, and English literature en route to cadetship in July 1892.

In the best navy tradition, Walcott's life remained on course. He joined the Royal Navy's prestigious Mediterranean Fleet, but in 1895 contracted scarlet fever at Salonika and required emergency hospitalization in Beirut. It was here that he earned the nickname "Trilby" because orderlies continually shaved the part of his head that only a Trilby hat, when stylishly worn, might cover. Then, when doctors discovered his weak heart, the young officer's naval career entered the doldrums for three months before he rejoined the fleet. Within seven years, Trilby was a full lieutenant, and by 1904 in command of HMS *Hunter*, a torpedo-boat destroyer built at the same yard as the *Asia*.[10]

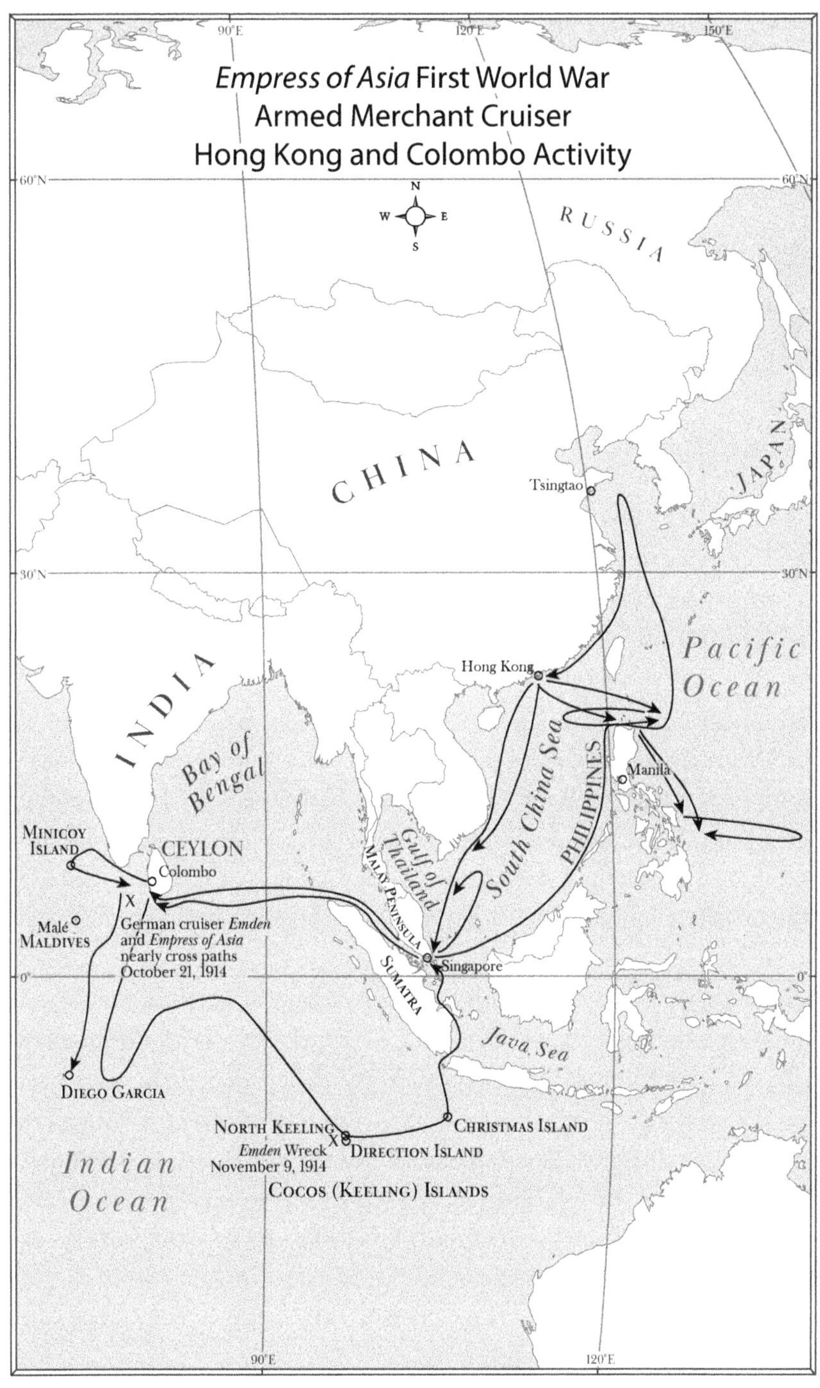
Empress of Asia First World War
Armed Merchant Cruiser
Hong Kong and Colombo Activity
RUSSIA
CHINA
JAPAN
INDIA
Tsingtao
Hong Kong
Pacific Ocean
Manila
PHILIPPINES
South China Sea
Gulf of Thailand
Bay of Bengal
MALAY PENINSULA
CEYLON
Colombo
MINICOY ISLAND
Malé
MALDIVES
German cruiser *Emden* and *Empress of Asia* nearly cross paths October 21, 1914
Singapore
SUMATRA
Java Sea
DIEGO GARCIA
NORTH KEELING
Emden Wreck
November 9, 1914
DIRECTION ISLAND
CHRISTMAS ISLAND
COCOS (KEELING) ISLANDS
Indian Ocean
90°E
120°E
150°E
60°N
30°N
0°

An incredibly rare photograph showing the *Asia* deploying her guns while serving as an armed merchant cruiser during the early days of the First World War in that capacity.

Walcott was a newlywed of 18 months when he took command of the repurposed passenger liner. Despite a mostly inexperienced crew and outdated guns, he put his ship through her paces, prowling for enemy ships, no matter the weather. As one of the navy's former top amateur boxers, Walcott was a fighter, but with the *Asia* he was being asked to punch above his weight.

The weather turned nasty on the night of August 11 when the ship encountered angry seas in the Balintang Channel between the Batanes and Babuyan Islands off the Philippines. In the galley, pots, pans, and dishes clattered and slid from their secure cupboards onto the floor, and there was the unmistakable smell of vomit and spoiled water gurgling in the drains.

As the barometer plummeted, the wind increased to hurricane force, producing mountainous waves. These briefly calmed in the storm's eye but were still choppy and unpredictable. The "wireless aerials came down, fortunately doing no damage, except smashing a table on the Bridge and a window in our bathroom," noted James.[11]

Robinson, whose oft-used words "nothing special" to describe most weather conditions, assigned significantly more weight to the monster hammering the ship's bow. It was a "howling gale" with "tremendous sea." Heading into the waves at 10 knots, it took half an hour to pass through the eye before encountering more violence on the other side.[12] Although this was a nightmare for the inexperienced crew, it was highly frustrating for old salts who understood how important it was to stay focused no matter the circumstances — alarming or mundane. "Many ... have never been to sea before ... and great difficulty being experienced in making the Pathans [Pashtuns] understand," noted James, who added "... almost driven crazy by the Chinese sailors, a lot of Rickshaw Coolies who know as much about a ship as they do about flying."[13]

Entering the archipelago southeast of Shanghai, *Asia* rendezvoused with the French cruiser *Dupleix*, the Royal Navy's *Triumph*, a cable ship, a collier, and two torpedo boats. Training their guns on appropriately named Barren Island (Hai Jiao) on August 15, the warships fired several practice rounds and likely killed or disturbed seabirds before valiantly steaming north into the Yellow Sea.[14]

Just 16 days earlier, Germany's Vice-Admiral Maximilian von Spee, commander of the East Asia Squadron, prudently moved his warships from the German-held port and leased territory at Tsingtao (Qingdao) into the Pacific to avoid blockade by Allied warships. His other intent, of course, was to sink as many Allied merchantmen and warships as possible while operating from Pagan Island in the Marianas before returning to Germany. That November, shortly before his squadron was largely annihilated by the Royal Navy, von Spee scored a major victory off Coronel, Chile.

These were the days before ship-borne radar, and not knowing exactly what naval assets the enemy had nearby, created a tense atmosphere aboard the *Asia*. "Darkened ship at night which is rather miserable, but at the same time very necessary," noted James.[15]

"Most uncomfy feeling," Robinson jotted in his diary. "With this poor coal ship [we] can only make 17 knots and German ships in Tsing Tau [*sic*] can make 24, 19, 17.... Seems like sending a camel out to do scouting, as everyone can see our huge bulk long before we could see them.... If any

German vessel is out and sees us unsupported & can call the other ships out we are trapped. Once again I say it is a very uncomfy feeling, especially after dark."[16]

Reporting nothing, Walcott received orders to return to Hong Kong. Incredibly, while the *Russia* had lost her grand piano during conversion, the *Asia* had kept the very same one Robinson had admired as it was swung aboard at Fairfield. So, with piano accompaniment that Sunday morning, hymns rose in the lounge during church service while the ship charted a southern course. This "sounds a bit off on account of all the furniture having been removed … but the Tommies and Seamen sing for all they are worth," noted James.[17]

While coaling at Hong Kong in preparation for the 1,460-nautical-mile voyage to Singapore, junior officers of the watch used a steam cutter to patrol around the ship in an effort to prevent more Chinese firemen from deserting. Then, on August 20, James wrote: "Proceeding towards Singapore, hoping to meet one of the enemy's armed merchant cruisers that are supposed to be in the vicinity."[18]

On the night of August 23, the crew of the Peninsular and Oriental Steam Navigation Company's *Arcadia* were no doubt startled when the *Asia*, running at some 20 knots without lights, stole up to the 6,603-ton cargo vessel before flooding her with searchlights. What the *Arcadia* lacked in speed she made up for in history, being the same ship that had carried Rudyard Kipling on his final return voyage from Aden to Brindisi, Italy.[19]

In the meantime, Japan had declared war on Germany. As the *Asia* made Singapore, the stealthy *Emden*, capable of 23.5 knots, passed through the Molucca Passage south of the Philippines and was soon in Sumatran waters, setting predatory eyes on the Indian Ocean. The next day, *Asia* commenced slow patrols through two nearby archipelagos — the Riau with its 1,796 islands scattered between Sumatra, the Malaya Peninsula, and Borneo, and the smaller Anambas Archipelago to the northeast. Both offered plenty of hiding places and so armed parties were dispatched in searing heat to search for enemy activity, usually returning with nothing more than mosquito bites, coconuts, and the canvas bags they had stuffed with sand for placement around the gun positions.

For a panoramic view above and around the jungle islands, the crew rigged an empty beer cask to the main mast well above the existing crow's nest. One slip, and the man climbing into it could plummet to his death, but it was worth the risk — if not for the man for the ship. Known as a "truck," the perch had two main disadvantages — it was out in the beating sun, and it smelled of stale ale with oak overtones. Still, it was cooler than the bridge and downright wintery compared to the stifling engine room.

Toward the end of August, following another man-overboard drill, Walcott and Robinson loosened their collars and settled into the humidity of the officers' mess to swap anecdotes and share a little about the important people in their lives. But foremost in both of their minds was the war and Japan's place in it, and whether that could lead to quicker peace.[20]

Fighting in Belgium, however, implied otherwise following the Battle of Mons, which had forced a British retreat not answered until early September. By then, the *Asia*, after annihilating or scaring more seabirds on another uninhabited island, had left the northern edge of the Anambas, heading south.

Back at Singapore, as the *Asia* coaled and Robinson opened letters from Jess, 23,000 Japanese troops landed 100 miles from Tsingtao in neutral China. It was the first of two waves of infantry, and in a matter of weeks, Japan knocked out and secured the German-held port and surrounding territory. The following evening, the *Asia* began heading east across the South China Sea to assist with the naval blockade of the Philippines.[21] An American colony possession, it had, like the United States, adopted a neutral stance. It was more interested in achieving independence than involving itself in a far-distant European war. But neutrality in the Philippines meant German merchant ships could continue to call and find safety from Allied warships. Indeed, a month before the *Asia* joined the blockade, nearly two dozen German ships were anchored there.

While anxious to fulfill the *Asia*'s next assignment, James, in typical fashion, was unimpressed by the ship's appearance after coaling, noting how she "proceeded to sea with no time to wash ship's side as usual, everything in a filthy mess."[22] It did not help that the air over the South China Sea was thick with dusty brown ash that Robinson attributed to a large brush fire, earthquake, or volcano. "Feeling rotten. Only can see about a couple of miles. Coast of Borneo, ten miles off but not visible."[23]

Spotting enemy ships and other suspicious movement through the haze was challenging and potentially reduced reaction time, so intelligence gleaned from communicating with or boarding friendly tankers and cargo vessels was vital, even if far from perfect. On one occasion, the boarding party learned about the heavy loss of a Canadian troopship in the Atlantic; in another, while off Manila Bay on the 7th, the *Asia* was informed by a Japanese vessel that there were 13 German merchant ships in the bay laden with coal or other supplies and one had recently left flying British colours.[24] Keeping the ships hemmed in was the top priority, even if it meant missing the excitement of pursuit and attack.

However, the crew still prayed for a scrap. "Cruising backwards and forwards off the Samar coast on the lookout for anything getting away from Manila, but no luck," noted James. "Rather exciting sometimes during the night when we sight a vessel, all hands at action stations, guns bearing on the stranger until we are satisfied she is not an enemy."[25]

A few days later as the *Asia* pulled alongside her sister ship, *Russia*, it was as though James was staring at a mirror image. "She did look an enormous vessel, a very fine sight, but the guns look like pop guns, they are lost almost, suppose we look almost exactly the same. Being painted grey from truck to waterline has made an awful difference in the appearance of us both," noted a wistful James. "She had captured a steamer … with a cargo of tea [supposed to be German goods] and has sent her to Hong Kong with a prize crew. One officer and two French bluejackets [enlisted men] sent here to be landed in Hong Kong for Hospital."[26]

From Hong Kong, the *Asia* commenced armed escort duties for a Colombo-bound convoy that would first call in Singapore. With Russia allied with Great Britain and France, the Russian-protected cruiser *Askold* ran point for the four troopships while the *Asia* brought up the rear. "One comfort is there is no cry of 'Darken Ship' this evening, which is nice," noted James.[27]

Providing armed escort for safe passage in and around Singapore and west to Colombo was essential not only for the transportation of arms and men but for the movement of oil, rubber, horses, and other supplies as well. Most of the international trade routes passed through the narrow Strait of Malacca between the Andaman Sea and Singapore. Off Ceylon, shipping traffic crossing the Indian Ocean converged in a tapestry of routes leading to and from ports, including oil depots along the Bay of Bengal, Southeast Asia, and beyond.

While the *Asia* was guarding the convoy off French Indochina en route to Singapore, the *Emden* was in the Bay of Bengal targeting and destroying more than just ships. On the 22nd, she shelled and set ablaze the large oil depot at Madras, India, and from there sank more ships before crossing south of Colombo and heading out to sea.

When not busy navigating waters strewn with islands and various hazards, Robinson satisfied his literary bent with back issues of the *Saturday Evening Post*.[28] With the embarkation of more troops complete, the convoy left Singapore on September 28. "Usual routine during the day. Convoy steering rather erratic and not keeping at all a good line," noted James on the 29th. "At 6 pm sighted a steamer's smoke, went to investigate, she proved to be a Frenchman, so our hopes were again dashed to the ground."[29]

Through the Strait of Malacca, then out across the Andaman to the Indian Ocean, the small convoy remained on high alert. James had sailed these waters before — in 1896 as third officer on the four-masted barque *Dowan Hill*. Calling at Colombo would refresh memories, and he was anxious to see how the old city had changed. Meanwhile, heat proved suffocating and relentless. Troops on the two transports were fortunate to find relief in the bow spray, a pleasure not afforded to those aboard the *Asia* due to her towering freeboard.[30]

After anchoring at Colombo on October 4, the ship turned to hunting duties with HMS *Hampshire*, a Devonshire-class armoured cruiser built 10 years before the *Asia*. On search-and-destroy patrols, they were to scour the ocean for enemy ships, including the notorious *Emden*. For the *Asia*, it was time to shine and hopefully play a role in destroying the ship that had back in August tried to hunt and kill the *Empress of Japan* "like a fox."[31]

Over the next two and a half weeks, the *Asia* would almost get her chance, although there was the quiet feeling among most that such an encounter would end badly for the modestly armed ship. For a week, she and the *Hampshire* patrolled off Colombo, then on October 13, they left for Diego Garcia, an atoll in the Chagos Archipelago far out in the Indian Ocean, seven degrees south of the equator. In his diary that night, James noted: "If we can spot her [the *Emden*] we shall most certainly engage her, and our Skipper says he will fight to a finish. If so we should most probably either burn up or get potted by a torpedo."[32]

After reaching the mouth of the atoll's blue lagoon on the 15th, the *Asia* dispatched her steamer and whaler with an armed landing party, led by naval intelligence officer Charles Montagu Scott and watchkeeper George Goold. Their mission: gather intelligence from a sparse human population, vastly outnumbered by coconut crabs and tiny lizards. As the party headed to shore, the men peered into shallow, crystal-clear water teeming with colourful reef fish and outcrops of coral.

Standing in the steamer, Scott admired the view. He was a man of the world. Born at Bombay, the son of a Royal Navy commander, he spent some of his childhood in England where he became a cadet, a midshipman on HMS *Revenge* at 19, and a lieutenant on HMS *Fantome* in 1911, assigned to survey work off Australia. He had fallen in love and married at Kowloon seven months before joining the *Asia*.

Goold, born at Broxbourne, England, had earned his first mate's certificate and joined the CPR Empress fleet in 1913, starting out as fourth officer on the *Empress of Japan*. During the 1930s, Goold would skipper both the *Russia* and *Asia* before assuming command of the *Empress of Canada*, sunk by torpedoes off Africa in March 1943.[33]

While the *Hampshire* withdrew to conduct another patrol, the *Asia* steamed around the island paradise and its blue rectangular lagoon. However, for the crew, there was not much to see except coral, sand, palm trees, pristine water, and the colossal crabs that when not engaged in climbing trees or cracking coconuts could strip any carcass within days.

Disappointingly, "at 9 pm, Scott and his landing party returned with the galling news that the *Emden* had called there ... with a collier and both had left on Saturday [five days earlier]," wrote James furiously. "The European [inhabitants] had no idea we were at war and had treated the Germans as well as possible which the Captain was good enough to acknowledge by a letter which the manager of the island gave to Scott."[34] From all reports, Captain Karl von Müller cleverly convinced islanders that his ship was part of a world-united fleet participating in manoeuvres.

As her crew shook their heads in disbelief, the *Asia* sailed north, but before catching up to the *Hampshire* on the 17th, Chinese stoker Wong Wa fell ill and died of beriberi within days. Wong, who was from rural China, was the first of several to succumb to the non-infectious but debilitating disease while the ship served as an armed merchant cruiser, and his body was committed to the deep. The history of the disease shows it is triggered by a lack of thiamine, or vitamin B1, in the diet, and it was especially prevalent among Japan's urban upper class and within its military. When brown rice is milled and polished into white rice, it becomes lighter and does not rot as quickly. However, the process also removes much of the thiamine, which had dire consequences for the Japanese military during the 1904–05 Russo-Japanese War when 27,000 died from the condition. Mixing the popular staple with barley, meat, or vegetables reduced cases, but most of the Chinese crew on the *Asia* preferred polished rice.

It was a terrible year for beriberi on the ship. Between the start of her commissioning in August and December 31, 38 cases occurred, five fatal. Of the 38, 34 were Chinese stokers, who comprised more than half the ship's company.[35] Symptoms among severe cases included extreme chest discomfort, edema, muscle weakness, and a rapid heart rate. Outbreaks occurred on other ships, including the *Empresses of Japan* and *Russia* and the Royal Indian Marine ship *Northbrook*.

On the 18th, the *Asia* rejoined HMS *Hampshire* to resume patrols. Three days later, on the day set aside each year to commemorate the Royal Navy's victory at the Battle of Trafalgar, both ships were running parallel to each other west of Colombo when it was later thought the *Asia* had come within only 10 miles of the *Emden*.[36] Robinson and James did not state this in their journals, but *Asia* gunner Edgar Mole noted the following in his diary on October 22, which suggests a much closer encounter: "A light was sighted ahead. We thought that it must be the *Emden*. This time we chased her for an hour and then lost her in the fog."[37] Bitterly, James later observed: “Instead of hearing of some glorious victory to commemorate the day, we hear that the ‘Emden’ has sunk six more merchantmen and captured a Government collier.”[38]

The *Emden*, which was heading east while the *Asia* steamed west, crossed into the Andaman Sea and on the 28th sank the Russian-protected cruiser *Zhemchug* undergoing maintenance and repairs at Penang. Eighty-one Russian sailors died, with another 129 wounded, some fatally. Only days before, the *Asia* had been patrolling the Maldives, another tropical paradise, and to seasoned sailors like James and Robinson it seemed more like a scientific rather than a naval mission in search of the *Emden*. “Still cruising about,” noted James, “the effect of the brilliant sun upon water covering coral reefs is simply splendid and seen from aloft, a sight never to be forgotten. From the deep blue of the Indian Ocean through all shades of blue, green, and yellow to the white surf breaking on the reefs and beach, all kinds

Gunners pose for a photo aboard the *Empress of Asia*, not long after their ship was requisitioned as an armed merchant cruiser for service in the South China Sea and Indian Ocean.

of weird fish can be seen many fathoms down owing to the absolute clearness of the water."[39] Not clear was the location of the *Emden*.

After coaling and maintenance at Colombo, the *Asia* resumed patrols; however, action stations blaring at 2:00 a.m., November 4, had James leaping from his bunk certain they had finally caught the elusive German warship. "We went for her full speed, everything ready and all anxious for the order from the Bridge to fire but alas she was a B.I. [British India Navigation Company] Steamer with no lights, and told us she was very glad when we told her who we were, as she had made up her mind we were a German cruiser."[40]

Incredibly, three nights later, the *Asia* found herself pursued not by enemy but by the *Askold*. "Shortly after midnight received wireless from 'Askold' to the effect she was chasing a suspicious looking vessel, we turned round and soon made out a cruiser which proved to be the 'Askold,'" noted James on November 7.[41]

Finally, it was His Majesty's Australian Ship *Sydney* that intercepted and destroyed the *Emden* on November 9 — a well-told event that includes the *Emden*'s last moments and the daring escape of survivors, including her captain. Less known are the supporting roles of the *Asia* and *Russia*.

On the 10th, as the *Asia* was off the southern tip of India with the *Russia* and *Hampshire*, she received a wireless report concerning the whereabouts of the *Emden* and ordering their immediate assistance. Steaming south at full speed toward Cocos Islands, the ships met a massive convoy transporting 25,000 Australian troops, horses, and stores bound for Gallipoli. The *Sydney* had been part of that convoy's escort force until HMAS *Melbourne*, a Town-class light cruiser, received a distress signal from Direction Island whereupon *Sydney* turned to investigate.

Four days after the *Emden*'s demise, the *Russia* came alongside the *Sydney* and relieved her of 230 prisoners while *Asia* sent a party ashore on South Keeling Island to search for anyone who had escaped the stricken warship. "At 7 a.m. passed quite close to the wreck of the 'Emden' as the beach is quite steep there," noted James on November 14. "She is on the south end of N. Cocos Is., very badly battered, foremast carried away, two of her [three] funnels knocked out and the afterend … burnt completely out. A number of dead are still lying on the forecastle head and amidships. At 9 a.m. we arrived off S. Keeling Is. and Scott landed and obtained an account … from the man in charge of the Eastern Telegraph Cable Station."

James noted a landing party from the *Emden*, composed of three officers and 44 men, had commandeered the *Ayesha*, a local schooner, "towing her with their steam pinnacle." Counting himself among those who could appreciate von Müller's achievements, James wrote: "The Captain of the 'Emden' is saved and we salute him and his crew as having played the game throughout."[42]

The *Asia*'s crew witnessed the grisly aftermath for themselves on November 16. "During the afternoon landed a party to search the island, they returned at 5 p.m.," wrote James. "The beach is strewn with wreckage

The tangled, burnt-out remains of the German light cruiser *Emden* was witnessed by the crew of the *Empress of Asia*.

and lots of corpses in various stages of decomposition. The land crabs will soon leave nothing but the bones. The dead were evidently thrown overboard and are now being washed ashore."[43]

Gunner Edgar Mole had joined the landing party:

> While we were busy cleaning up the guns, the captain shouted down from the bridge for the Gunners to get ready with rifle, bayonet and ammunition to go ashore and search Cocos Island as he thought some of the Germans had escaped and were hiding there.… It was an awful job getting through the jungle. The under-growth was so thick and the ground strewn with coconuts. We opened several with our bayonets and had a drink.… After a couple of hours struggling through the jungle we came into the open and before we knew [it] we were surrounded by a flock of fierce-looking birds more like young eagles.… We had to keep them at bay with our rifles. I think they had never seen a human being before.… The mast of the *Emden* was sighted so we made for it as best we could. After we got clear of the trees we could see the wreck quite plainly.… As we were walking along the beach we came across several dead bodies that had been washed off the wreck. Some must have gotten ashore alive as we came across a shelter made up of blankets and leaves, there must have been some wounded as there were a lot of bandages.[44]

Mole's diary also notes that the party could not board the *Emden* "on account of the waves breaking over her … and the sea full of sharks."[45]

While still looking for the *Ayesha*, the *Asia* steamed northeast to Christmas Island, then north into the narrow Sunda Strait between Java and Sumatra where she passed the infamous Krakatoa caldera. Continuing northwest, the voyage brought a smile to James as he recalled a moment in 1896 when he and a shipmate on the *Dowan Hill* jokingly christened their skipper "Deadlight Dick, terror of the Java Sea."

Seventeen miles into Bangka Strait, the ship encountered a wall of smoke from wood fires that kept her whistle blowing for more than an hour. Reaching Singapore on November 23, the armed merchant cruiser had logged 25,841 nautical miles since joining the Royal Navy. She had not yet claimed any kills or prizes but had at least been in on the hunt for the *Emden*. Like many warships, her steady, monotonous patrols had helped keep ships safe during a very dangerous time.

4

GUNS AND SHUTTLECOCKS

We made a clean job of it, cut her right in two, and the bow and stern ends both floated.

— Assistant Executive Officer Herbert James

Yet again, Assistant Executive Officer Herbert James was not pleased with the *Asia*'s appearance at 5:00 a.m., November 27. Four straight days of intense coaling in Singapore had produced swirls of grimy black dust that penetrated everything from nostrils to steel. It had smeared her top and sides and even crept into the cracks between the wooden deck planks. If Major Kersey, the man who had overseen her construction, had been alongside, he likely would have turned away in disgust.

After receiving a heads-up, James quickly ordered all hands to scrub the decks and spray down the superstructure, but even four and a half hours later the *Asia* was, he noted, "far from shipshape — according to my idea" when Vice-Admiral Sir Thomas Henry Martyn Jerram and his staff came aboard. After "a very nice speech," the vice-admiral complimented Robinson

Assistant Executive Officer Herbert James of the *Empress of Asia* was an experienced sailor and a stickler for detail.

on his "zeal and efficiency" and then "complimented me upon the trimness and cleanliness of the ship."[1]

With recent patrol areas now relatively safe from enemy depredations, the *Asia*'s sailors feared they could be "paid off," but that was prior to the loading of more coal and four months of stores. Rumours were they were heading to the Red Sea to fight the Turks.

If familiarity breeds contempt, especially among men who live weeks on end in close quarters, then it likely sparked the cryptic comment from James when Robinson received a promotion. "Robinson is promoted to the rank of Commander, R.N.R. [Royal Naval Reserve], despite several very broad hints, has not yet done the right thing."[2] James's comment points to a serious lack of trust, but it does not identify the cause, and his expression of animosity was limited to his journal, because openly criticizing a fellow officer or member of the crew went against code.

Frustrations aside, it was relatively smooth sailing to Colombo where, on December 4, 1914, she embarked for Aden, a tiny British protectorate on the Arabian Peninsula near the Red Sea. Squalls and ocean spray removed some of the recent coaling filth, which pleased James, and overall, it was clear blue skies, starry moonlit nights with flying fish, and a wake of bioluminescence.

In the shimmering heat across the Arabian Sea, objects near and far lost definition while physical exertion required appreciatively more effort. For recreation, badminton on the afterdeck proved popular, but only late in the afternoon and played with wooden bats, not rackets, produced by the ship's obliging carpenter. For James, the bats made "for a much faster and better game," although wind and sea claimed shuttlecocks, later replaced by celluloid Ping-Pong balls that also bounced and sailed into the ocean.

On December 9, the *Asia* pulled into the scorching oven of Aden where she joined Britain's auxiliary cruiser squadron. For the British, the enemy was the Ottoman Empire that had for economic and strategic purposes long since conquered and occupied a strip of inhospitable desert along the west side of the Arabian Peninsula. But in November, the Turks had entered the war as a German ally, and with that came an immediate threat to Aden from Turkish forces in the province to the north — Yemen.

The Ottoman Empire's population in 1914 was 25 million, including 10 million Turks and six million Arabs, although the empire itself was in decline. What had made Aden important to the British was its location close to the eastern entrance to the Red Sea. It provided a strategic, though undermanned, military base for protecting British interests and cultivating influence among Arab sultanates. Protected by a thin isthmus, the seaport was also a vital coaling and water station for steamships.

While other provinces bordering the Red Sea had ceased or not acknowledged allegiance to the Ottoman Empire, Yemen was an outlier. To the north was the province of Asir where the native chief, Muhammad ibn Ali al Idrisi, declared himself an enemy of the Ottomans. Wedged between the protectorate and Idrisi's followers, the Turks had trouble replenishing weapons and ammunition via the desert, so they relied heavily on dhows and other vessels under sail on the Red Sea.

For the Allies, though, there was more at stake than securing the protectorate and supplying arms to Idrisi. The longer game was to formulate wartime agreements that would lead to the partitioning of the Ottoman Empire. Consequently, British policy grew in Arabia to include the signing of an alliance treatise with sultans and Arab leaders aimed at encouraging or assisting Arab insurrection against the Turks, a strategy that helped launch the Arab Revolt of 1916 and involved, rather heavily, a fellow named T.E. Lawrence.

Meanwhile, the delivery of Allied war supplies, soldiers, and other goods between Australia, New Zealand, the Far East, Europe, and into Africa depended on secure travel through the Gulf of Aden, the Red Sea, and the Suez Canal, the last being under British control and thus closed to the enemy. Oil, of course, was especially vital to Britain's war and economic interests. With oil-fired ships and its army becoming more mechanized, there was growing interest on Persian oilfields developed by the Anglo-Persian Oil Company.[3]

At the southern end of the Red Sea in the narrow strait between it and the Gulf of Aden is Perim Island, strategically important to any force intent on controlling maritime traffic to and from the Red Sea. A couple of miles from the eastern mainland, the tiny volcanic island was not far from the village of Sheikh Said and a fort that could easily fire on vessels. Turkey, which had invaded Yemen in 1872, occupied these mainland positions on the edge of the loosely defined British protectorate, and not long after entering the war sent additional troops, including Arabs into the area, further alarming Allied interests.

A month before the *Asia* reached Aden, a large Allied troop convoy, bound for France, landed Indian soldiers and deployed naval guns to destroy the fort. While roughly 400 Arab and Turkish defenders fled across the desert on horses, at least six died in the attack. Allied losses were four killed, 15 wounded.[4]

What followed for the *Asia* was the first part of a 10-month assignment entailing monotonous patrols through baking heat with a frustrated crew

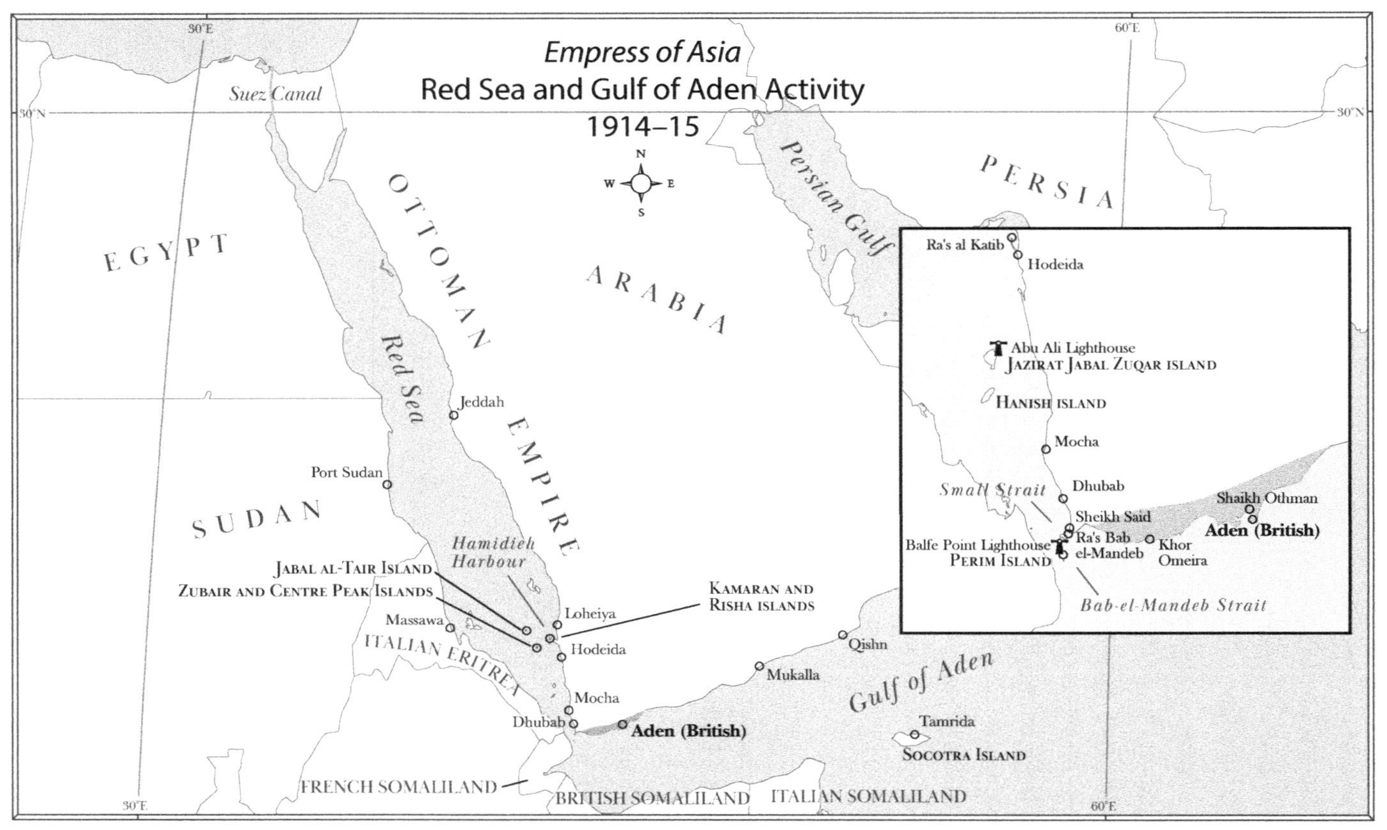
Empress of Asia
Red Sea and Gulf of Aden Activity
1914–15
N
W
E
S
30°E
60°E
30°N
Suez Canal
EGYPT
OTTOMAN EMPIRE
ARABIA
Persian Gulf
PERSIA
Red Sea
Jeddah
Port Sudan
SUDAN
Hamidieh Harbour
JABAL AL-TAIR ISLAND
ZUBAIR AND CENTRE PEAK ISLANDS
Massawa
Loheiya
KAMARAN AND RISHA ISLANDS
Hodeida
ITALIAN ERITREA
Mocha
Dhubab
Aden (British)
Mukalla
Qishn
Gulf of Aden
Tamrida
SOCOTRA ISLAND
FRENCH SOMALILAND
BRITISH SOMALILAND
ITALIAN SOMALILAND
Ra's al Katib
Hodeida
Abu Ali Lighthouse
JAZIRAT JABAL ZUQAR ISLAND
HANISH ISLAND
Mocha
Dhubab
Small Strait
Sheikh Said
Shaikh Othman
Aden (British)
Balfe Point Lighthouse
PERIM ISLAND
Ra's Bab el-Mandeb
Khor Omeira
Bab-el-Mandeb Strait

whose main wish was to engage the enemy with firepower. For the longest time, it seemed as though the armed merchant cruiser ran on testosterone as well as coal but was repeatedly fated for long patrols without firing a single shot in anger.

Aden, though, remained the hub of operations for Allied warships, including two other Empresses, the *Russia* and *Japan*. Along with other warships, they worked to prevent dhows from smuggling weapons or supplies into Ottoman-held territory north of Aden, which James described as a "God forsaken place," but one he would like to visit. In addition to boarding and firing on dhows, the squadron bombarded enemy shore positions and blockaded ports.

However, it was news from away that sparked celebration on December 10, 1914, when the *Asia*'s crew learned how most of Germany's East Asia Squadron had met its demise on December 8 in the Falkland Islands. Two days later, with her crew covetous of the actions of other warships, the *Asia* embarked on yet another Red Sea patrol. This did not begin well. "Rotten job getting away [from Aden] on account of stern swinging round with wind into the bank while getting anchor on board," grumbled Robinson.[5] Then, while attempting to hoist her steam cutter on board, 25-year-old Chinese stoker Lee Tuk lost his balance and plunged into the water where he drowned. Walcott deployed a search party, which turned up nothing. However, the fellow's name now appears on a Hong Kong war memorial.

While Turkish forces assembled in Yemen, there were sketchy reports of a planned amphibious assault on Perim Island. This was the next port of call for the *Asia*, but due to her troubles at Aden, she did not reach the island until early evening on December 1.

After transferring bags of mail to the *Russia*, she returned to Aden where the ship's company had a little fun at the expense of another Allied warship. With her officers assembling in the wardroom, the captain delivered a tongue-in-cheek speech before pinning a tin cross with two bars inscribed "Askold" and "Aden" on Charlie Scott's inflated chest. "The joke is that the Russian cruiser 'Askold' nearly fired on us on the night of November 7th, but the 'Empress of Russia' and another of our ships … were both fired on while entering Aden though not displaying the correct recognition signals," noted James. "As Scott is in charge of [our] signalling staff we considered he saved the ship & so is suitably decorated."[6]

Anchored in the relentless heat off Perim on December 30, the *Asia* welcomed aboard 42 ratings from the *Russia*, including Royal Navy Reserve and Royal Navy Volunteer Reserve men, replacements for the Pathans and British gunners who had signed on in August. Escorted on board hours later were several Arab and Turkish prisoners of war who, noted James, "seemed very cheerful" when landed at Aden the next day.

Less than a month later, James was promoted to "Acting No. 1," second in command to Walcott, replacing First Lieutenant Q.B. Preston-Thomas, Royal Navy, who had dislocated a leg. The next weeks were a monotony of patrols, armed boardings (with little to show for it), and crucial water deliveries to personnel guarding lighthouses on Abu-Ail, Centre Peak, and Jabal al-Tair. Hauling kerosene tins of water uphill on rugged, sun-scorched tracks through oppressive heat left men sunburned, exhausted, and dehydrated with blistering lips. Worse, the thought of becoming a climate casualty rather than a combat casualty did not sit well with some.

Coaling at Port Sudan in late January, there was "enough breeze to lift the coal dust and put it over everything," noted a disgruntled James, whose quick thinking three days later averted disaster at Perim. An underwater wreck at the entrance to the small harbour caused a dangerous riptide, and the *Asia* became "unmanageable," recorded Walcott in a January 30 report to the commander-in-chief, East Indies. "The reversing of the Port propeller had no effects, and the Ship was only saved from grounding by the smart manner in which Lieut. James, R.N.R., let go both anchors, the ship being finally brought up within 100 feet of the rocks on Lee Point."[7] This quick thinking earned him a mention in Dispatches.

Routine replaced excitement, though on February 2, James noted: "Sent a boat off to board a dhow, but nothing happened. We do not seem to have any luck, but suppose we should be thankful that we are alive and enjoying comparative comfort & luxury, far different to the fellows in Destroyers in the North Sea. Here the weather is perfect, just cool enough to keep us in whites [uniforms]."[8] However, his tone quickly reverted the next day. "Still cruising about the Red Sea, wish we could get near the fighting, as everyone is getting rather fed up with acting as a sort of Police Patrol."[9]

It was galling for all aboard when the *Asia* delivered mail to her sister ship and learned she had "fired a few shots in anger" at the enemy along the shore. Even worse, when on March 8, the *Asia* collided with a dhow shortly after commencing a patrol from Aden: "I awoke at 3:45 a.m. to the signal of 'Away Sea Boat' to find we had run down a Dhow. We lowered boats and picked up all the survivors, 21, stayed around for an hour and then proceeded. We made a clean job of it, cut her right in two, and the bow and stern ends both floated. I am sorry to say that the survivors claim that 13 were drowned."[10]

While reports of the lost men and cargo piled up, the *Asia* welcomed a new master — Captain Philip Howard Colomb, the son of a retired vice-admiral, who arrived in mid-March.[11] As the Red Sea patrols continued, a week later,

Although time has severely damaged this rare photo, it shows some of the crew posing for posterity aboard the *Asia* while the ship was engaged in the Red Sea.

several Arab chiefs and envoys joined the crew en route to Kamaran, bringing with them a treaty guaranteeing Idrisi and his followers independence as long as they assisted Britain against the Turks. The days of monotony abruptly ended on April 10, 1915, when the *Asia* finally fired on the enemy.

Colomb was in pursuit of a suspicious dhow running toward shore near Kamaran Island. From 25,500 feet, a shell pierced the air but missed the dhow; neither did it slow her down. Another, from 24,000 feet, sailed past and went unheeded before a third missile landed close by causing the dhow's intimidated crew to lower sails. The *Asia* raced in and her armed, adrenalin-fuelled boarding party, led by the courageous Scott, took control and triumphantly towed the dhow alongside the warship like a dead whale. "We took the crew on board as prisoners and took some of her cargo out," noted James. "After being eight months in commission we have at last fired a shot in anger, as the last round fired, the Captain gave the order to hit her, but she presented a very small target being end on."[12]

Philip Howard Colomb, Sr., had served during the China Expedition and as a commander in the Crimean War. After marrying Ellen Bourne, a British subject from New Brunswick, Canada, he fathered seven sons, all heavily influenced by his strict military discipline. At least five boys entered military service, including Philip Jr., who was packed off to a naval cadet training ship at 13 where he mixed with 126 cadets of similar age. While the boy impressed Colomb Sr. by earning top marks, the old commander was more than likely annoyed at reading the comments. "Contentious and disposed to question orders," wrote one commanding officer. "Zealous and painstaking. Impatient of control. Very Good Organizer."[13]

Entering manhood, Colomb Jr. progressed from lieutenant to captain of HMS *Encounter*, then flag captain on HMS *Pembroke* before coming aboard the *Asia* where, noted James, he imparted "a few words of comfort and joy" for the ship's tired and bored company pestered by clouds of mosquitoes.

Unsurprisingly, malaria found Sub-Lieutenant George Goold, who despite "looking very seedy," rejoined the ship after being "frightfully disgusted" by the "very crude" hospital at Aden. In mid-April, as the *Asia* weighed anchor for Bombay, James also found himself on the sick list, sweating profusely and doing everything to avoid hospitalization while dosing himself with the pain-relieving, fever-reducing drug Phenacetin. A much-needed ship overhaul followed at Bombay where she would remain for two months before returning to Aden on June 29 for the second part of her Red Sea mission. The situation, however, was not good.[14]

Two weeks earlier, the Turks had deployed at least a dozen dhows to attack Perim, but the island's garrison easily thwarted the assault, which saw only a few unlucky men land.[15] In the interim, Aden had become even more vulnerable when the *Asia* steamed into port, anxious to assist with her big guns. However, instead of remaining at the seaport, she deployed to Perim for added defence while the Turks were getting ready to attack Aden. This lack of knowledgeable command and coordination in boosting the defences of the protectorate would also lead to weakened support from important Arab allies. In addition, the Sepoy Mutiny, which had originated in Rangoon and Singapore, had now reached Aden, and while arrests followed, the Sepoys rioted and fired on their officers.[16]

Two months later in early July, the Turks began their advance toward the Sultanate of Lahej (Lahij) and Shaikh Othman (Shaykh Uthman), the latter being the vital fresh headwater source that sustained the protectorate. British ground forces, however, failed to reinforce both locations, which came under attack on the 4th and were abandoned the next day.[17] In addition to combat casualties, the oppressive heat claimed British soldiers through serious cases of heatstroke and dehydration. It was a massive logistical nightmare that left troops without water, food, or other crucial supplies.[18]

As for the *Asia*, she was ordered to leave Perim and proceed to Aden at full speed. While Lahej was 18 miles north of the seaport, Shaikh Othman was only five miles away. Along with other Allied ships, the *Asia* took up defensive positions close to shore where her guns became an additional deterrent to any thoughts Turkish forces might have had of crossing the narrow strip of land leading to the seaport.

Meanwhile, Herbert James's temperature had shot up. He was losing weight, down to 170 pounds after contracting erysipelas, a bacterial infection accompanied by a bright red, raised, and well-defined rash on his face. It was rough, but he forced himself out of his bunk, got dressed, and tried to hold down food while focusing on the ship's supporting role in a crucial operation. "I have had a lovely face. My ears have been like those of an elephant … however it has gone on very nicely."[19]

Referencing the seaport's defence, James noted: "All available troops are out, holding them back and the 'Russia' and we are on the eastern side of the Peninsula to assist with shell fire if they get close.… We do hope they will … make a dash for it so we can drop a few 4.7 lyddite [a high explosive containing picric acid] among them. So far our troops seem to be holding them in check a few miles out."[20]

On July 18, Robinson took charge of a party that coordinated the movement of several thousand reinforcements recently arrived from India. As the men flooded into Aden on four transport ships, Robinson and other sailors worked to load soldiers, horses, mules, ammunition, and field guns onto lighters bound for shore. "They evidently mean business now," noted James soon after a sandstorm blasted the port. "The ship looks almost khaki today with all the sand, as fine as flour, simply everywhere.…"[21]

Landing east of Aden two days later, the infantry waited until the next morning before pushing north to Shaikh Othman. Offshore, the armed merchant cruisers, with no lights showing, stood ready with their guns. "It was quite hazy owing to a fresh breeze, and the sand blowing so when daylight arrived we could hardly see the houses at Shaikh Othman," scribbled James in his diary.[22]

By 6:00 a.m. the following day, the dust had settled, and from the *Asia*'s bridge it appeared the Allied forces on the attack's right flank had not only secured their objective but had also rejoined the main thrust. Another entry reads: "At 8 a.m. we heard the reports from the H.A.C. [Honourable Artillery Company] and R.G.A. [Royal Garrison Artillery] batteries. This continued up to 11 a.m. [and] then all was quiet. Of course, it was not clear enough for us to see any of the other troops."[23]

At 4:00 p.m., the *Asia*'s crew learned the attack had driven the enemy back, causing heavy losses, even beyond Shaikh Othman, but for now the settlement, its supplies of fresh water, and Aden were secure. All ranks of British casualties numbered more than two dozen while between 50 and 60 Turks were killed or wounded.[24] While James knew men had died, he wrote that the attack's success "was very good news, but at the same time we were very sick at not having taken any active part in the operations. One of our signalmen landed with a party from the 'Empress of Russia' [and] is now a hero on the lower deck being the only member of our crew who has so far been in action."[25] In that adrenalin-rich moment, it was easier for James and others to gripe than to appreciate the contribution they had made, even without firing a single shell.

Both the *Asia* and *Russia* returned to Aden for coaling, but the dirty, laborious task lasted longer than usual because military authorities had commandeered many of the labourers assigned to the work. On the plus side, there was a congratulatory message from the general officer commanding for splendid work and co-operation in landing troops. Robinson, however, was still restless. In his view, the ship was burning valuable coal without anything to show for it. "Messing about not seeming to get anywhere or do anything. Must really start to do something definite — seem to have lost interest in things."[26]

Several Pashtun soldiers line the deck of the *Asia* during deployment off Aden.

Even if cultures did not tend to mix naturally, they travelled closely together in late July when the *Asia* carried 350 Indian soldiers and eight British officers from Aden to Perim before returning with 250 soldiers and an officer.

This temporary troopship duty spawned more complaining among the crew; however, it was something else entirely that raised Robinson's ire as the ship approached anchorage at Aden. "Arr[ived]. just before dark," he jotted in an angry hand. "For the first time in my life I think I had a job taken out of my hands for no reason by Colomb. Arranged to swing ship to anchor but found berth too small so was going to turn ship without it when he took it out of my hands. A dirty mean trick in the way it was done, making me look small before everyone."[27]

That evening, the ship called again at Kamaran where the air was fresher, but the sea registered 90 degrees Fahrenheit. On July 30, orders were received to be on the lookout for a Dutch steamer, the *Djebres*, reportedly loaded with an Austrian cargo, "which we will confiscate if we can bag her," speculated James. "It has been simply boiling today." The *Asia* caught up to the vessel the next day, and beneath the cruiser's threatening guns, the *Djebres* had no choice but to surrender. "As we have not communicated with her I bet she feels a bit sick to see us clinging round her stern like a bulldog after a toy terrier."[28] Two days later, any promise the "bulldog" had of steaming into Aden with her prize evaporated when she lost the honour to another warship, which freed her up for more patrols and "dhow snatching."[29]

"Hotter than two Hells & that's not swearing, it's plain fact," grumbled Robinson on August 2, while the ship chased dhows along the west coast of the Red Sea, north of Jabal al-Tair.[30]

In late August, there was hope the *Asia* would offer naval support to Idrisi and his followers who were planning an advance on Loheiya. Colomb, who had joined the HMS *Minto* for a closer look at the coastal town of Medi, wanted to consult the chief, but when he returned to the *Asia,* he announced the ship would weigh anchor for Aden. Intelligence from Scott, and confirmed by Colomb, estimated enemy strength at Loheiya at 500 Turkish, 2,000 Arabs, and two German officers. In his journal, James presumed Idrisi was not ready to attack the city. "All our hopes dashed to the ground again," he lamented.[31]

The blazing heat, though, remained the number one enemy, and while those in the engine room dripped sweat, grew more ugly blisters, and lost weight, there was no relief. Within a month, the temperature in James's cabin hit 97 degrees Fahrenheit. "The weather is simply stifling and we are all suffering from prickly heat, and many people have been greatly troubled with boils.... I am glad to say my weight is keeping down. I still only weigh 170 pounds, but what the dickens will my blue uniforms, and shore togs look like on me?"[32]

Naval intelligence officer Scott also felt the heat but continued to earn more than the humorous gongs bestowed on him. On September 18, he left the *Asia* to command the 2,198-ton armed boarding steamer HMS *Lama*. Measuring 276 feet long, the *Lama* was small, but equal to the task. "She looks a nice little ship, and far more suited for this patrol than these enormous ships," observed James with mounting envy as she took on ammunition and stores. "She has three 4.7-inch guns and two maxims [machine-guns]."[33]

On October 15, with the temperature rising to 95 degrees Fahrenheit in the shade, the *Asia* left Aden for Bombay. While the ship sailed east, Scott and his crew were busy in the Red Sea where they encountered a dhow that fired on the *Lama*'s steam cutter and whaler crews, killing one sailor and wounding another.[34]

Bombay marked 63,604 nautical miles since the *Asia*'s wartime commissioning at Hong Kong. Joining her on an officer swap was the very familiar L.D. Douglas, who had travelled to Liverpool with Robinson in 1913 and would assume command of the *Asia* in 1921.

Although not suited for armed patrols on the Red Sea, mainly due to her enormous size, deep draft, and large turning circle, the *Asia* had given her best. Dismantled and landed on the wharf at Aden before she left were her eight large guns. After paying off surplus personnel, a skeleton crew with Colomb in command and Robinson as navigator embarked for Hong Kong for a major refit under the supervision of Captain Edward Beetham, the CPR's marine superintendent.

James, who left the ship at Bombay on October 21 with the rank of first lieutenant, joined the armed boarding steamer HMS *Suva*, towing supply barges into Basra in the northwestern Persian Gulf. He then became port intelligence officer at Colombo, continuing in that capacity while his wife, Ethel, raised their daughters, Marjorie, Doris, and Barbara, in Vancouver. "He was a stern, formal person, and always immaculately dressed," recalled his grandson, Robert Gibson. "He had a sense of humour, but he never had a button undone."[35]

5

ATLANTIC AND PACIFIC

... we climbed on board and were led down many stairways till I was sure we were actually near the bottom of the sea.

— **Chinese labourer Li**

In April 1916, after being seared in the Red Sea, the *Asia* returned to North Pacific passenger service under the command of Captain Arthur Wellesley Davison, formerly of the *Russia*, who exchanged ships with Robinson on January 29. By then, the ocean routes between North America and the Far East were considerably safer thanks to these two ships, the *Empress of Japan*, and many other vessels that worked to achieve this, often at a terrible cost.

Six hundred and thirteen passengers of all ages arrived in Victoria, British Columbia, aboard the *Asia* on May 6, 1916. This crossing marked her first inbound voyage since her armed merchant cruiser days, and the list boasted 308 first class, 34 second, and 271 steerage. As the war ground on over the next several months, the *Asia* resumed her regular mail and silk runs with manifests full of tourists, missionaries, business executives, and diplomats.

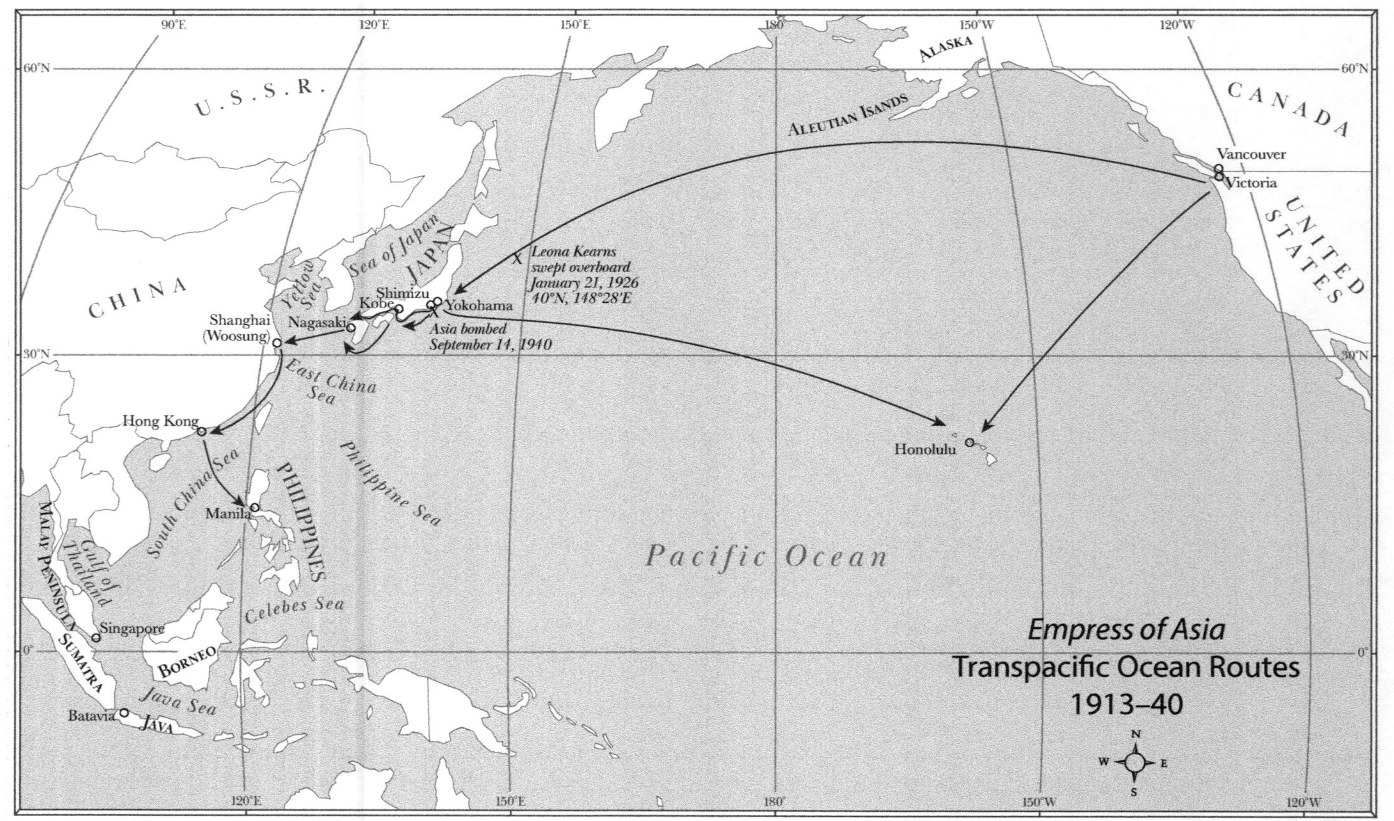
Empress of Asia
Transpacific Ocean Routes
1913–40
U.S.S.R.
CHINA
ALASKA
ALEUTIAN ISANDS
CANADA
Vancouver
Victoria
UNITED STATES
Sea of Japan
JAPAN
Yellow Sea
Shimizu
Kobe
Yokohama
Nagasaki
Shanghai
(Woosung)
Leona Kearns
swept overboard
January 21, 1926
40°N, 148°28'E
Asia bombed
September 14, 1940
East China Sea
Hong Kong
Philippine Sea
Honolulu
South China Sea
Manila
PHILIPPINES
Pacific Ocean
MALAY PENINSULA
Gulf of Thailand
Singapore
Celebes Sea
SUMATRA
BORNEO
Java Sea
Batavia
JAVA

On May 14, 1917, for example, the diverse passenger list included Hudson's Bay store commissioner H.E. Burbidge of Winnipeg, whose father ran Harrods of London, as well as a bearded Count Ilya Tolstoy. Author and second son of Leo Tolstoy, he had just concluded a successful U.S. lecture tour that, to no one's surprise, focused on his father.

Ilya had picked a tumultuous time to return home. It was shortly after the first of two violent Russian revolutions that forced Tsar Nicholas II to abdicate and be replaced by a provisional government. Ilya would mark his 51st birthday shortly before the ship reached Yokohama, and as a child had been described by his father as "always thinking about what he is told not to think about. Invents his own games.... Hot-tempered and violent, wants to fight at once; but is also tender-hearted and very sensitive."[1]

One of the youngest passengers arriving at Victoria on April 30 was nine-year-old Charles Woodruff Yost and his mother, Gertrude. Born at Watertown, New York, Master Yost had a bright future as a U.S. ambassador,

Passengers loved acquainting themselves with the officers, including the obliging Captain Arthur Wellesley Davison, May 1916.

but in that moment, his priority was making sure his toy soldiers answered roll call. During the past six months, those little paper and lead men had "campaigned" all the way from California to Hawaii, the Philippines, China, Manchuria, and Japan.

Gertrude had planned to travel just as far as Honolulu until an intrepid female traveller enticed her farther with the "world is your oyster" approach. "The fact that half mankind, including the Japanese, were engaged in the First World War and that China since the overthrow of the Manchu dynasty had been involved in almost constant civil war seems not to have entered their calculations," noted Yost in his unpublished memoirs.[2]

Encounters with passengers aboard the *Asia* and other steamers throughout his adolescence and early adulthood surely left an indelible impression, perhaps shaping his views of people and events that altered the world from the Red Sea to the Somme. It is also conceivable that they led to his pursuit of a diplomatic career with early Foreign Service postings that included

Bundled up against the cold while crossing the North Pacific, a female passenger poses in front of a lifeboat in 1916.

Egypt, Ceylon, and Thailand — and on to permanent U.S. representative to the United Nations in 1969.

For Gertrude, it was a time for healing. "This tour was my grandmother's effort to deal with her grief at the sudden loss of her husband," explained granddaughter Felicity Yost. "Even if her family thought it foolhardy, it was truly a brave decision on the part of a Victorian widow and her young son."[3]

A few weeks after Gertrude and Charles reached Japan, the United States declared war on Germany on April 6, 1917. Appreciating what she and her son had seen and grateful for a forecast of exceptionally calm weather across the North Pacific, Gertrude felt it best to return home with Charles via Yokohama and Victoria.

During their nine and a half days at sea on the *Asia*, the boy retrieved his little soldiers and strategically arranged them into fighting columns on

Gertrude Yost and her son, Charles, enjoy a moment together while returning home from a long vacation.

the thick wooden tables in the writing room or library. Previously, when he was on another steamer passing through the Straits of Shimonoseki, the boy's martial deployments had caught the discerning eye of a Japanese army officer. "Having spread my lead soldiers out on a table I was surprised and pleased to be accosted by the officer who, having politely asked my permission, joined me in a friendly war game for the rest of the voyage. I often thought of this episode years later — after Pearl Harbor — and [while] remembering my courteous companion, [I] was never able to feel quite as savage about the Japanese military as most of my compatriots did."[4]

Two months after their return at 4:30 p.m., July 6, the *Empress of Asia* sounded three short blasts and reversed her engines before one longer blast announced she was underway from Vancouver. Settling in for this first leg to Victoria was a mixed party of 29 united by a single purpose: the American Red Cross Mission to Russia. Prominent among them were Lieutenant Colonels Frank Billings and William Boyce Thompson — the latter an American business magnate — and Dr. William Sydney Thayer, chairman of the mission's medical and scientific group.[5]

Thayer, born in Milton, Massachusetts, was a professor of clinical medicine at Johns Hopkins University, while Billings, a prominent Chicago physician, was an infectious disease expert. Like Herbert James and Captain Robinson, Thayer, who was in his early fifties, was a diarist who filled 164 pages describing the mission from its inception to its return to the United States months later. Their goal was to deliver 70 tons of medical and surgical supplies to the Russian army and coordinate relief efforts among Russian agencies.[6]

From Vancouver, it would take the expedition three weeks to reach Vladivostok via Yokohama, and another two to arrive at Petrograd via the Trans-Siberian Railway. For Thayer, who shared an upper cabin with Billings and Philadelphia doctor D.J. McCarthy, the ocean crossing offered opportunities for personal reflection, Russian language lessons, and speculation on what might happen to Russia after the fall of the Romanov dynasty.

Besides keeping a diary, Thayer wrote to his wife, who although in very poor health, had encouraged him to join the expedition. While saying their goodbyes, he knew it could be their last moment together, and he was right, since she did not have long to live.[7]

As they sailed out of Victoria that July 7, Thayer stood in the balmy breeze on the Promenade deck, taking in the view and breathing the fresh sea and land scents along the Strait of Juan de Fuca. It was "a beautiful afternoon, cool, but clear" with a "glorious outlook," he noted in his diary. A few hours earlier, the professor had been busy observing his fellow passengers, assessing various personalities and working on a possible approach to at least one or two Russians. Several Dutch were on board, and various members of the Society of Friends — Quakers — heading to Russia to assist refugees. "I am trying to find a likely Russian to talk to," he noted, aware there were 200 Russians returning home, however, "I am afraid the party is too large to allow the sort of intensive study I should like to give."[8]

Mission chairman Billings, and Thompson, who had offered to cover expenses, earned early favourable reviews. "Billings grows on me as one sees more of him, a really fine figure, simple and strong, and right-minded with good clean emotions. T. is an interesting figure, fat and smiling, not one who would impress you as a man of any particular force and yet a man of large affairs, evidently very able and looked up to by his associates."[9]

His assessment of McCarthy of the Phipps Institute for the Study, Treatment and Prevention of Tuberculosis was also good. The doctor had spent a year as American inspector of German prison camps and so had much to share. "McC is a good fellow, capable, rather sensitive and quick, I fancy, but we get along, and shall get along well. He vaccinated me today in a way of which I don't approve — but he is forgiven."[10]

A vaccination parade on July 8 lasted all afternoon. Using the cabin shared by Thayer and McCarthy, all members of the party were administered a typhoid shot.

On the 13th, following his daily Russian-language lesson, Thayer settled into an agreeable spot on deck. It was warm out of the wind, and a welcome shaft of sunlight had poked through the fog. A Belgian, only identified as P. May, sat with him. "A nice fellow, but of rather the exquisite type; wrinkles his forehead with a look of great suffering every time the fog horn blows, a muscular effort which in the end must amount to a good deal."[11]

The next day, while members of his party collected cash for the Red Cross, a sharp blast from the ship's whistle interrupted conversations and seized the crew's attention. The *Asia* reduced speed and began turning around. "The Chinese crew rushed to the deck and our Dutch friends gave us a good performance of an incipient panic crying 'submarine' and running to and fro most actively.... I fancied it was a boat drill and continued on my way to my stateroom to remove my uniform and coat [as] we had just had a picture taken. When I came out, we were going slowly backwards, a boat fully manned hanging over the side ready to go out and pick up the poor Chinaman who jumped overboard. 'Twas no use. We were going twenty miles an hour and he is gone; poor chap."[12]

That afternoon, Thayer, Billings, Thompson, and another member of the group met to discuss a rather embarrassing development concerning one of the colleagues in their party — a handsome 29-year-old who seemed less interested in the mission and its members than the attentions of a female passenger. Thayer was disparaging: he "has attached himself to a rather silly married woman with two little children, and spends all his time with her. In addition, he has begun bragging to some of the younger attachés of the party of certain amorous exploits which are of small credit to him. Such an individual must be called to account sharply and B [Billings] wanted our advice."[13] After much deliberation, the group felt it best to ship the man home from Yokohama if he did not mend his ways.

Although the United States had entered the First World War three months previously, the professor was keen to recite a few recent lines from American poet and Harvard graduate Alan Seeger, who had enlisted in the French Foreign Legion soon after the outbreak of the war and was killed in action three days into the Battle of the Somme. Sitting nearby in the lounge during the recital was a large group of passengers from the Netherlands, a country that remained neutral during the war. It was Thayer's observation that some in the Dutch crowd were "obviously — no apparently — *Les Bosches* sympathizers. Twice I made several of them sit up very sharply with obvious indignation.... At the end we all rose and sang the *Marseilles*, a Dutchman of the right sort leading with vigour."[14]

Thayer could also not stop thinking about the Chinese man who leaped overboard. "He was the doctor's boy and had been caught stealing several hundred dollars. He was taken before the captain, tried and convicted and was being given over to the men at arms to be imprisoned in the forecastle when he broke away and jumped over. There is something very pathetic in the thought of the poor little frightened wretch jumping to death in a moment of despair."[15] The man's name was Tam Cheong, 30 years old from Canton, China, employed as a saloon boy.[16]

On the 17th, the Red Cross Mission to Russia disembarked at Yokohama and carried on with a journey that would pass through an historic timeline leading to further division in the world.

○ ○ ○

As the toll of war rose on the Western Front, the Canadian government requisitioned several ships, including the *Asia* and *Russia*, to transport non-combatant Chinese labourers from Shantung (Shandong) Province in north-eastern China to Canada. The requisitioning was part of Canada's response to a British scheme to find additional labour to work behind the lines, thus freeing up more British men to fight. Roughly 80 percent of the Chinese who were organized into parties or "battalions" for the Chinese Labour Corps (CLC) came from Shantung, and their duties included unloading ships, building and repairing roads, and stockpiling ammunition. After the war, they assisted with clearing debris and recovering human remains from the battlefields; however, for a long time after, their services were largely left out of the First World War narrative.

Boarding ships at Weihaiwei and Tsingtao, approximately 84,000 crossed the Pacific to the William Head Quarantine Station west of Victoria. Travel east across the North Pacific was stormier, but safer than following the Indian Ocean, Red Sea, and Mediterranean route. One transport, the *Athos*, carrying civilians and Chinese labourers under a French scheme, was lost by a torpedo off Malta.

Once cleared at William Head, the labour battalions continued to Vancouver where under armed guard they boarded special CPR trains to eastern ports. Between April 1917 and March 1918, the *Asia*, under Captain Davison, transported 13,989 men in six Pacific crossings.[17]

While written accounts from Chinese labourers are rare, the Kautz Family YMCA Archives at the University of Minnesota holds an undated article written by W.W. Peter, a medical missionary who quotes a man identified as Li. Whether Li existed or not, the account rings true and reflects the fact that most of the men had never been to sea, let alone crossed an ocean.

The account states Li boarded the *Asia* on April 16, likely at Weihaiwei, where lighters conveyed the men to steamships. The date coincides with the *Asia*'s sailing schedule, as she left Hong Kong on April 12:

> The boat that we were to take was so large it could not even get to the shore. We had to take smaller boats out to where it lay anchored. On approaching the big boat my heart beat faster in fear for I saw with dismay that the boat was already full. Yet we climbed on board and were led down many stairways till I was sure we were actually near the bottom of the sea. When I climbed back up the shore was no longer in sight. I had left home. My courage almost failed me. What journey into the unknown was I beginning? I looked about me. Here were two thousand more like me, and yet not like me. I never saw what made the ship go. There were no sails.[18]

When British lieutenant Hugh Lowder arrived in Canada on March 24, 1918, he was in charge of a CLC battalion that would not see anything more of Canada than the quarantine station. On May 23, this officer with the 3rd Battalion, the King's (Shropshire) Light Infantry, led 1,783 members of MM Battalion on board the *Asia* alongside the barnacle-encrusted pier near the hut used to fumigate passenger luggage. Instead of travelling to Vancouver, the lieutenant's men, along with another CLC contingent, sailed south to the Panama Canal and on to New York and Liverpool.[19]

While at William Head, Lowder, somewhat of a progressive, had hoped to assign his men to paid work gangs outside the station's CLC encampment.

However, his proposal failed at the outset when the camp commandant insisted the men be employed at a nearby farm that could take no more than 20 men.[20] The secret nature of the CLC movements and the fact all other Chinese entering Canada had to pay a head tax most likely affected this decision, plus there was fear some men might escape. Some did break free briefly, but most were just trying to stay healthy, since mumps, conjunctivitis, and other illnesses were rampant. The sickest of Lowder's contingent — eight men — were sent home while another man died and was buried at William Head. In all, 105 labourers from MM Battalion were too sick to board the ship.[21]

Neither the station nor the *Asia* impressed Lowder. The former was filthy and crowded when he arrived, and the latter's lighting and ventilation was subpar for the densely packed men on board. "During the voyage from Canada to New York, the daily average of sick was 27, the men being kept on deck all day, and sleeping on deck at night owing to the heat."[22]

Lowder's preference to keep his men working was clearly engrained: "On the 29 of May I placed 92 men in the ship's stokehold at the request of the Chief Engineer [James Adamson]" with the understanding the men would do this without pay because the ship had a full complement of men in the engine room. Lowder noted the existing firemen from southern China were destined to leave the ship at New York, and so if his men "did well enough the Master would recommend their employment as firemen across the Atlantic, paid at firemen's rates."[23]

The lieutenant was correct regarding what awaited the ship's existing Chinese crew. China had been at war with Germany since mid-1917, and so there were no neutrality issues. Nevertheless, instructions received from the CPR's marine superintendent stipulated the Chinese crew was not to proceed to Liverpool. With Canadian Pacific Ocean Services (CPOS) covering the cost, these experienced men disembarked and travelled by rail from New York to Vancouver for Hong Kong–bound ships. Rushing eastbound from Vancouver to New York was a replacement crew of 431 obtained from the *Empress of Japan*.

Lowder was an intelligent, practical fellow, quite familiar with the strong differences between northern and southern Chinese communities. While he welcomed Chief Engineer Adamson's request, he was reluctant to mix his men with firemen from southern China. He also admitted his men were "overburdened with fat owing to lack of exercise" while in transit and "unaccustomed to the work and that the heat in the tropics might overcome them."[24]

Despite these concerns, 140 members of Lowder's group toiled as trimmers while the ship coaled at Colón on the Caribbean side of the Panama Canal. By June 18, the *Asia* was at New York where severe depression took hold of Chinese cook Tsang Fong, who grabbed something heavy and jumped 59 feet over the side into the oily harbour. He sank immediately, and half an hour later his body was recovered and turned over to harbour police.

The *Asia*'s logbook for this period also records the deaths of two CLC men. Wei Jeng Wen, 28, with his CLC registration number recorded as 134062, died of an embolism on June 28, eight days after the ship left New York; and Wang Feng Chui, 37, with registration number 133389, died two days later — cause not specified. Their remains were committed to the deep.[25]

○ ○ ○

The *Asia*'s last acts of the war involved transporting thousands of American troops to Europe in 1918, repatriating Canadian personnel two months after the armistice, and returning thousands of the CLC to China in 1919–20.

Wearing rectangular dazzle to camouflage and distort her lines, making it more difficult for the enemy to track course and speed, the *Asia* was ready to commence her duties as a troopship when her reliable chief engineer contracted the Spanish flu and required hospitalization in New York. One crew member who would not be there to wish Adamson better health was his old friend McMurtrie, the easygoing, pipe-smoking boilermaker who had finally taken leave at Liverpool.

In October, while those troopship duties were underway, steerage steward Edward Fisher from Vancouver died of pneumonia after entering Liverpool's Walton Hospital. Then, the death on November 7 of 40-year-old Wong Wong, a greaser in the noisy engine room, was attributed to the "effects of smoking opium."[26]

The *Asia* and her unique "dazzle" paint scheme.

Altogether in 1918, the *Asia* completed seven perilous voyages from New York to Liverpool and New York to Brest, France, and after each crossing, she returned empty, usually without escort.

With tens of thousands of military personnel waiting to go home, arrangements for the ship's repatriation voyage to Victoria and Vancouver began in late December 1918. On New Year's Day, Major William Henry Macdonald of the Canadian Army Medical Corps (CAMC) was happily awaiting departure from London's Euston Station on a Liverpool-bound train. Originally from Newcastle, New Brunswick, he had served in a stationary hospital and casualty-clearing station on the Western Front.[27] While he looked forward to seeing his wife, Victoria, and their two adolescent boys at Medicine Hat, Alberta, the major appreciated an opportunity to sit for a while. The last two days had been hectic. Ordered to serve as senior medical officer for troops embarking for Canada on the *Asia,* Macdonald had arranged for additional medical supplies on the voyage to Victoria and Vancouver via the Panama Canal. He had also met and briefed his medical team, which included six nursing sisters, two orderlies, and a couple of corporals assigned to escort duty.

Among the nurses was Margaret Cecile Kennedy of Ingersoll, Ontario. Unmarried, Kennedy had joined the CAMC in September 1914 and sailed

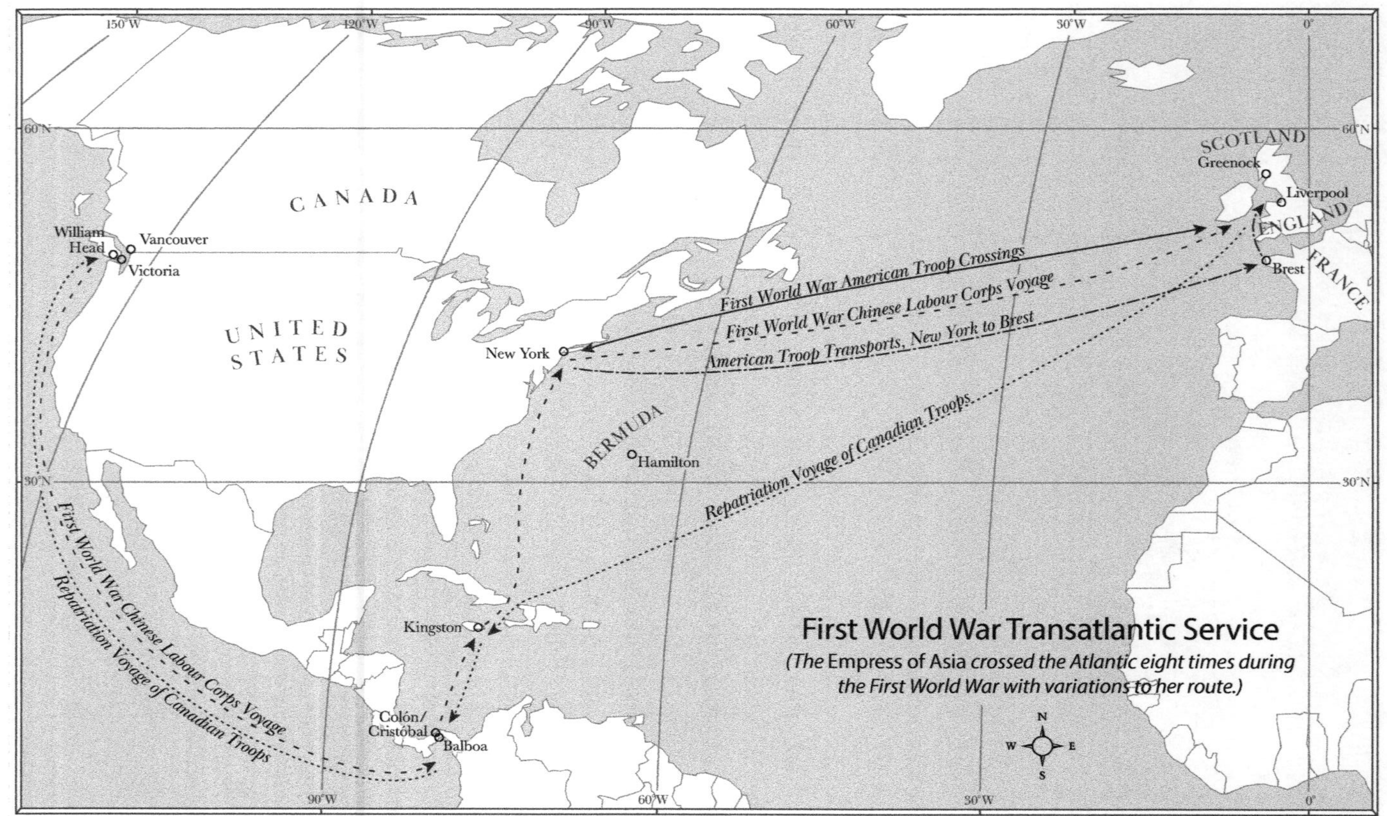
First World War Transatlantic Service
(The Empress of Asia crossed the Atlantic eight times during the First World War with variations to her route.)
First World War American Troop Crossings
First World War Chinese Labour Corps Voyage
American Troop Transports, New York to Brest
Repatriation Voyage of Canadian Troops
First World War Chinese Labour Corps Voyage
Repatriation Voyage of Canadian Troops
CANADA
UNITED STATES
SCOTLAND
ENGLAND
FRANCE
BERMUDA
Greenock
Liverpool
Brest
Hamilton
New York
Kingston
Colón/ Cristóbal
Balboa
William Head
Vancouver
Victoria
N
W
E
S
150°W
120°W
90°W
60°W
30°W
0°
60°N
30°N

to Europe that October from Quebec with the First Canadian Contingent. Shortly after her arrival, she accidentally slammed her fingers in an elevator door and spent seven days in hospital with badly bruised digits.

Kennedy had spent much of the war in England, France, and at sea on a transport bound for Canada in August 1916 and then twice on the British hospital ship *Araguaya* in 1918. Incredibly, she had dodged the Spanish flu but not diarrhea, measles, and mumps. Like Macdonald, she was exhausted when she took her seat on the Liverpool-bound train but looked forward to applying her training on the homeward troopship.[28]

Early on January 2, 1919, as the *Asia* prepared to set sail, besides the medical team, she carried 1,359 passengers, including several children. One of the youngest was five-year-old Derek Johnston, drawn to the black tails (ribbons) hanging from the caps worn by the Seaforth Highlanders. Sneaking up on an unsuspecting soldier, the youngster discovered he could easily snatch the cap by grabbing the tail and then beat a hasty retreat to his parents' cabin. "A couple of visits from the soldiers asking for their caps and stern paternal admonitions" ended the hijinks, but it was fun while it lasted.[29]

Overall, it was not a healthy ship. Many troops were sick or nursing physical and/or psychological wounds. Macdonald's team made a thorough inspection of the vessel and ruled favourably on the sanitary conditions, but more importantly, there was no influenza detected among the civilian passengers they examined at Liverpool. At the time, the world was moving into a third wave of the deadly pandemic.

Among the soldiers swinging into hammocks were Privates John Beston and Jonathan Clifford, both quite finished with the war and just wanting to move on with their lives. Beston, from Tipperary, had fought at Vimy Ridge with the 102nd Canadian Infantry Battalion, but while near Lens on June 7, 1917, the former ranch hand, who had enlisted in Ottawa in 1915, suffered a gunshot wound to his left thigh. Treatment and months of convalescence followed before he returned to the battlefield where enemy fire again found him in September 1918 — this time on his right side.

Clifford, born in Wiltshire, England, was returning to Mount Lehman in British Columbia where his wife, Mabel, three sons, and five daughters anxiously awaited his arrival. While her husband was overseas with the

Canadian Forestry Corps, Mabel found it arduous to support the family on his assigned pay and paltry separation allowance awarded to spouses or dependents of service personnel.

After reaching Panama's Colón Coal Pier on January 14, Beston and Clifford were among hundreds of military personnel on the *Asia* who began formulating the beginnings of the Land Settlement Committee with the objective of approaching the BC government with a plan to establish co-operative farming communities for returned soldiers.

At its core were four highly decorated officers, plus a sergeant, a gunner, and a corporal. Lieutenant-Colonel William Latta, Major Arthur Mills, Lieutenant Walter Kirchner (committee secretary), and Lieutenant Alexander Walker (committee chairman) had much to consider, including possible locations in the Upper Columbia Valley, Southern Okanagan, and Peace River districts. While these men had been fighting in France and Belgium, discussions on formulating provincial and national policy on how best to reintroduce returning soldiers to civilian life were well underway in Canada. In 1916, the British Columbia Returned Soldiers Aid Commission had already recommended that the provincial legislature implement a co-operative land settlement plan for returning service personnel. While the federal government and provinces worked toward revising the federal Soldier Settlement Act in 1918–19, John Oliver, British Columbia's premier, moved forward with an Act to Provide Lands for the Use and Benefit of Returned Soldiers.

From a tentative idea developed independently on a repatriation ship, the *Asia* Land Settlement Committee was proof of how important the issue was to soldiers who favoured their own community settlements. Acting Sergeant John Bird, who had been temporarily blinded by mustard gas while observing the enemy in November 1917 and was returning home to Cranbrook, British Columbia, certainly favoured the idea. "I was interested and attended the meetings," he explained in a 1964 CBC interview.[30] "Lieutenant Kirchner seemed to be the spark plug in getting the meeting started. There was also Latta who became a director of the Land Settlement Board and the idea was that it would be nice if men who had served on the battlefield would carry on into civil life into a farming community. That was the whole idea."

Born in Caithwaite, Cumberland, England, Bird had enlisted in October 1915 at Vernon, British Columbia, with the 54th Kootenay Battalion. After bidding farewell to his wife, Marian, and two-year-old son, the CPR store clerk shipped overseas on the *Saxonia*. He "first noticed smell of mustard, then started coughing, eyes started to smart & water, vomited several times and finally could not see," details his wartime medical sheet corresponding to his service with the 11th Canadian Infantry Brigade. "Blinded for about 10 days. Vision much improved. Can read but not advisable. Wearing dark glasses."[31]

After 12 days at sea, the ship's coaling at Colón and time at Gatun Locks presented Bird and others with the opportunity to visit noisy, smoky bars, sample food, and partake in other enjoyments. They were "given a royal welcome all day and most of the night" with "dances, sightseeing tours, and concerts arranged for their benefit. At Gatun, people showered them with oranges, bananas and … magazines."[32]

Hawkers were out in force at Gatun, enticing sailors and soldiers with souvenirs or colourful trinkets for their sweethearts. Enterprising photographers and their assistants turned snaps of the ship into postcards. One joyful shot shows every inch of the *Asia*'s upper decks crammed with returning men, the ship's crew in white, with a scattering of women — one in a large hat near a lifeboat and another at the stern above the ship's name and port of registry. Somehow Vancouver had lost the letters *C* and *U*. Far below the portholes, water used to cool *Asia*'s condenser can be seen splashing into the canal, while cables tether the ship to electric locomotives known as mules.

Lost in the image, of course, is any sign of the reported inferior food, although from later stories told to the press in Vancouver, the grub was far worse than what officers received. One story noted the porridge was lumpy, half-cooked, and served without milk; the coffee was filthy; and the sausages were raw. The headlines prompted investigation by the Department of Militia and Defence's Colonel E.E. Clarke, director general of supplies and transport, who immediately called to task the CPOS and ship's captain.

Clarke's demand for a full account yielded a stinging rebuttal from Captain Davison, who alleged the commanding officer of the troops "had no control over the men." The troops' quarters, he wrote, suffered from a serious lack of regular cleaning. "We did everything in our power to make the men comfortable, giving them practically the whole of the ship for their use.... None of this appreciated by most of the men, who retaliated by making as much trouble as possible, and writing all kinds of rude and obscene remarks all over the ship." In regard to the food, the CPOS stated the cooking was good with the exception of one or two instances.[33]

It is worth considering that the complaints by soldiers on the *Asia* not only reflected a pushback against inferior food and superior officer status and entitlements but also a broader postwar realignment flowing into the 1920s. The rise of radical political parties, a growth of labour militancy, unions, and the co-operative movement was colliding with the old social order.

Well before the soldiers' complaints reached Ottawa, Major Macdonald and his medical team were on the go day and night. Along with the usual broken bones, sprains, dressing festering war wounds, and pulling teeth, they dealt with eye and groin injuries, venereal disease, a pierced foot, penis ulcer, lice, recurring seasickness and sunstroke, acute bronchitis, mumps,

Thousands of repatriated Canadian soldiers and a number of civilians crowd the outer decks and superstructure of the *Asia* as she transits the Panama Canal bound for Victoria, British Columbia.

pneumonia, and on the civilian side, four children with chickenpox. Five days before reaching Colón, Macdonald, with sanitary conditions in mind, directed Davison to "erect a large swimming tank on B Deck," which "proved very useful and popular."

Two days later, with yet more food complaints, Macdonald investigated and "found a mixture of vegetables and boiled meat, in which the vegetables were sour and the whole unfit for consumption." Macdonald wrote a sharp letter to the troop commander complaining about the ship's filthy condition, urging him to get his men moving on sanitary fatigues.[34]

The mood lifted considerably on January 24 at Victoria where a massive, cheering crowd, estimated at 15,000, surged past, over, and through barricades to areas reserved for families. Song and prayer animated joyful tears, a band played "God Save the King," and speeches honoured the men and their sacrifices. But before stepping off the ship, where they had just spent 23 days, the returning men, all excepting 19 who had slipped off without permission, received $5 in ones and twos, celebration money delivered by Military District No. 11.[35]

A happy and boisterous scene greets the troop transport when she finally arrives in Victoria.

o o o

Thirty-seven Canadian Pacific steamships with a gross tonnage of 329,960 did war service as transports or armed cruisers, with 12 lost to enemy action. These ships transported approximately a million troops, four million tons of war supplies, plus thousands of horses and mules.[36] All told, the *Asia*'s Atlantic duty during the war and not long after the armistice saw her complete eight crossings.

On shore, the *Asia* Land Settlement Committee kept its word, sending a delegation to meet provincial government representatives. "They appeared to consider the matter sympathetically and then the committee decided to go to Ottawa," recalled Bird.[37]

To cover expenses, the interested veterans, including Bird, contributed a dollar each. The committee met with future prime minister Arthur Meighen, then minister of the interior, who Bird recalled judged the plan financially impossible. "My opinion of that was there spoke the apostle of big business," added Bird. "It was the dollars that counted. Sufferings that soldiers had gone through counted for nothing. I never had a very good opinion of the Honourable Arthur Meighen after that. Well, [the committee] came back and the British Columbia cabinet decided they would start soldier settlements and the first one was at Merville."[38]

On Vancouver Island, nine miles north of Courtenay, the Land Settlement Board purchased 14,000 acres from a logging company. By May, soldier settlers, including many from the *Asia*, had begun clearing Merville for homes and other community buildings, and by late September, 400 people were living in temporary shacks and assisting with construction. Clearing the land and farming were challenging, but by 1922, in addition to farms, Merville had a church, community hall, post office, and store. This was not the only such community that had grown from discussions on the ship; Bird, who would live to 81, and others had successfully settled into a similar community at Lister in southeastern British Columbia, not far from Cranbrook.[39]

The families of John Beston and Jonathan Clifford were among those who called Merville home. Blasting or pulling stumps with teams of horses and steam engines was hard, dangerous work, but there was shared purpose as they endeavoured to put the war behind them.

That summer, smoke from a forest fire burning for two weeks hung over the valley, and on July 6, 1922, wind sent the flames racing toward Merville. The conflagration swept through town, killing livestock, incinerating buildings, and claiming the life of Clifford's 15-year-old-son, Jack. Clifford also lost his home, as did Beston and just about everybody else in Merville. What remained was scorched earth and blackened tree stumps. At least three men caught in the firestorm found refuge in a stream as flames raced over them. For the veterans, it felt as if they had never left the Western Front.

PART TWO

BETWEEN WARS

6

BOYS TO SEA

If you consent to go, you may be away from home for one, two or three years and will have to go to the danger zone.

— **Marine superintendent, Canadian Pacific Steamships Limited**

It is something of a coincidence that 66-year-old Walter Oliver is finally that stranded "old man" at home in a Vancouver apartment complex named after one of the world's greatest explorers. The irony is that even from his church-sponsored single room on the sixth floor of the Columbus Tower, the former merchant mariner lacks an ocean view. It is 1976, 34 years after the *Asia* burned, more than a decade after his marriage ended, and two years after a stroke removed any chance of returning to the *Pitt Lake Express*, the tender he operated between Port Coquitlam and the BC Forest Products wharf at Pitt Lake. The solo work, although a far cry from what he had been used to, had been steady — 186-mile round trips per week on a vessel that transported acetylene tanks, oil drums, retreaded tires, and slabs of meat wrapped in cheesecloth.

Oliver had missed the scenery along the tidal river and into the narrow mouth of the lake. The channel and its large mud flats — nasty in fog — demanded attention, but on the lake he could relax while mindful of strong winds.

On the morning of May 14, 1971, as his diesel-powered tender approached Goose Island, 12 inmates from a correctional institution were engaged in a recreational canoeing adventure when they encountered a squall on the west side of the lake. Two of the men drowned while others clung to a rockface. Oliver reached the scene in time to pull the men aboard and land them safely. The perishables he had been transporting north all spoiled, so he lost money on the trip, but never sought compensation. The men were young and pleasant — almost like having a deck crew again. Besides, it was considered bad taste for a skipper to charge for an act of mercy or profit from the misfortune of others. "In such cases," noted his son, Nelson, "usually after a period of time a couple of bottles of rum might be presented for assistance."[1]

Oliver's recovery from the stroke had gone well, so he sat at his small Arborite kitchen table, making good on a promise: punching keys on his Smith-Corona, taking him to sea again, to a time before he signed onto the *Empress of Asia*.[2] He could easily relate to lads dreaming of the sea, and recall the 1920s, heyday of the big passenger liners even though his early sea time was born of necessity on a cargo ship.

○ ○ ○

What he remembered most from his first working voyage was the Strait of Magellan: stark, breathtaking scenery; narrow, meandering channels, plunging rocky walls with a thin covering of dried grasses or stunted evergreens; and the endless icy fjords surrounded by snow-topped mountains. Splendid, yes, but with innumerable place names such as Port Famine, Useless Bay, Mercy Bay, and Desolation Island, sailors would be foolish in undertaking the journey unprepared.

The last thing 15-year-old Walter Oliver and the crew of the *Anglo Colombian* wanted to see in the 350-mile strait was a "williwaw," a sudden squall that could strike with gale force and lead, at best, to confused seas and poor visibility. It was challenging enough without gyrocompass or echo sounder to follow the narrow, meandering channel. All the *Anglo Colombian* had to work with was her crew, a magnetic compass, and "deep-sea lead" — an anchor tied to a rope, marked off in fathoms.[3] By avoiding treacherous Cape Horn, the 8,407-ton ship would be somewhat more protected in the strait separating mainland South America from the Tierra del Fuego Archipelago and dividing the Atlantic and Pacific.

For Oliver, a life at sea was the only good choice, one that would lead to regular employment on more than a dozen ships. A harder option facing boys from his hometown led to a life of "black lung" and the collieries of Durham County where some 170,000 miners worked the pits.[4]

His escape began on November 4, 1925, when he signed Articles of Agreement calling for up to two years' service on the same vessel. Battened down and laden with coal, the 479-foot ship *Anglo Colombian*, originally launched in 1915 as SS *Afrika*, cleared the port of Cardiff, Wales, and steamed from the Bristol Channel to open sea, bound for Rio de Janeiro. It was the first port of call on a 50,000-mile circumnavigation of the globe, a journey lasting more than 11 months with no certainty he would see his family again.

He definitely had no way of knowing that within a year his parents, Walter Sr. and Eliza, would move again from England to Canada and settle in New Westminster, British Columbia. Emerging wounded from the First World War, Walter Sr. had endured a painful recovery in the home of his in-laws at Easington Colliery. It was there that young Walter witnessed how the coal pits shortened or irrevocably altered the lives of both miners and their families. But it was a lively place — an epicentre of trade union and Labour Party activity. "He began viewing the world through somewhat of a class lens," noted his son, Nelson Oliver, of Port Moody, British Columbia. "He was more than just sympathetic to working people. He saw himself as one."[5]

In those first hours out of Cardiff, as pellets of sleet pricked his young face, Walter Oliver was unaware of the icy dangers beneath the creaking steel hull. The Atlantic's name, meaning "Sea of Atlas," came from Greek mythology. Foreboding and vast, it was part of the planet's mysterious saltwater realm covering more than 70 percent of the Earth's surface. Ocean currents, trade winds, high seas, and exotic bustling ports with peculiar and colourful customs lay ahead on voyages tied to legend, lore, and of course, superstition. Did he know, for instance, that it was bad luck to sail on Thursdays, Thor's day, the god of storms? Or Friday, the day widely believed to be when Jesus was crucified? Did he know that a storm could be conjured by whistling or singing into the wind, and what about those monstrous creatures he had seen depicted on maps? A lot to learn for the novice sailor.

For the time being, the practicalities and contingencies of good seamanship were far more important, and for his own protection, heeding the mercantile marine code of silence when it came to admitting fear or calling out the shortcomings of fellow crew.

While crossing the Atlantic, at roughly 30 degrees north latitude, the *Anglo Colombian* entered the horse latitudes where, Oliver had heard, Spanish mariners bound for the West Indies jettisoned horses because the feed ran out while their ships languished on windless equatorial waters.[6]

After clearing the treacherous Strait of Magellan, the *Anglo Colombian* steamed west into the Pacific, then took a northerly route, keeping parallel to the Chilean coast. She took on a load of nitrate, passed through the Panama Canal, and then steamed to Charleston, South Carolina, where sailors could not find a single drop of rum or whisky; however, there were ice-cream parlours aplenty. Galveston, Texas, was her next port for sulphur bound for New Zealand. Even with such a dangerous cargo, owners considered it cheaper to send the ship around the cape instead of paying canal passage. This time while rounding the Horn, the *Anglo Colombian* encountered a gale shipping 49- to 52-foot-high waves, including one "greybeard" that washed a chicken coop off the boat deck 39 feet above the waterline.[7] Sensing possible death, the men down below prayed and held fast in the humid air reeking of body odour and bilge.

On October 17, 1926, after completing his first round-the-world voyage at the port of Immingham in northeast England, Oliver decided to stay with the ship for six more trips. It was not until July 3, 1929, that he signed off — more than three and a half years after first stepping aboard.

○ ○ ○

Lying on the kitchen table next to his typewriter was the little dark blue book issued to Walter Oliver by the Board of Trade — otherwise known as the Continuous Certificate of Discharge Book, a valuable document in identification of service for merchant sailors recording the ships they signed onto throughout their careers. Included was a column bearing a stamp from the ship's master that graded the sailor's ability and general conduct. In addition to capturing Oliver's service on the *Anglo Colombian*, his book listed, between July 3, 1929, and January 1941, nine different merchant ships from cargo vessels to oil tankers. Throughout, Oliver's character was consistently rated "Very Good."[8]

What was slowly coming out of his typewriter was, of course, for family. It was important for his grown children to know about the *Asia*, but within the context of where he had been and what life was like at sea during the 1920s and 1930s. He wanted to convey the momentum of experience that served him into the Second World War and his first encounter with the ill-fated ship at Vancouver, and how service in the British Merchant Marine made him different from many Canadian Pacific Steamship officers, especially those who had not gone to sea during the Great Depression.

Although there had been nastiness and incompetence over the years, his intent was to prepare a travelogue that still respected the mariner's code of silence. So, with index fingers poised above the keys, Oliver recalled how much older he had been than some lads who signed onto the *Asia* in 1941. The year he joined the *Anglo Colombian*, some of the *Asia*'s future crew members were still in childhood.

○ ○ ○

Maurice Atkins would barely have been walking. Born on August 6, 1924, young "Morris" was 497 miles from the closest ocean, residing in southern Alberta at a dusty crossroads called Gem. The only waves were in the grass. "I guess you could call it a settlement," Atkins explained. "It was roughly twenty-five kilometres northeast of Bassano and there were just four corners with a two-room schoolhouse, but no church. Travelling north you went into God knows where, and going south you ended up the same way. There was plenty of farmland and people raised horses and hogs."[9]

Besides his father, Maurice Sr., and mother, Christabel, Atkins had a sister, Barbara, three years older. "Most families within fifty miles had around eight children, so ours was relatively small." The other mouth to feed was an orphaned pony that strayed onto the property and was adopted by young Maurice. He named her "Cutie."

Maurice Sr., a stoic but enterprising fellow, ran a general store, and for a sideline relied on his beat-up truck to deliver hogs from ranches to the abattoir in Calgary. "My father was a businessman and from the store he sold anything he could get his hands on. Mainly we were a grocery store, although we sold ice in summer."

When the boy was older, he accompanied his father on hog runs, which began after midnight. With a heavily loaded truck exuding ripe livestock smells, father and son settled in for the 78-mile trek to the city, Maurice curled up with his head on the seat next to the rattling gearshift and his father's warm thigh. These were not quiet journeys — besides the clatter and grind of the engine and various animal snorts, the youngster could hear the protesting springs as his father deftly navigated the rutted roads, leaving behind a tail of dust. "The truck was a wreck, but it ran."

Reaching Calgary before sunrise was vital. "Once the sun comes up the hogs lose weight and the farmer selling the hogs loses money, which of course made them rather unhappy," Maurice recalled. "It was a great day getting the hogs to Calgary on time, because afterwards we had breakfast in a restaurant. That was really something." For an impressionable Atkins, the hog runs proved how punctuality mattered.

When the Depression hit, the family, and Cutie, piled into the truck and drove west through the mountains, settling at Brentwood Bay on Vancouver Island's Saanich Peninsula, 11 miles north of Victoria. Maurice Sr. purchased another store, but times were tough as predictions of war unfolded, and in early 1939, while Maurice was in grade 9, his father pulled him aside. It was time, he said, to quit school. "I had no say in the matter. My instructions were to meet with the high school's headmaster on a Saturday and write the entry exams for HMS *Conway*, the venerable naval academy on the Mersey in England. I was fourteen and by then Hitler was raising hell. Not far from *Conway* were the Cammell Laird slipways and the führer's bombers came over and mined the Mersey one night, sinking several ships at anchor, and we had to evacuate." It would not be long before Atkins, as a 16-year-old cadet, signed onto the *Asia*.[10]

o o o

Geoffrey Tozer was nearly three when Walter Oliver went to sea. His parents had moved from England, settling in British Columbia's Okanagan Valley at Coldstream, southeast of Vernon. Born December 10, 1922, he, too, became a child of the Depression, but the beautiful valley hugging the shores of Lake Okanagan offered opportunity for escape at a time when families remained poor. The boy's curiosity led to invention — always making something out of nothing. Through trial and error, he created tools from discarded objects, using them to build or repair other things.

Coldstream, then home to just over 700, was on the northern end of the much smaller Kalamalka Lake, but the myths and reported sightings of a serpent-like monster living in Okanagan Lake captured the boy's imagination. Like others, he claimed to have spotted the head and neck of the creature in August 1936 while on a four-day fishing trip with a fellow sea cadet. At 13, he never doubted what he and his pal saw, even if it was for only a few seconds. Whatever it was, it caused them to pull in their lines and row to shore.[11]

Whether to seek out the fabled beast or have fun in the smaller lake named after the Okanagan First Nation chief, Tozer and his pals built a diving bell out of a discarded hot-water tank and a bicycle pump. Applying globs of tar, the boys stuck on a window and took turns walking along the silty lake bottom.

This, of course, led to an interest in larger bodies of water and a larger sense of adventure, leaving him restless and wanting to make something of his life. So, in 1939, when Canada declared war on Germany, independently from Britain, the young man took matters into his own hands. Convinced he was too young to enlist in the army or navy, the tall, lanky boy with brown hair and hazel eyes set his sights on the Merchant Marine, assigned the crucial job of delivering wartime supplies.[12]

On November 25, 1940, the precocious 17-year-old posted a letter reminding the marine superintendent at Canadian Pacific Steamships that he had submitted an application to join one of the company's ships as a boy seaman. While the response was not good, it nevertheless constituted a reply. "I have had your name on my list, waiting for you to make another application," stated the superintendent. "I will now keep it in front of me. Unfortunately, there is very little hope of a vacancy … most of our vessels being away on active service."[13]

The superintendent was telling the truth. By spring 1940, 18 of the company's 20 ocean liners and freighters had been taken over for wartime troop and supply duty. One freighter — the SS *Beaverburn* — was lost on February 5 to a torpedo in the North Atlantic.[14]

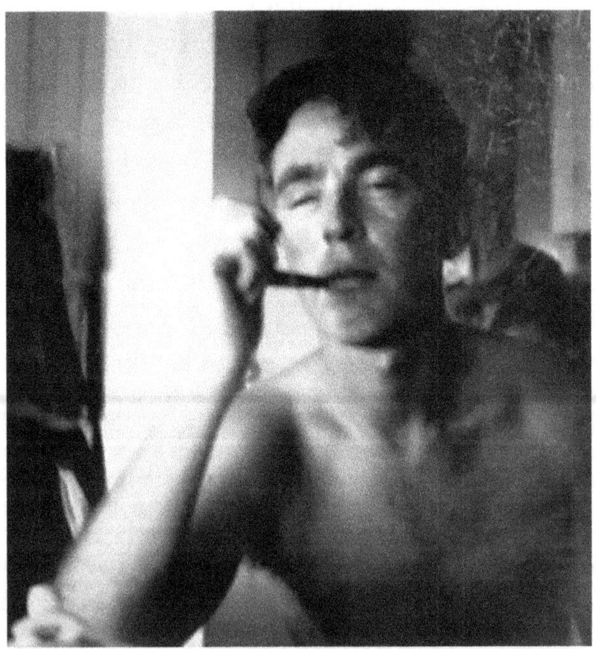

Geoff Tozer lived for adventure on the high seas but, more importantly, wanted to be of service to his country during the war.

Good news finally arrived 72 days after he had posted his appeal. "I think I will have a vacancy on one of our vessels sailing for England at the middle or end of next week," wrote the superintendent. "If you consent to go, you may be away from home for one, two or three years and will have to go to the danger zone...."

The letter, coincidentally dated exactly a year before the loss of the *Empress of Asia*, promised wages of $15 per month, plus a war bonus. It stated he could expect to be messed with six other boy sailors and "not have to mix with the crew generally." Additionally, if he worked hard and showed promise on the *Asia*, "every endeavour will be made to teach you as much as possible of the seafaring business." It was exactly what he wanted to read, and the last line, warning him to "be ready to report in Vancouver on telegraphic advice," was especially exciting.[15]

○ ○ ○

Jack Ewart, Ernie Higgs, Douglas Elworthy, and Geoffrey Hosken were not much older than Geoffrey Tozer and Maurice Atkins.

Ewart, nicknamed "Rusty" for his red hair, was born on October 26, 1921, at Point Grey, a separate municipality until amalgamated with Vancouver in 1929. Competitive and gregarious, he breathed sports, especially rugby and baseball. He loved the smell of the sea, followed the shipping news, and admired the steel leviathans resting on their chains in English Bay. By the summer of 1936, he had a summer job earning $30 per week working for his Auntie Lil at a store in nearby Roberts Creek and even got to drive a delivery truck. He was only 15, but knew he still had to finish school.[16]

With high school finally behind him, Ewart headed north to the Blubber Bay Quarry on Texada Island where the lime dust clung to his clothes and lungs. He worked the quarry for six months, and the pay was good, augmented by tough, dangerous overtime work on the green chain at a nearby sawmill.

By Christmas 1940, Ewart was back home where talk of Hitler and the need to stop him was everywhere. Like Maurice Atkins's father, Ewart's

Like many his age, young Jack Ewart wanted to contribute to the war effort. His opportunity came on the *Empress of Asia*.

dad, Norman, was worried about the state of the world. Nevertheless, he encouraged his son to take the next step toward a career at sea. The First World War army veteran had suffered a gunshot wound to the arm while scouting out German trench positions in northern France, so he understood the dangers, but heard through rumour that the old *Empress of Asia* was due in port. She was the same ship he had arrived home on following the armistice. Part of the rumour had the government replacing her Chinese crew with an all-white one for wartime service, even though it was widely understood the Chinese crew had been loyal for more than 20 years. But it had also become apparent that many Chinese crews on foreign-registered ships were reluctant to serve in foreign war zones. When the *Asia* arrived at Vancouver on January 11, 1941, only four Chinese remained with the ship. Ewart applied and within days signed on as captain's steward.

○ ○ ○

Ernie Higgs and his non-identical twin, Cyril James, nicknamed "Buster," were born in Victoria on November 9, 1920, the sons of Daisy and Frederick Higgs. Ernie, who later became a master mariner, left school in grade 8, and enrolled in the sea cadets while the family was in Vancouver. "He was a bright, but restless child and had fallen in love with the ocean," explained his daughter, Joan Higgs, of Roberts Creek. "It came from early exposure through his father and uncle, who built and sailed their own boats. I know Dad was happy in that life and so was Buster."[17]

Ernie and Buster, who had a younger brother, Gerald, were over six feet and muscular. They were the kind of boys to call if you needed an arm's length of anything at the hardware store or something on the top shelf. Socially, they were alike, although if there was one thing Ernie could not stand, it was a braggart.

In 1937, after turning 16, the boys signed onto the 24-year-old *Asia* but did not work aboard at the same time. Buster joined as a messenger in the purser's office on March 25, 10 days before commencing a Pacific crossing. He signed off on July 12, seven days before Ernie signed on as a messenger on the bridge. The reason for Buster's leaving remains a mystery, but it happened on the day the *Asia* returned to Vancouver from Hong Kong and Yokohama.[18]

A photograph from September 1937 shows Ernie and Buster wearing their navy blue uniforms. Standing side by side, arms draped over each other's shoulders, pipes clenched between their teeth, they are a cocky picture of brotherly love. A handwritten note on the back, written by Ernie, describes the pair as "The Salty Pipe Smokers!"

Sadly, tragedy struck the family on the afternoon of Saturday, November 11, 1939, Remembrance Day. Working as salmon fishermen on the 33-foot-long troller *Shannon J*, Buster and another lad volunteered to seek help when the vessel's engine quit at the mouth of the Fraser River. There was a fresh southeast wind of 19 to 25 miles per hour, so taking advantage of a flood tide and wearing oilskins and rubber boots, the two set out in the boat's 10-foot dinghy, hoping to reach Sand Heads Light at the mouth of the Fraser in the Strait of Georgia. Both, along with the dinghy, disappeared and a search found nothing. Nineteen-year-old Buster Higgs and 24-year-old Bobbie Pinkerton had drowned or died of exposure.

Twin brothers Ernie Higgs (left) and Buster Higgs enjoy the pipe.

Ernie's daughter distinctly remembers her father telling her how he overheard his mother say, "Oh, why did it have to be Buster?" Joan Higgs said her father could have taken his mother's words the wrong way, but what he heard or chose to hear left him not only with the loss of a brother but also with insurmountable survivor's guilt.[19]

Between signing onto the *Asia* and Buster's death, Ernie Higgs had completed 14 round trips between Canada to Japan and China and was on board when the ship once again cleared for Asia on October 18, 1939. At the time his brother went missing, he was in Hong Kong, and the *Asia* did not return to Vancouver until December 1, approximately four weeks after the tragedy.

Sea time kept the young bridge messenger busy into early 1941 when *Asia* entered war service. Ernie had become familiar with every section of the upper decks, and years afterward, when asked to describe what it was like running messages between bow and stern, Higgs could, even after

Alzheimer's set in, quote the exact number of steps on the various ladders and stairways. "He recalled very specific details," noted Nelson Oliver. "It was like he was standing on the ship, looking at things — pointing things out. I sensed a weakened memory, especially in recalling individuals, but noted nothing when I talked with him in 2002 to suggest dementia. He remembered how the silk trains were waiting near the pier with their engines running and how they left — fully loaded — before he got off the ship. His recollections were accurate and if he did not know the answer he would make that clear."[20]

○ ○ ○

Fourteen months before Walter Oliver joined the *Anglo Colombian* at Cardiff, a six-year-old steerage passenger from that same port city stepped off the 13,867-ton passenger ship *Antonia* in Quebec City. It was September 28, 1924, and his name was Douglas Richard Elworthy.

Accompanying him was his mother, Gladys, and two younger siblings, Terence and Mary. Their father, Thomas, had arrived three months earlier on the same ship; as a shipwright and a carpenter during three transatlantic crossings to Boston and New York, Thomas certainly knew his way around a tub.[21]

Between 1926 and 1931, Douglas only caught glimpses of his father, who was away working on the *Empress of Russia.* On May 25, 1930, while Douglas was completing grade 6, his father jumped off the *Russia* into the ocean to rescue a female passenger who had gone overboard, and in recognition six months later, R. Randolph Bruce, British Columbia's lieutenant-governor, presented Thomas with the Bronze Medal of the Royal Humane Society of Great Britain.

Soon after leaving high school, Douglas and a partner staked a claim on a small gold mine in British Columbia's Cariboo district, but the outbreak of war quickly shifted his focus from land to water. On November 9, 1939, he signed onto the 6,406-ton tanker *Albertolite,* built in 1912 and having undergone three previous name changes. Purchased in 1929 by Imperial Oil, she frequently carried crude from California oil-tank farms to British

Columbia refineries. Undoubtedly, transporting massive amounts of crude through all kinds of weather was risky business, but it did not compare to the dangers ahead. At age 22, on January 23, 1941, Elworthy signed onto the *Asia* as a member of her catering crew where he would work below decks as a pantryman. Earlier that month, the 28-year-old liner had just completed her 307th crossing, and despite showing patches of rust, she was still considered a well-made ship. Of course, there were old salts who thought it a crime that her furnaces were still dependent on coal as she set off for war, meaning keeping good steam would rest with the engineers, trimmers, and firemen, otherwise known as the "black gang."

Elworthy figured he was contributing to the war effort. What he did not know was that within a year his service would reach the height of wartime sacrifice.

○ ○ ○

Geoffrey Hosken would have been nine in 1925, but just seven years later, shortly after Walter Oliver signed off his fifth ship, he embarked on a Pacific treasure hunt the likes of which young Geoffrey Tozer could only dream about.

At age 15, thanks to his father's reputation and strong shipping connections, Hosken became the youngest member of an expedition to Cocos Island, approximately 310 miles southwest of mainland Costa Rica. The goal was to locate buried treasure, and the 4,350-mile voyage to the small island began at Vancouver's Pier D on February 21, 1932. Crewing the motor schooner and arctic trading vessel *Silver Wave* was an odd assortment of two dozen adventurers comprising various nationalities and trades. There were war veterans, sailors, engineers, loggers, miners, "old salts," and boys "fresh from school."[22]

For decades, Cocos had attracted treasure seekers, including two previous expeditions based in British Columbia. The 1920s brought growth and opened the world to further exploitation and adventure, and the Depression had led some to find fortune through bolder plans.

Wishing bon voyage on this excursion were family and friends, the press, and a city councillor who delivered a speech on behalf of the mayor. The

Geoff Hosken went looking for treasure well before signing onto the *Empress of Asia*.

day before the *Silver Wave* sailed, a report in the *Vancouver Sun* had little to say about Hosken other than describing him as a "plain treasure hunter."[23] The flamboyant leader of the expedition, Lieutenant-Colonel John Edwards Leckie, and more seasoned mariners on board captured the limelight, including the schooner's second-in-command, Captain Bob Adams, nicknamed the "Wolf of Alaska."[24]

Also on board was aviator Dennis Rooke, formerly of the Royal Flying Corps, who in 1927 crashed his De Havilland Moth in India while attempting to fly from England to Australia.[25] Another notable was 79-year-old Captain Augustus Whidden, a Nova Scotian who had already participated in two previous expeditions to the island.

Two days from his 60th birthday, Canadian-born Leckie was a veteran of the Second Boer War, the First World War as commander of the 16th Battalion, and the Russian Civil War in command of a unit in the Archangel-Murmansk theatre. He had earned a Distinguished Service Order and the French Croix de Guerre and was made a Fellow of the Royal Geographical Society. However, he was best known for mineral speculation and mining activity in Canada, Mexico, Venezuela, and Africa.[26] So the search for Spanish gold on a remote island surrounded by sharks seemed custom-made.

Despite the lack of mention in the press, the Hosken family had credibility too. Geoffrey's father, Captain A.J. Hosken from Cowes on the Isle of Wight, had served as an officer on various transpacific ships, including the *Empresses of Japan* and *Russia*, and was skipper of the *Monteagle* during a daring rescue mission (see Chapter 10). Geoff's mother, Veronica, born at Port Hope, Ontario, was the daughter of highly respected R. Edward Gosnell, who in 1893 became British Columbia's first permanent legislative librarian.[27]

He also served as private secretary to a number of premiers, including Theodore Davie and Richard McBride. But likely it was A.J. Hosken's connection with the *Empress of Canada* captain Ernest Philip Green that secured a place for his son on the *Silver Wave*. Green had reportedly been involved in an earlier expedition to the island and had attended a demonstration on a fancy metal-detecting device. Fitted into a box, the Metalophone featured a long coil played out horizontally over the ground, and while wearing earphones, the hopeful operator bided his time.

The expedition ended in early 1933 and disappointingly produced no buried treasure, although young Hosken benefited from sea time. Instead of returning on the *Silver Wave*, the 16-year-old and four others sailed to Costa Rica where, after a six-week layover, they boarded a freighter to San Francisco. When he finally reached home, Hosken had been away for 13 adventure-filled months but suddenly had to adapt to a far less exciting life — punching the clock as a machine operator at the American Can Company. All that changed, however, on February 11, 1941, when Able Seaman Hosken joined the deck crew of the *Asia* and met Walter Oliver for the first time.[28]

7

SALT OF THE SEA

He had a grip of steel, but never ever used it with a handshake because he was too much of a gentleman.

— **Norm Hutton**

Walter Oliver was at the Pitt River Government Wharf in Port Coquitlam, British Columbia, when he suffered the stroke in 1974. He was lucky that the sudden loss of blood to his head occurred on a day he was not making another supply run upriver to the logging camps. The outcome would have been quite different had it happened while he was at the helm of the *Express* because everything went dark, his face felt numb, and he could not form proper words.

Unable to control his feet, Oliver stumbled and fell onto the wharf, but fisherman and shake-splitter Dave Thompson, who had been working the wharf, got to Oliver before he joined his much-admired older *Asia* mates who had "gone before" — namely, rugged Second Officer Cecil Crofts, sanitation engineer John Drummond, Chef Archie Gee, and trimmer-turned-storekeeper James Jackson. More recent was the death of Ernest Raymond

Punter, *Asia*'s "iron-grip" quartermaster who died in 1975, the year before Oliver began chronicling his time at sea. Among the older good fellows still carrying on were First Officer Leonard Johnston and Third Officer James Donnelly. All of them, like his younger mates, had been aboard the *Asia* during her final moments and to a man had logged plenty of sea time.

○ ○ ○

Cecil Crofts from Leicester was 15 when he and his mother, Elizabeth, boarded the *Megantic* at Liverpool and reached Montreal in October 1911. His father, John Crofts, who had resided in Canada for two years, was there to greet them when they stepped off the train at Vancouver. After passing his civil service exams at Vancouver Normal School, Crofts began working for the post office and shortly afterward joined the 6th Regiment, Duke of Connaught's Own Rifles, where he learned to fire a rifle. By 1916, while his father was in France with the No. 13 Field Company, Canadian Engineers, the now 19-year-old was already a member of the Royal Naval Canadian Volunteer Reserve at Esquimalt. Just two years later as he was working his way up from ordinary seaman to leading seaman, he had served on the East and West Coasts, as well as several stints aboard His Majesty's Canadian Ship *Rainbow*.[1] At 22, the quick-thinking Crofts was in possession of the General Service Medal and Victory Medal when discharged at Halifax from HMCS *Niobe* as a yeoman of signals.[2]

After returning to Vancouver, the young man resumed work at the post office, but there was no doubt in his mind where he needed to be. He had found time in 1919 to write his papers to become a coasting mate, and later that same year, signed onto the *Prince Albert* as third officer. Over the intervening years, Crofts gained considerable sea time on a half-dozen ships and went on to obtain his master's certificate in 1927.[3] During this time, he travelled up and down the West Coast; sailed across the Pacific to Yokohama, Osaka, and Auckland; traversed the Indian Ocean to Bombay; and while aboard the *Canadian Skirmisher* as third officer, visited New York in August 1925.[4] It seemed every time Crofts turned around he was, in nautical terms, watching yet another harbour "go astern." It was remarkable that he found time in June 1926 to marry his sweetheart, Gertrude Adamson. Signing on as the *Asia*'s third officer in August 1934, Crofts cycled through a series of Depression-era promotions and demotions but eventually settled into the duties of ship's second officer in September 1940.

○ ○ ○

Crofts and Oliver were significantly younger than Scottish-born John Drummond, whose extensive ship-borne knowledge was nicely balanced by a good measure of artistic talent. Born in Edinburgh in 1882, Drummond spent his younger days on the Isle of Man, working as a plumber on coastal vessels running between the British Crown dependency — located in the Irish Sea — and the Scottish mainland. For a time, he resided at Saltcoats, a town on the west coast of North Ayrshire, Scotland, deriving its name from the salt-harvesting industry in the Firth of Clyde. In 1922, Drummond and his family relocated to Canada aboard the CPOS liner *Victorian* and settled in Vancouver where he returned to the sea.

Of average height, somewhat portly, and with the face of a pugilist, the easygoing Drummond became *Asia*'s sanitation engineer in 1926. His first voyage departed Vancouver on February 8, just 10 days after she arrived in port with details of a raging storm that had swept a young American baseball player overboard to her death (see Chapter 10).

Drummond experienced first-hand the domino effect delays created on ocean travel after the *Asia* was forced to postpone departure by several hours, owing to an Atlantic storm that resulted in the late arrival of European passengers and mail, with connections through Halifax to the West Coast. In addition to embarking 300 passengers, the ship left with a cargo of automobiles, motorcycles, tires, 500 tons of asbestos, and 1,200 tons of flour. More unusual was a shipment of trout eggs that Charles Elliot, the British ambassador to Japan, planned to introduce to lakes and streams at the prestigious Tokyo Angling and Country Club.

The job of sanitation engineer was crucial on a passenger liner, or indeed, any large ship. Drummond had to ensure the *Asia* took on an adequate freshwater supply at each port, since without it the vessel would be unable to function. Still, it could be easy for some passengers to overlook the significance of his work or form impressions based on his title or the dark stains and smudges on his white coveralls. There is a photograph showing Drummond leaning against the ship's railing next to his good friend and fellow Scot, Tom McMurtrie, who was still with the ship years after her maiden voyage. Drummond in his battered peaked merchant sailor's cap, work boots, and coveralls is in contrast to the grey-haired boilermaker, McMurtrie, who sports a pipe and is spotless in his summer whites against the fog.

When not checking the water supply, the sanitation equipment, or the vast labyrinth of pipes used to move cold, hot, and wastewater throughout

John Drummond (left) and Thomas McMurtrie aboard the *Empress of Asia* in the 1930s.

the ship, the affable Scot spent free time pursuing his passions as a marine artist and amateur music composer. His favourite spot was in his cabin astride a stool beside the small porcelain sink — a portable, flip-up easel wedged in place. On the wall opposite him was a collection of already completed paintings, each edged in handmade wooden frames. It was relaxing, some would say meditative, to be so engrossed in an endeavour that added tiny strokes of blue-grey ocean or a dab of white to the forecastle.

Music, inspired by the sea, deeply moved the man who sometimes slipped into the first-class lounge to sit at the grand piano beneath the circular skylight and pencil in the notes or accidentals of his latest composition. A huge admirer of Scottish novelist and poet Robert Louis Stevenson, his particular favourite was "Requiem" with its poignant line "Home is the sailor, home from sea," words ultimately inscribed on his headstone at Vancouver's Mountain View Cemetery. Fortunately for historians, Drummond kept a detailed journal of that time, which remains one of the best accounts of the *Asia*'s final days (see Chapters 15 to 17).

○ ○ ○

Arthur (Archie) Ernest Gee was an old man of 32 when Walter Oliver first went to sea. Born September 12, 1893, he, too, started early — before his 18th birthday in 1911 and prior to the *Titanic* disaster.

One of seven children born to William and Annie Gee of Kirkdale near Liverpool, his birth certificate shows his father's occupation as ship's cook, which no doubt left his mother the task of raising five boys and two girls in their working-class neighbourhood. Pressured into helping make ends meet, young Archie started work in 1908 earning a dollar a week, but when it became apparent he could earn three to four times that at sea as a larder cook, it only made sense to follow in the wake of his father.

What Gee lacked in height, he made up for in strength with a stout, muscular body and bollard-like legs. The young man with wavy light brown hair also loved to box. A photograph shows him bare-chested in tight dark shorts with his left arm extending into a fist and his right hand coiled above his hip, ready to strike.

Still single in 1913, the 20-year-old sea chef suddenly fell in love during a family visit. His sister, Rosa, was planning a party, and she asked her mother if her piano teacher could attend with her sister, Flora. A courtship, kindled by earlier glances, blossomed between Archie and Flora, and wedding plans were soon underway, but so, too, was Archie's sea time during the First World War.

Survival was top of mind, but Gee was one of the lucky ones. He weathered the loss of two ships by enemy action, one containing the decorated white cake his mate had baked in the galley for his upcoming wedding. Archie Gee's grandson, Dave Thackray of Vernon, British Columbia, clearly remembers his grandfather explaining how he got off the vessel:

> It was the second time he had to abandon ship. People were panicking. Many jumped into the water, but my grandfather had some experience with this and so after assessing the situation soon realized the ship was not going down fast. He calmly went below to his bunk, packed a few belongings into a little suitcase, and grabbed his greatcoat off the hook. Back on the main deck, instead of jumping into the North Sea, he asked the deck crew to lower him in a lifeboat, presumably along with others who had not jumped. Once alongside the doomed ship, he helped pull crew out of the water into the boat and as they sat there shivering they were impressed to see the chef fully dressed and dry with his greatcoat and suitcase.[5]

Archie and Flora were married in the summer of 1918 in West Derby, Lancashire, and shortly after the war, Archie commenced work as a cook on transatlantic passenger ships. Meanwhile, his parents and siblings had moved to Canada and settled in Vancouver where in 1920 his father, William Gee, signed onto the *Empress of Asia* as chief cook and chef. Interrupted by two brief stints on the *Empress of France* and *Empress of Canada*, William held that job until 1930 when he handed the galley over to a highly creative Italian-born chef (see Chapter 9).

Archie and Flora Gee wed during the First World War.

In May 1928, just before the world plunged into economic depression, Archie, who had graduated to grill cook, Flora, and their two children, Olga and Archie Jr., left Liverpool and joined the rest of the family in Vancouver. It was difficult for newcomers to settle and find work, but on August 1, Archie signed onto the *Canadian Farmer*, a 2,410-ton ship built by Collingwood Shipbuilding in Ontario for the Canadian Government Merchant Marine. However, months later he was back on terra firma as cook at Hotel Georgia, one of Vancouver's premier hotels. Like many families during the Depression, the Gees watched every penny and always paid up front. Avoiding debt was one of Archie's golden rules, but missing a mortgage payment on their small wooden-sided bungalow was unavoidable when Archie Jr. required an appendectomy.

Bringing a third child into the world was out of the question, yet little Beverley arrived in 1933, six years younger than Archie Jr. and eight years

Archie Gee signed on as chef of the *Empress of Asia.*

younger than Olga. "I was a mistake. There was no question about that," recalled Beverley, who turned 88 in 2021. "I never felt like I was a mistake or anything, and it didn't bother my upbringing. Mom and Dad were struggling like everybody else and so when the chef's job on the *Empress of Asia* came up, he promptly took it. I'm sure it had something to do with him getting a bit of help from his father."[6]

On June 22, 1939, Gee signed Articles of Agreement to join the *Asia*'s large catering crew, and while acutely aware of how big and famous the ship was, what mattered most was prepping meals for hundreds of passengers that equalled or even surpassed similar dishes offered in posh restaurants and hotels — not realizing the ship's catering department would soon be feeding hungry troops, not tourists.

Between June 1939 and January 1940, Gee was seldom home, completing 10 round trips from Vancouver to Asia, with layovers in Vancouver lasting anywhere from seven to 12 days. During his time in the galley, he worked under three different captains and was on board the *Asia* in September 1940 when Japanese aircraft bombed her off Japan, more than a year before Pearl Harbor (see Chapter 13).

While still at Hotel Georgia, young Beverley liked to surprise him. "I was about three and we had a towel rack on the back door and I remember hiding there. The towel would cover only my head, but, of course, I thought Dad couldn't see me, and so he would come in saying, 'Oh, where's Bev? Where's Bev?' Then he would say something like, 'Oh, she's gone! Nobody knows where she is!' Meanwhile, I'm thinking I have a good hiding place. I always remember how he did that and how he would often — on payday — arrive home with a 'diamond' ring for me from Woolworth's."[7]

Mesmerizing for Beverley was the butterfly tattoo on the inside of her father's left forearm. "I'd be sitting on the back porch with him and he'd move his hands in such a way that it would make the muscles in his arms move. It would look like the butterfly was flying, and, of course, he would tell me it was flying away."[8]

Not a man of many words, Gee would sit and listen while Flora did most of the chatting during social visits. He had a good sense of humour, though, and enjoyed whistling "bits of tunes" while he worked in the small garage or out back with his beloved German shepherd, Fritz. Moreover, the patient and loyal dog seemed to enjoy listening to Archie when he played the banjo next to Olga at the piano.

Flora ran the household, and despite her husband's culinary talents, donned an apron and cooked most of the meals because, as Beverley noted, her father "was fed up with cooking by the time he got home, and he always praised her for what she had baked. They were very much in love. They didn't show a lot of affection, either one of them, but you just knew they were in love. He never mentioned another woman's name — ever. It was always about Mom."[9]

Life got more serious for the Gees when Archie, at age 45, decided to remain with the *Asia* after the outbreak of war. Despite working hard and living responsibly through the darkest days of the Depression, the Gees soon faced two paralyzing wartime losses (see Chapter 18).

o o o

Ernest Raymond Punter was not especially tall, but he was solid. He never pumped iron for exercise — his heavy lifting was on the job or in assisting friends or relatives. Norm Hutton of Powell River, British Columbia, says his uncle was naturally strong. "He could pick up a full-sized sofa, flip it on his back and either carry it down a flight of stairs and place it in the back of a van or run up a flight of stairs and put it down gently in an apartment, then ask, 'What's next?' He had a grip of steel, but never ever used it with a handshake because he was too much of a gentleman."[10]

With his thatch of thick sandy-coloured hair and broad features, Ernest looked as if he belonged in a rugby scrum when, in fact, he was a mellow individual who loved to sing as a bass-baritone and was always impeccably attired. His dress shirts had detachable Windsor collars accentuated by ties displaying a neat Windsor knot, and his suits always appeared freshly dry-cleaned and perfectly hemmed above black dress shoes polished to a military shine. Ernest's nephew remembers him as patient and generous, someone who would step in as peacemaker between "unbending types."

The second youngest of five children, Punter was born on April 30, 1911, at Luton, Bedfordshire. His father, Edward, was an accountant, and his mother, Annie, a seamstress, which may explain why he dressed so well. His parents were both quite musical. Annie sang soprano and played piano, while Edward performed in a men's quartet. On May 31, 1927, not long before the Punters relocated to Canada, son Ernest enrolled as boy seaman, second class, assigned to a Royal Navy training establishment. His nephew recalled him saying how he "cried for his mother after joining up" and leaving home.

Punter rose from boy seaman to ordinary seaman and served on three warships, including the old Iron Duke–class battleship HMS *Benbow* and the heavy cruiser HMS *Shropshire*. When he left the navy, he joined his mother, father, and siblings in Canada where during the Depression he participated in prospecting trips into the BC interior. He also found work as an electrician before returning to sea as a deck hand on merchant ships. In February 1941, Punter signed onto the *Asia* as her quartermaster, and naturally, because of his physical strength and previous experience, the junior ratings respectfully called him "Mister Punter."

As ship's helmsman, he worked alongside the captain or officer of the watch and local port pilot to safely bring the vessel in and out of harbour and navigate each port's unique channels, shallow water, active tides, and other ships. As such, the upkeep and use of the *Asia*'s navigational equipment were also part of his remit.[11]

o o o

First Officer Leonard Johnston and Third Officer James W. Donnelly were prominent in Walter Oliver's memory as very reliable officers who, like Crofts, were among the last to leave the burning ship. Johnston was five years older than Oliver. He had grown up on Vancouver Island in Esquimalt and lost his father in 1920 when he was 14. Three years later, before his mother remarried, he signed onto the *Asia* as a cadet. As his career took off and he went from one ship to another, Johnston found himself moving so often between officer levels that it was a wonder he could remember what rank he held at any particular time.

In January 1932, while serving as the *Russia*'s fourth officer, he married Elizabeth Mary Toye, a stewardess he had met during previous voyages. Before leaving the *Russia* in 1934 to serve as second officer on the new 26,032-ton *Empress of Japan*, Johnston assisted in rescuing four sailors on a waterlogged Chinese fishing junk southeast of Hong Kong. The water was so rough that the *Russia*'s commanding officer, A.J. Hosken (Geoff Hosken's father), positioned his ship to shield both the junk and the *Russia*'s No. 3 lifeboat from the wind. When the junk's distraught captain declared his intention to go down with his boat, members of the rescue party physically manhandled him into the lifeboat. Another rescued crew member, suffering from exposure and shock, underwent treatment while the ship completed her voyage to Hong Kong.[12] As Pat Terry, marine editor of the *Vancouver*

Leonard Johnston was a highly competent first officer.

Sun, noted, "It was another act in the long drama of the sea" featuring life-saving measures initiated by one of the Empresses (see Chapter 10).[13]

Promotions followed Johnston onto the *Empress of Canada* at Hong Kong in February 1939 when he became first officer, then chief officer, a step below captain. In mid-October 1940, the highly experienced Johnston signed on as the *Asia*'s first officer.

o o o

James Donnelly, a native of Liverpool, was 21 when Walter Oliver went to sea, and was third officer by the time the *Asia* embarked on her final voyages. At the age of four, Donnelly accompanied his mother, Hannah, and a brother across the Atlantic to Quebec City aboard the *Empress of Ireland* on their way to join their father, William, in Vancouver. By the time James turned six in 1911, the family was residing in Nanaimo on Vancouver Island where William Sr. worked as a labourer, although his wartime enlistment papers also state he was a firefighter.[14]

The boy was 10 when his father headed to Victoria to enlist in the Canadian Expeditionary Force. After saying goodbye to his family, William crossed the country by train and proceeded overseas from Halifax on the SS *Lapland*. He would never see them again. While serving in France with the 72nd Battalion, William went missing on March 1, 1917. However, it took until October to list him among the dead; his name was later inscribed on the ramparts of the Vimy Memorial alongside more than 11,000 other Canadian soldiers posted as "missing and presumed dead" in France.

Two days before Christmas 1920, 16-year-old Donnelly signed on in Vancouver as apprentice on the *Harold Dollar* of the Dollar Line, a position held until at least April 1922. Three years later, while Walter Oliver was circumnavigating the globe, Donnelly was an able seaman aboard the Canadian Government Merchant Marine *Canadian Rover* before returning once more to the Dollar Line as deck officer. Then, as chief officer on the *Robert Dollar* and on the *Chief Capilano*, he added more sea time making regular calls to Asia.

Six days before he and his wife, Violet, celebrated their third wedding anniversary, Donnelly signed onto the *Asia* on June 8, 1936; less than six years later he refused to abandon ship until her final moments and days later had a key role in the escape from Singapore.

○ ○ ○

When it came to deciding whether to falsify his age or forgo an opportunity for employment, James Jackson by necessity settled on the former. Signing onto the *Asia* on February 10, 1941, as part of the delivery crew to Liverpool, he listed himself as being in his forties when, in fact, he was 56.

Born November 1, 1884, at the market town of Chester-le-Street in County Durham, south of Newcastle upon Tyne in England, Jackson's hometown was close to where Walter Oliver spent his youth at Easington Colliery. Both were Geordie boys, but Jackson's childhood was tougher than

Oliver's, especially after his mother died of tuberculosis and his father remarried. He and his brother, Bob, were shipped off to an orphanage where they remained until claimed by their maternal grandparents. At age nine, Jackson entered the coal pits in Bebside, Northumberland, until 1929 when he relocated to Canada and resumed life as a miner at age 45.

Incredibly, in 1941, Jackson was not the *Asia*'s oldest crew member when the ship entered the war. In fact, more than a dozen other men were older, a few by several years. At five foot six, but tough, Jackson relied on his strength. His job as trimmer demanded resilience as he delivered coal from the mountainous heaps in the bunker to the firemen at the blazing furnaces. It was dirty, brutal work, and portraying oneself as a younger man was possible with displays of physical strength and a face and hair smeared with coal dust.

Like other members of the engine room crew who signed onto deliver the ship to Liverpool, Jackson had the option to return to Canada before the ship commenced her wartime voyages. A wage of $90 per month was good, and so it was not a decision taken lightly.

○ ○ ○

Leaning over his typewriter, Walter Oliver knew he, too, had made the most of his time throughout the Depression when work for many was non-existent or difficult to find. Before signing onto the coal-burning *Canadian Beaver* at the height of the Depression in September 1933, he worked three other merchant ships and joined the Royal Canadian Naval Volunteer Reserve at HMCS *Naden*, a shore-based training facility on Vancouver Island. He remembered feeling quite at home with barrack life and squad drill, with ratings from the Prairies, many of whom had never been to sea, but he especially enjoyed firing a rifle and the six-inch naval gun. He racked up two weeks of "sea time" on HMCS *Vancouver*, although the old destroyer never left the quay, so he considered it a paid holiday.

Prior to joining the minesweeper HMCS *Armentières* on the west coast of Vancouver Island at Bamfield, he worked the inland coastal waters, feeling at times like an itinerant longshoreman while signed onto the Union Steamship Company's *Catala*. A round trip lasted six days, transporting loggers, cannery workers, and supplies between Vancouver and Prince Rupert, returning with a hold full of canned salmon, and an occasional corpse — carried double fare.

It was the *Beaver*, though, that gave Oliver his more memorable Depression-era lesson on problem-solving. When she ran out of fuel in rough, dangerous water, he and fellow crew members used fire axes and crosscut saws to cut deck cargo, mainly pallets and wooden shipping crates, into five-foot lengths to feed the furnace.

On the evening of September 3, 1939, Oliver was aboard the tanker MS *Henry Dundas* at Tandjoeng Oeban, nearly 280 miles north of Palembang, Sumatra. The ship was loading "motor spirit" bound for South Africa. That was when the 29-year-old heard Britain and France had declared war on Germany at 11:00 a.m. London time. For merchant sailors, this spelled the end of routine ocean voyages.

At nearby Singapore, the ship's master received wartime instructions, sailing orders, code books, and a weighted steel box from British naval

While working ships in Southeast Asia, Walter Oliver figured it made sense to keep a cabin near the oil refineries at Palembang, Sumatra.

authorities. This box, along with its classified documents, was to go over the side before the ship fell into enemy hands or was lost or abandoned. Wireless communication became severely limited, and as Oliver recalled, there was no time to apply dull grey paint to the British-registered ship before she departed for Durban, a decision that might have saved her from attack.[15]

Meanwhile, the German pocket battleship *Admiral Graf Spee* was about to wreak havoc in the South Atlantic. After leaving Wilhelmshaven, Germany, on August 21, 1939, the warship began hunting Allied merchant shipping through the early fall and into mid-December.[16] From Cape Town, the *Henry Dundas*, still sporting peacetime colours, commenced a dangerous crossing to Trinidad in the West Indies. On the evening of December 1, an unidentified warship passed the tanker, heading in the opposite direction. The next day, south of the island of St. Helena, another merchant ship, the *Doric Star*, fell victim, sunk by the *Graf Spee*, which also captured the crew. Oliver was sure the lack of grey paint had spared the tanker because the Germans might have mistaken her for a neutral ship.

From the Caribbean, the *Henry Dundas* steamed to Halifax where she joined a small transatlantic convoy. Although scattered by storm and deliberately sectioned off, the battered ships reached Thames Haven east of London in early January where the lucky *Dundas* delivered her petroleum. She was later re-enforced with cement and fitted with guns, black screens, and degaussing equipment to protect against magnetic mines. Then after sampling the sights of London, Oliver rejoined the ship for another tense round trip to the West Indies, sitting on, for half the voyage, 12,000 tons of high-octane fuel.[17] Assisting with North Atlantic escort duties during her return voyage via Halifax was the battle cruiser HMS *Repulse*, later destroyed off Malaysia, northeast of Singapore (see Chapter 15). There was one more return trip to the West Indies before Oliver signed off the *Henry Dundas* in August 1940 and boarded a Swedish ship, the *Axel Johnson*, bound for Montreal in late September. By then, the Netherlands had fallen, Belgium had surrendered, British and French forces were driven back at Dunkirk, France had fallen, and the Battle of Britain was well underway. "While [aboard the *Axel Johnson*] at anchor on the Mersey," recalled Oliver, "in came German planes attacking the ships, during daylight, and dropping their bombs."[18]

Oliver certainly had to tip his hat to the many fine men he had sailed with up to then. They — and the ones he eventually got to know on the *Asia* — were the last of a fine breed with oceans of experience. Crofts, Donnelly, Drummond, Gee, Johnston, and Punter, for example, had all come together by circumstance. Most left loved ones behind, but others had nobody. The good ones worked hard, knowing a well-run ship was the sum of her crew and — crucially — dependent on clear orders from the captain and chief engineer. Most, too, understood the occupational hazards they faced, and the grim prospect of being lost without a trace.

As Oliver contemplated those interwar years, which were both difficult and rewarding, he recognized there was a huge piece of the *Asia*'s story he had not experienced. His time on cargo vessels and tankers had coincided with the *Asia*'s voyages during the Roaring Twenties and Depression when the great ship, in addition to transporting freight, silk, and mail, was fulfilling what she was meant to do: conveying vast numbers of tourists, organizations, and business executives across the Pacific.

Worldwide, travelling by steamship had become popular and exciting again, and although visiting Egypt was a top choice owing to its pyramids and newer discoveries just west of the Nile River, many were drawn to the hand-crafted offerings of China and Japan. People had to visit, see it, and purchase it for themselves, and so the big steamship companies on the West Coast had quickly capitalized on the heightened "lure of the East." The liners and their heavily promoted offerings in the 1920s became extensions of shore-based entertainments, with excellent food, silent pictures, full orchestras, and of course, jazz — a huge hit in the Prohibition-era speakeasies.

Oliver also recognized that the 1920s ushered in more affordable and comfortable travel for the middle class. Suddenly, there was a microcosm of society shuffling up the gangways exhibiting the latest styles and a greater sense of personal freedom, especially among women. The war was over, and it was time to dance and perhaps, if not too frightened or shy, make friends and fall in love with the ocean again.

8

RISING TWENTIES

In the afternoon, we went to a reception given by Mr. and Mrs. John D. Rockefeller, Jr. and met members of their party.

— **Passenger William F. Russell**

The *Empresses of Asia* and *Russia* emerged from the First World War wearing their dull grey paint scheme with parts of their camouflage, known as dazzle, still showing. They were less than 10 years old, in excellent shape, and during an overhaul at Hong Kong, *Asia* received fresh paint, including a deep black hull with green "boot-topping," a kind of pinstripe along the top of her hull. She also acquired a brilliant white superstructure, and for her funnels, a light brown-orange colour known as buff.

To Walter Oliver that was when she looked her best — a proper Empress by the time Henry and Mary Frances McNeal and their 21-year-old daughter, Helen, stared up at her from Vancouver's Pier D on June 22, 1921. As the crow flies, the three Americans were 1,650 miles from home in Joplin, Missouri, and shortly they got the opportunity to peek into their cabin — No. 139, portside, Bridge deck — a day before sailing.

o o o

Skippered by Captain Alfred James Hailey, the *Asia* had been alongside since Monday the 13th after knocking an impressive two hours off her fastest steaming time across the Pacific. Apart from fog along the Japanese coast, the weather had been fine for "fast passage and the *Asia* ploughed the main at a good 20 knots for the greater part of the route."[1] When she arrived in Vancouver, her passenger list was 732 with 429 in steerage, including many who hoped to make a fresh start in Canada (see Chapter 11). While the Victoria press heralded the ship's speed, it noted there was "nothing to be gained … in making this side of the Pacific ahead of schedule," owing to train timetables. After departing Victoria, the *Asia*, carrying another precious cargo of silk, had loafed around on her chains in Burrard Inlet prior to reaching Vancouver where separate passenger and silk trains awaited.[2]

The temptation to smash transpacific speed records remained strong in the 1920s and breaking them produced more bragging rights along with marketing opportunities. Three years later in July, the *Asia* did not break the existing record — set by the much newer *Empress of Canada* on June 17, 1923 — but she came close at eight days, 14 hours, 48 minutes.[3]

With heavy postwar passenger and cargo volume returning to the transpacific routes, there was renewed optimism, even before the *Canada* and *Empress of Australia* joined Canadian Pacific Steamships (CPSS) — the new name of Canadian Pacific Ocean Services (CPOS) as of 1921 — in April and June 1922. The company emerged from the war with terrible losses but had aggressive plans to promote its long-standing All-Red Route. This enthusiasm and key business decisions led to a profitable decade that bid farewell to the old but reliable *Empress of Japan* and the venerable *Monteagle* that year. E.W. Beatty, the CPR's president, reported that by 1921 traffic had increased by more than 200 percent over the fiscal year ending June 30, 1914. Between June 30, 1920, and the time the McNeals stepped aboard a year later, the service had carried 35,555 passengers across the Pacific, including 9,761 first class, 1,854 second, and 23,940 steerage. The volume for 1914, all classes, was 14,575.[4]

These increased traffic volumes boosted competition for both passengers and freight, much of the latter being raw materials bound for Japan, in a region experiencing tremendous manufacturing and export growth. With most of that competition now coming from the United States and Japan, the CPR was quick to emphasize this in its 1922 annual report. The United States Merchant Marine was operating 10 passenger and cargo ships from San Francisco and Puget Sound to China and Japan and other East Asian countries — they were the "largest, most expensive merchant ships built in the United States."

On the heels of a major announcement for a new dry dock in Vancouver, CPOS general freight agent W.T. Marlow reported in June 1920 that cargo space used to transport repatriated members of the Chinese Labour Corps was now available for commercial freight.[5] Meanwhile, the All-Red Route continued to shine, supported by mail delivery contracts and passengers intent on "seeing the world." Modernized accommodations throughout, with cabins and open bunks in steerage, proved a vast upgrade over the age of sail, while in first class, the number of berths rose from 296 to 377.[6]

For those in first class, expectations went beyond comfortable, safe passage. Privacy, a well-appointed cabin with fine linens, hot running water, and workable switches for the bed lights were paramount. The company also understood that courteous and prompt service in the dining saloon, lounge, and smoking room was not a luxury but an expectation, as was the high demand for popular entertainment. The *Asia*'s first screening of a motion picture occurred on February 19, 1920, with five comedies featuring Hollywood stars and five documentaries extolling the Canadian landscape as viewed from CPR trains and hotels.[7]

○ ○ ○

On the morning of the McNeals' departure, Mary and Helen went shopping for a hat with a brim. After all, they were just hours from boarding the ship, and a fashionable hat would look splendid while keeping the sun off Helen's face during the crossing.[8]

For those who could afford the latest fashions, the challenge during most pre-voyage shopping excursions was more about selection. The 1920s released waves of new looks that spilled into shop windows with colourful, statement-making styles. Cotton and wool remained popular, but nothing matched the soft, flowing feel of a silk gown, the creation of which depended on ships like the *Asia*.

The CPR's transcontinental railway and fast transpacific ships were perfect for moving raw silk, mostly from Yokohama to Vancouver.[9] Postwar competition from Japanese and American lines, particularly the Seattle-based Admiral Oriental Line, upped the ante. However, insurance rates and competition meant only those who could guarantee fast, secure delivery would continue to profit. The heavily guarded silk trains were equipped with lined, airtight freight cars, and once the 60-pound bales were loaded and secured, the trains had priority as they raced across the country.

Vancouver remained a leading North American silk port, and two months before the McNeals went shopping, the *Asia* had arrived with $8.5 million worth of product, 90 percent of it raw. Beneath the headline "Cargo of Silk Set New Record," the *Vancouver Daily Province* noted the "Empress of Asia, instead of being satisfied with creating a dock-to-dock record for speed, also set a high-value figure in importing silk."[10] It noted that while some of the cargo was bound for Europe, most was destined for eastern North America.

A *Victoria Daily Times* story emphasized priority given to the shipments, then valued at over $7 per pound. The CPR, in conjunction with the New York Central, delivered 3,697 bales from Yokohama to Vancouver to New York in just 13 days, 16 hours — and nine days on the *Asia* and four days crossing the continent.[11] Still, it depended on the stevedores, who raced on board and stacked the bales onto a conveyer belt or wheeled the product into railway cars, using hand trucks.[12]

When the *Asia* came alongside on the afternoon of Monday, May 17, 1926, there was enough raw silk to fill 13 railway cars, all of it destined for France. It took 14 minutes, sometimes less, to load one car, and the train departed three hours after the *Asia* docked.[13] However, the largest shipment transported by the ship arrived in August the following year. Valued at $12 million, it weighed 1,000 tons and filled 20 railway cars.

○ ○ ○

Henry Harrison McNeal did not join his wife and daughter in their quest for the perfect hat. As the head of the family, he had decided to take his family on a six-month world tour. Although there was never a good time for a vacation, "Papa Mac" could afford it and perhaps might actually enjoy it — a good decision in retrospect, since angina pectoris claimed him at age 63 in 1929. Born just months after the U.S. Civil War, Henry was an American success story, the son of Ohio- and Kentucky-born parents who owned a prosperous farm in Naples, Illinois. He split his time between school and farm until age 16 when he started work as an engineer on an Illinois riverboat. By the early 1890s, the young man was in Kansas City, employed by a building contractor who owned a zinc mine northeast of Joplin.

As the mine's property manager, Henry spotted a business opportunity. Mines needed reliable machinery to extract the resource, and soon McNeal & Company was born, incorporated as McNeal Machinery Company with Henry as president and general manager. The company turned a profit building and repairing high-demand machinery for commercial and industrial enterprise, a success leading to other manufacturing achievements.[14]

The launch of the McNeals' world tour depended on coordinating railway schedules. With business in Chicago, Henry left by train a day ahead of his wife and daughter, who headed north with a stop in Kansas City where they shopped, ate, attended two shows, and had their fortunes read. Reunited at St. Paul, Minnesota, the family boarded another train and crossed the Dakotas into Canada, through spectacular Banff and Lake Louise, reaching Vancouver on the summer solstice. By then, Mary had assumed the duties of scribe, noting on Wednesday, June 22, how well pleased the family was after viewing the beautiful ship and gazing into cabin 139.[15] They were among 850 passengers when they strode aboard the next day, each of them holding second-class tickets, suggesting that Henry, even with his wealth, was no spendthrift.

The first-class cabins on the Promenade and Bridge decks included many notables who, depending on their itineraries, might rely on half a dozen currencies by the time they had completed their ports of call.

Among them were Lo Wen Kan, vice-president of the Peking Supreme Council, returning to China after an international conference in Europe; J.J. Abbott, vice-president of the Continental Bank of Chicago; Dr. R.H. Torrey, "an evangelist of international repute"; and influential Dr. John Ferguson, financial adviser to the Chinese president and renowned authority on China.

The Canadian-born Ferguson was famous for speaking his mind. In 1919, during the repatriation of the Chinese Labour Corps from war-torn Europe, he objected to the poor treatment the Chinese labourers received during their cross-Canada railway journey (see Chapter 5).[16]

Earlier in the week, Dr. Torrey, dean of the Los Angeles Bible School, delivered keynote addresses at the "commodious tabernacle" near Vancouver's Canadian National Railway station. There the faithful heard staples ranging from "The Resurrection of Christ" and "Why I Believe the Bible to Be the Word of God." Looking forward to delivering that same message in China, the preacher was also excited about seeing his missionary son. "Dr. Torrey again proved his ability as a speaker by the manner in which he handled the subject and drove home his points to a large audience," praised the *Vancouver Daily Province*.[17]

If Torrey read the article, he might have noticed the story on page 17 entitled "A Sabbath Disturbance." It described how a "Scotch Presbyterian C.P.R. constable called down fire and brimstone" on a group of Chinese sailors playing cards on the *Empress of Asia* that Sunday. Shocked by such behaviour on the Sabbath, the "brawny officer" marched up the gangway and attempted to halt the gambling until one of the Chinese crew "tangled" with him. Equally resistant, the other players "buzzed down on the Scot who beat a strategic retreat to the deck," where the ship's officers "made a human barrage of themselves and held back the sailor mob until the policeman had made the gangway and escaped."

The story noted how any dignity associated with the constable's retreat from the *Asia* was "marred by the Chinese making him the target for wooden shoes which they took from their feet and heaved down from the steel cliffs of the Empress." It concluded by stating: "Calm Sabbath peace rested on the scene after the bluecoat had finally disappeared."[18]

Meanwhile, for the McNeals, the evening included a three-course meal in the second-class dining saloon and the likely sharing of bon voyage telegrams before the muzzle-loaded Nine O'Clock Gun at Brockton Point in Stanley Park marked the hour, its boom echoing across Coal Harbour. Simultaneously, the "bo's'n blew his whistle and up went the gangplank on the big *Empress of Asia*. She pulled out to sea wafting back sweet strains from the Filipino orchestra across the water," noted the *Vancouver Sun*. "Several hundred people stood on the Pier D promenade as the big ship moved away, waving bon voyage to their friends bound for the Orient."[19]

"We sailed at 9 P.M.," wrote Mary. "Watched steamer out to harbor and went to bed about 10:30 P.M." The family had marked the end of a perfect day, save for the new dress left behind in "Van Couver [*sic*] through the fault of purser."[20]

When the McNeals slipped out of bed the next morning, the *Asia* was at Victoria. By 9:30 a.m., she was underway again, steaming northwest through the Strait of Juan de Fuca, past Race Rocks, and into the unpredictable grey ocean. Receding off her port side was the northwestern tip of Washington State, the last bit of America the Missouri family would see for some time.

Promenade strolling and ocean-gazing filled the morning before the family retired to the parlour's warmth. A nap followed lunch, but unfortunately so did leaden skies and cooler air. The barometer dipped and rain came on Saturday, June 25, just in time for the *Asia*'s first fire drill. The next day, with the ship enveloped in a fog thick as wool, her master became more circumspect.

○ ○ ○

If one had to rank the *Asia*'s captains from 1913 to 1942, the cautious but pragmatic Commander Hailey would be close to or at the top. Fourteen months earlier he had returned the ship to Victoria after a rope became entangled in a propeller. While his white beard presented the look of a quintessential sea captain, his nickname "Bricks" was a reminder of the fiery red beard he once sported.

Well liked by both crew and passengers, Hailey, the son of an Anglican minister, had learned from experience to rely on his good judgment, especially in fog. The inquisitive Mary McNeal had no way of knowing that, for Hailey, fog revived painful memories of his beloved mother, Elizabeth.

In late May 1914, Hailey had been on a run across the Pacific when the ship received a wireless message stating the *Empress of Ireland*, bound for Liverpool, had sunk within minutes in the St. Lawrence River — a collision in heavy fog with the Norwegian collier *Storstad*. Elizabeth, then a widow and listed as Mrs. David Trail Hailey, had recently visited her son and his wife, Marion, in Vancouver and was among the 1,014 passengers who consequently died in Canada's worst maritime disaster. Hailey's granddaughter, Nelle Picard, said it was an awful situation because when he received the news, he could not return to Canada to be with his family.[21]

Also on board the *Ireland* that fateful day were four men who had worked in *Asia*'s engine room during her maiden voyage. While Henry Briard, James Rankin, and Albert Smith survived, Walter Scott, who had served as the *Asia*'s third electrical engineer, disappeared in the frigid river, his body never recovered.[22] All four were returning home as repatriated supernumerary crew members from the *Asia*, which meant that while on board the *Ireland* they received nominal pay with little work required as they travelled free. Scott was 26, a native of Montrose, Scotland. He had apprenticed as an electrical engineer before joining the shipbuilding yard at Fairfield that had built the *Ireland*. It had been Scott's hope to surprise his mother with a homecoming visit, so she had no idea he was on board.[23]

Captain Hailey had, noted his granddaughter, "swept Marion off her feet." The couple married at Vancouver's St. James Anglican Church in 1910, a parish rooted in Anglo-Catholic tradition. At the time, Hailey was captain of the *Empress of India*, and fittingly, the church's interior swayed with decorative palm fronds. He remembered his mother looking on, dressed in a violet charmeuse gown and hat, as Marion came down the aisle wearing an embroidered Indian silk gown. Around Marion's neck hung the lustrous pearl pendant Hailey had given her as a wedding gift.[24] For the honeymoon, the skipper and his bride stayed on land and travelled through the Rockies to eastern Canada, eventually raising five children with the help of a nanny and a Chinese houseboy.[25]

○ ○ ○

Mary McNeal's succinct diary entries became noticeably shorter the morning of June 27. "Feeling mean," she wrote. "Had lunch in room. Water very rough. Remained in room all day, but ate all of my meals. Boat made 461 miles. Turned back 1 hr." Up on the bridge, Captain Hailey adjusted the time based on the speed and direction of his ship. This ensured the liner would arrive on time at her port of call. It also allowed passengers to adjust their watches and keep track of the ship's progress. With the North Pacific shipping large swells the next day, Mary stayed put. "Sea very rough, boat swaying & rocking. Felt very mean ... but better late in afternoon. Made 466 [nautical] miles on a very rough sea. In bed all day, but missed no meals."

As the weather cleared, weary passengers tentatively emerged from cabins, and on July 1, the ship's staff captain Cecil Claxton led a tour from the upper to lower decks. Keeping pace with the officer, Mary scribbled notes while Henry's gaze was likely drawn to the equipment and machinery, including the bow winch, cable holders, and large electric derricks above the cargo hold. Mary deftly jotted down the *Asia*'s specifications, from length, width, and height to the diameter of the propellers, amazed that "100 tons of fresh water is used daily, carried in a shallow tank in the ship's hold" and that the "boat is equipped with a distilling [desalination] system in case of emergency." With Claxton's help, she pegged the number of souls at 1,509, including a crew of nine Filipinos, 58 Europeans, and 509 Chinese.

From the upper decks, the tour passed through the enticing smells of the galley, which left an indelible impression on Mary, who described it as "a wonderful kitchen & bakery" with "five bakers busy all the time" and it included an "electric bread mixer, dish washer, and ice cream freezer." Deeper into the hull, the McNeals became aware of a change in skin colour, language, accommodations, and amenities. She observed three "Hindues, [*sic*] rest all were Chinese" and "36 Bolsheviks being deported to Russia, some put in second- & some third-class to avoid trouble. Gambling was going on in steerage among the Chinese." In the engine room, tour guests covered their ears and held their breath while staring at the massive turbines, tightly riveted boilers, and roaring furnaces where "good steam" on outbound voyages still depended on quality coal from Vancouver Island and the underappreciated labours of firemen, greasers, and trimmers.

The next day, while those men poured on more coal, the *Asia* crossed 466 miles of ocean, and while doing so, entertained passengers with a masquerade ball. Swirling with colourful costumes and the uplifting sound of the orchestra, the event was symbolic of the contrasting nature of ocean travel. On the surface, the festivities emphasized the company's efforts to combine travel and fun, but beneath was the multilayered responsibility of keeping the ship safe. During the early hours of Sunday, July 3, Mary luxuriated in a warm bath before taking breakfast and helping Helen, who was under the weather, pack for Yokohama. The evening included a church service and lecture by the incomparable Dr. Torrey, followed by the most glorious sunset Mary had witnessed at sea. Land rose on the horizon on July 4 and after "breakfast we went on deck & watched all the passengers being lined up for [health] inspection. Inspection officers came on aboard about 10 A.M. We stood in line 1 hr. & nearly suffocated it was so hot."[26]

While Henry and a companion watched the unloading of their trunks and baggage by hotel porters, mother and daughter boarded a *jinrikisha* to the Victorian-styled Grand Hotel where guests could sink into plush chairs in a lounge overlooking Tokyo Bay. Surrounded by luxury and attentive servants, European and North American guests could admire the ocean while down on the wharf porters and sweaty stevedores continued to unload the ship as more *jinrikishas* jostled in the dust. Within easy reach was the Grand Hotel's guidebook, full of tips and assurances that promised a continuum of luxury from ship to shore:

> Now, what is that little Japanese boy standing around for? Ah, if you please, he is your boy. Yes, your own particular boy, so long as you remain a guest at the Grand. In some parts of the world he might be called a "valet" but in Japan he is more: he is a "valet" and "maid," combined for he does more than either or both: he anticipates your wishes

> before you, yourself, are fully conscious of them. He is part of the Grand Hotel service![27]

That evening, the McNeals joined 600 guests in the hotel's elegant dining room. "Nearly everyone from the *Asia* was there," noted Mary. Before calling it a night, the family relaxed on the veranda, enjoying the breeze and "a gorgeous display of fireworks" featuring "large floral pieces & the last one a large American flag" shimmering above hundreds of small boats hung with Japanese lanterns. It was the Fourth of July.[28]

Within 26 months, the Grand, which had hosted thousands, including Rudyard Kipling, W. Somerset Maugham, and William Howard Taft, vanished along with most of Yokohama and Tokyo in the Great Kantō earthquake. In the aftermath, the *Asia*, but more so the *Empress of Australia* under the command of seasoned Captain Robinson, formed part of a massive rescue and relief effort (see Chapter 11).

○ ○ ○

When William F. Russell, dean of the University of Iowa, his wife, Clo, and young sons, Bill, Jimmy, and Bobby, boarded the *Asia* at 4:00 p.m., Thursday, August 18, 1921, the Vancouver waterfront was abuzz because the passenger list featured a large party led by John D. Rockefeller, Jr., "son and heir of the richest man in the world."[29]

Reporters had been diligently filing stories ahead of the party joining the ship, which was scheduled to call at Victoria around midnight. Beneath the headline "Son of Famous Oil King Sails by Empress of Asia," the *Victoria Daily* Times ran a photograph of the suave, stylishly dressed man whose 82-year-old father owned Standard Oil. Rockefeller, who would return to Vancouver that November aboard the same ship, was sailing to China with members of his world-famous Rockefeller Foundation, eager to participate in the grand opening of the Union Medical College in Peking (Beijing). In condescending fashion, the press noted that "Rockefeller millions have been lavished to improve the health of the hundreds of millions of human beings in China and rid the world of some of the scourges which originate in that country."[30]

Clo Russell and two of her children, along with other passengers, pose next to Captain Alfred James Hailey in 1921.

Rockefeller happily informed newsmen that his father remained active, played golf daily, and hoped to live to a hundred. When pressed about his own work, he explained that while he continued to head a number of enterprises, he devoted "most of his time to directing the philanthropic institutions of his father."[31]

Joining the party was biochemistry professor Dr. Archibald Byron Macallum, who during the First World War headed the Honorary Advisory Council for Scientific and Industrial Research (later Canada's National Research Council). The scientist was in the midst of a seven-month assignment to help organize the council and lecture at the Peking college. His work on behalf of the Rockefeller Foundation commenced soon after he became chair of biochemistry at Montreal's McGill University.[32]

Like Macallum, William Russell was a world traveller, so when he received an invitation to join an Anglo-American body known as the China Education Commission, his wife and their young boys decided to accompany him. The family boarded a train at Iowa City and journeyed west to meet the *Asia*. It helped that Russell was familiar with ocean travel, having visited Yokohama in 1918 en route to Siberia. At the time, he had been studying educational techniques and teaching U.S. educational methods, although his granddaughter, Sarah Russell Spray, believes there was an intelligence-gathering component to his work.[33]

The stated objective of the China Education Commission was to conduct a survey of the foreign mission schools developed under British and American missionary societies, all of it unfolding while China was developing its education system.[34] "It was a hearts and minds effort to which he devoted his life," Spray noted. "He spent his career researching what forms of education fostered democracy and how best to support those principles."[35]

William Russell also kept a diary. On August 17, the day before sailing, it was raining and his seven-year-old son, Bill, "had a very bad attack of asthma and wheezed badly. We got the doctor and it kept us pretty busy.… In the afternoon, we went to a reception given by Mr. and Mrs. John D. Rockefeller Jr. and met members of their party.… That night, Bill's attack of asthma broke.…"

At 5:30 p.m., August 18, the *Asia* reversed engines to pull away from the pier, and a long blast of her whistle served as the exclamation point on her departure. Russell's diary for the next day notes: "Fine, clear day with the seas calm. One would hardly realize he was at sea. The boys found an inside place on the back deck where they could play." After referencing the daily time adjustments of setting the clock back each night, he noted: "Mr. Speer led the conference in the morning. Mr. Wallace made an excellent talk on the West China situation in the afternoon. Dancing in the lounge this evening. Clo danced with John D. several times."

Tossed by rough seas on the 22nd, Russell commented: "We were feeling pretty peaked, but we got the better of it quickly. Bill has proved to be the best sailor, apparently never noticing … any motion to the boat. Everyone else was dizzy."

It remained nasty four days later when Russell spoke with Fred W. Stevens, who was among those representing American-banking interests. "He showed that China was in a very bad way financially due to corrupt government; presented a discouraging outlook for the country."

On the morning of the 28th, the *Asia* was on "the edge of a typhoon" but avoided the worst, encountering fog and rain prior to reaching Yokohama. Japanese doctors boarded first, conveyed on lighters from the harbour with the job of ensuring passengers were free of contagions. Next, a coterie of police and immigration officials dispersed on deck to examine travel documents and make arrests if necessary. "We were all interested. Investigated, had our passports visaed [*sic*], and then went on deck to see the landing. It was lots of fun and the boys had a great time."[36]

The party, including the Russells, checked into the original Imperial Hotel in Tokyo, which was being replaced by a new Imperial designed by master architect Frank Lloyd Wright, who in the course of site planning and construction crossed the Pacific five times on the *Empress of Asia*. The ship's speed, coupled with her magnificent architecture and Crawley's stunning designs, helped make it Wright's favourite liner, the one he recommended to his 78-year-old mother, Anna, when she visited Japan in 1920.[37] Unlike Yokohama's Grand Hotel, Wright's Imperial survived the September 1923 earthquake.

Accompanied by his first wife, Catherine, and two clients from Illinois, the Wisconsin-born architect first visited Japan in 1905. His three-month sightseeing tour commenced in Vancouver on the *Empress of China*, and as Kathryn Smith notes in her article "Frank Lloyd Wright and the Imperial Hotel: A Postscript," Wright was "subject to seasickness, and, as a result, preferred the fastest means of transport."[38]

For the master architect, the CPR's coordinated land-and-sea network made travel convenient and comfortable. He could board a train in Chicago and steam through North Dakota and into Canada where he was fond of Glacier House near the Illecillewaet Glacier. "One can only speculate about Wright's interest in the phenomenon of frozen water," writes Smith.[39] Indeed, one of Wright's most inspired designs was Fallingwater, a residence built over a waterfall in southwestern Pennsylvania during the Depression.

By 1909, Wright was facing bad publicity centred on his personal life. While estranged from his wife, he had a relationship with Mamah Borthwick Cheney, wife of a former client. Then, on August 15, 1914, while Wright was in Chicago, a servant at his Wisconsin home murdered Cheney along with her two children and four others.[40] From this horror, Wright, now 49, returned to Japan on the *Empress of Asia*, departing Vancouver in late December 1916. Four years later, the H-shaped Imperial, built on a floating foundation with reinforced steel, had still not opened. But as Smith notes, the architect arranged to ship a Cadillac for the hotel's manager.[41]

Wright's fifth voyage on the *Asia* commenced at Hong Kong on March 31, 1921, arriving in Victoria in April. High hopes for a grand opening, or at least a partial opening of the new Imperial the following May quickly evaporated when a 6.8-magnitude earthquake damaged much of Tokyo on April 26, leaving Wright's new hotel — still under construction — badly shaken but intact. The next planned opening date of September 1, 1923, proved equally ill-fated as one of Japan's worst earthquakes devastated the city. Instead of welcoming tourists, Wright's Imperial, like the *Empress of Australia* alongside at nearby Yokohama, opened their doors and offered desperately needed emergency relief to countless survivors of the Great Kantō (see Chapter 11).

○ ○ ○

Roughly four months earlier, another prominent American, Richard Gimbel, on April 19, was preparing to sail on the *Asia* from Vancouver. Gimbel was the son of Ellis A. Gimbel, Sr., and grandson of Adam Gimbel, a Bavarian-Jewish immigrant who, along with his sons, founded the Gimbel department store chain, a business empire that had mushroomed to include flagship stores in Milwaukee, New York City, Philadelphia, and Pittsburgh. In 1923, Gimbel Brothers, Incorporated, was busy purchasing rival Saks in New York, turning it into a subsidiary and household name.

Combining business and pleasure, Richard Gimbel, accompanied by his older brother, Ellis Jr., packed a camera and used it to great effect while crossing the Pacific. Most of the photos are tagged with short captions that help illustrate the 1920s gaiety as glamorous first-class passengers pose beside the rail or unwind on deck against a backdrop of ship, sky, and sea. One features a bespectacled passenger identified as Jack Colton, sitting with legs crossed in a wicker lounger on the Promenade deck. In striped suit and

Comfortably seated on the Promenade deck, American passenger Jack Colton enjoys the journey in 1923.

checkered tie, he looks dressed for church or a business meeting, but this was customary attire, even while at sea.[42]

In another dated April 30, 1923, a young woman identified merely as "Mrs. Wood" stands next to the rail on the Promenade deck. Exactly five months earlier, also outbound from Vancouver, the widow Edith Vanderbilt and her 22-year-old daughter, Cornelia, must surely have enjoyed the same view. Both Vanderbilt women were extraordinarily wealthy. Edith had given birth to her only child in the opulent 178,928 square-foot, 250-room mansion her husband, George, had built as part of his Biltmore Estate in the rolling Blue Ridge Mountains of North Carolina, reportedly the largest privately owned home in the United States.

When she turned 21 in 1921, Cornelia suddenly inherited a $2 million annuity, left to her by her father, who died in 1914 of complications from an emergency appendectomy. Within a few years of embarking on the *Asia*, she

A cool ocean breeze plays with the stylish dress of a Mrs. Wood in the spring of 1923 while she journeys to Yokohama, Japan.

was a married woman and received an even larger inheritance before heading to New York City in the 1930s and then on to Paris where she divorced and flamboyantly dyed her hair pink. No doubt while on the *Asia*, the two reflected on George's opportune decision to switch the family's Atlantic crossing from the *Titanic* to the *Olympic* in 1912, although the disaster did claim the life of George's valet.[43]

Gimbel's photos in Yokohama show the ship alongside and moments before resuming passage to Kobe. In one, a web of streamers criss-crosses the space between passengers lining the rail and those waving goodbye. It was a joyful moment in May, but everything within the frame and beyond would be lost during that devastating September quake.

In the intervening months, the *Asia* completed two further round trips, and in early June, a few days before she reached Vancouver, the distinctive

A Richard Gimbel photo captures the web of farewell streamers as the *Empress of Asia* departs Yokohama en route to Hong Kong.

smell of ink emanated from the printing press three levels below. The press was used to crank out a ship-borne newspaper, but on this occasion, it was printing invitations for an exclusive first-class social gathering. Organized by a group of passengers and aided by the always-efficient chief steward, the themed dinner party was just one of many hosted during the 1920s in the private dining saloon. These beautiful invitations listing guests and culinary delights often became prized souvenirs autographed by participants.

An article in the *Victoria Daily Times* in late March 1925 championed the advantages of ocean travel when the *Asia* arrived with the season's first holiday seekers. "Steaming across an ocean over which the sun shone in all its springtime glory during the day, and through nights of stars and soft winds, the Empress of Asia arrived here today from the Far East with the vanguard of the tourist and holiday rush...."[44]

Less newsworthy was the dangerous work in keeping the ships supplied and moving. On May 23, Angelo Conto, a 34-year-old Italian immigrant, suffered several broken ribs while loading coal onto the liner at Vancouver.[45] Two years later, the names of three Chinese crew members did

The autographed cover of a menu for a 1923 private dinner aboard the *Empress of Asia*.

not make the press, nor did their deaths delay the *Asia*'s departure on June 9, 1927. All three died of meningitis, one on the ship and the others at St. Paul's Hospital.[46] The papers did name world-famous polar explorer Roald Amundsen, who joined the *Asia* that evening on her 70th voyage to Japan and China, little realizing that just a year later, the strapping Norwegian, along with five others, would perish attempting to rescue a fellow explorer north of Norway.[47]

Two weeks after Amundsen's departure from Vancouver, the *Victoria Daily Times* published a report on how vital the province's West Coast ports were to foreign and domestic commerce. Statistics from 1926 showed 29,887 ships with total tonnage of 50 million had arrived at the province's four principal ports, with Vancouver and Victoria recording 19,767 and 6,303 vessels, respectively. Such volume boded well for the economy. In 1927 alone, 3,227 passengers arrived on the *Empress of Asia*, while 2,857 embarked for Japan and China. Incredibly, this outbound traffic on just one ship occurred at a time of violent unrest as the National Revolutionary Army (NRA), under Chiang Kai-shek, marched north to defeat warlords and unite China.

The NRA, the military arm of the Kuomintang (Chinese Nationalist Party) faced its own power struggles in March when it entered Nanking in search of enemy forces from the Northern Coalition hiding in the treaty port along the south shore of the Yangtze River. Hostility and distrust toward foreigners were high, and in Nanking led to rioting against foreign interests, precipitating a British and American navy response to protect citizens and consulates.

Foreign correspondents who relied on the big ocean steamers to reach their bureaus in Hong Kong, Shanghai, and Canton risked everything in reporting not only the Nanking Incident but violent clashes in foreign concessions at Hankow (Wuhan) in April, and massacres at Shanghai in June. Despite the risk, many foreign missionaries, consulate workers, and business types remained or continued to travel east on ships like the *Asia*, and as the situation worsened along the Yangtze, hotel lobbies in Shanghai became refugee dorms for fleeing American and British nationals.

Transporting a party of refugees, the *Asia* was the first Canadian Pacific ship to arrive in British Columbia from China in January 1927 — most were missionaries and their families who had boarded at Shanghai. Returning in late May was Bishop William C. White from Kaifeng in Honan Province, who told the press that approximately 4,000 missionaries had fled China since the start of the year.[48]

Correspondent Frederick Moore of the *New York Times* had been covering China since his days with Associated Press during the First World War. However, he and his wife decided to return home via Victoria so their children could attend American schools. The ship's 401 passengers that Sunday in October included 10 convalescing British soldiers from the Shanghai defence force, who like everyone else on board were more than a little weary because the ship had battled five days of powerful headwinds off Yokohama.

Pressed by reporters to comment on the worsening situation in China, Moore said no country "can have a stable government with such poor communications as exist in China." He went on to state that 98 percent of China's 400 million people were illiterate, with more than 90 percent starving or on the verge of starvation. "Things are going from bad to worse. China can be likened to a jellyfish. She has no backbone and no feet to stand upon."

Moore further commented that the word "patriotism" did not exist in the Chinese vocabulary, emphasizing how it was a sentiment invented by Chinese students educated at North American universities and colleges and used to "blaze the Nationalist propaganda." He said: "Sentimentalists in the U.S. had thought China could become more stable." However, they "are now realizing their mistake. China today is more decadent than she has ever been, but we must not forget that she led civilization 2,000 years ago and she may easily do so again...."[49]

Before leaving Peking for Shanghai, Moore hired fellow American Hallett Abend to fill the *New York Times* North China and Manchuria post. "He [Moore] gave me such sage advice, and sailed home, and I was left alone to make good if I could," recalled Abend, whose early assignments for the newspaper took him from Peking into Manchuria, then over Japanese-controlled and Russian-owned railways to festering Harbin with its displaced and impoverished White Russians.

Abend also covered the widespread famine in Shantung Province, and then in April 1928, while reporting on the NRA's advance, word came that Japan was sending military forces from its concession garrison at Tientsin, with another 5,000 arriving by sea at Tsingtao.[50] Although well-placed individuals continued to wave flags of optimism, there was no end to the violence that bled into the Second Sino-Japanese War in 1937.

Like Moore's, Abend's life intersected with the *Asia* on more than one occasion when he no doubt used part of his time aboard to write drafts or prepare for interviews. His first trip occurred while he was rushing from New York to China via Victoria on May 29, 1930, to cover China's Feng-Yen Rebellion. He was back on board again during an inbound trip in June 1939, followed by a voyage coinciding with the outbreak of the Second World War. Remarkably, on the first voyage, Abend landed in Shanghai only 60 days after he had sailed from Shanghai to Los Angeles on the Dollar Line's *President Lincoln*.[51]

Meanwhile, arrivals and departures continued to stir imaginations among those who could not afford the fare. When commodity prices crashed and the world entered economic freefall in 1929, ocean travel still held mass appeal even if attaining it was impossible for many.

Earnest Aden and James Van Hatten, two sea cadets from Vancouver, turned dreams into reality in 1930 when they earned temporary postings as *Asia* bridge messengers.[52] Two years later, a *Victoria Daily Times* columnist, driving to work, gained a glimpse into youthful ambition when he offered a ride to a seven-year-old boy anxious to witness the *Asia*'s arrival. The lad walked four miles to town and four miles back every Saturday morning to see the ships. "When I'm a little older I'm going to be a sailor," he said. "I want to see the rest of the world. I want to go to China and Brazil and everywhere."[53]

9

COLLAPSING THIRTIES

Up and over the rim of the World — strange, exotic places — queer outlandish customs ... Babel of conflicting tongues ... meet them all on your World Tour.

— **Newspaper ad for Canadian Pacific Ocean Tours, 1934**

The table was set for 30: 14 place settings along each side and one at either end. Before the photographer arrived, eyes had been drawn to the striking centrepiece, a beautiful fresh arrangement of magnolia and asparagus fern nearly blocking the view across the narrow table. It was October 13, 1933, and although the Depression had shattered confidence and destroyed lives, there was no indication of that among the guests who had arrived from the *Empress of Asia* to dine with several others in Manila.

Invited to the party was the *Asia*'s genial sanitation engineer, 51-year-old John Drummond, who had already deployed his soup spoon before the photographer's first flash exploded. It was a humid night in the Philippines, and several male diners, including the burly engineer, had removed their jackets and donned miniature pillbox hats, the kind that might adorn the head of an organ grinder's monkey. Looking more like a New Year's Eve party in July, any Friday the 13th superstitions were replaced with gaiety and sumptuous food. One thing seemed clear: the dinner guests were intent on recalling and celebrating the optimistic 1920s rather than the collapsing 1930s, if only for a few hours.

On the serious side, there was plenty to ponder besides the world's economy. In January, the U.S. Congress voted in favour of Philippine

Against the backdrop of a papier-mâché ship, dinner guests from the *Empress of Asia* prepare to dine in Manila in October 1933. Sanitation engineer John Drummond is on the left in white shirt and tie.

independence, and weeks later, Nazi Party leader Adolf Hitler became chancellor of Germany, followed less than a month later by the suspicious burning of the Reichstag. By March, Franklin D. Roosevelt was in the White House and immediately introduced the Emergency Banking Act. That spring, however, Nazi focus took a far more sinister and disturbing route with the persecution of Jews — opening of the first concentration camp, creation of the Gestapo, and legalization of eugenic sterilization. In a lighter vein, those who could not afford to see a show or dine in a club or restaurant might stay home and listen to ironic hits such as Al Dubin and Harry Warren's "We're in the Money" on radios purchased with an installment plan.

Back in Manila, though, the clatter of knives and forks quickly replaced polite conversation as diners tucked into the delectable and elaborate entrées that could have easily been replicated in the *Asia*'s galley. Like other Empress ships, she maintained top-rated chefs even during the Depression. Pat Terry, the *Vancouver Sun*'s plucky marine editor, did not have to be convinced after sampling an offering from the breakfast menu. Terry wrote how the master chef "focused upon me all the wealth of his craftsmanship and art, employed all the organizations of his vast kitchens in the *Asia*, and produced for me that poem in colour, texture, and perfume, the perfect omelette."

The chef was Pietro Colombo from Candelo, Italy, who had impressed the palettes of the Prince of Wales and Prince Henry while working at a hotel in Sheffield, England. Terry observed how the Vancouver waterfront was "cold, dismal, and dispiriting" when he was invited aboard in December 1934. On the Shelter deck at the centre of the ship, abaft the pantry and scullery, he entered the galley where he met the rotund chef, one of the "great personalities" of the dining saloon — a man who had an endless supply of anecdotes of ocean travel and culinary achievement. "I found him, this man of great hulk — he weighs, as a good chef should, over the 230 lb mark — deep in the mysteries of designing the immense cake set-piece which is to grace the dining saloon on sailing day."[1]

Colombo, then 34, had begun working on ships shortly after the First World War, employed in the galleys of well-known liners such as the *Majestic*, *Mauretania*, and *Leviathan*. Along with saloon steward Massimo

Guggiari, he had signed on in July 1930, when in addition to his engaging charisma, he rapidly became known as a master of artistic cake design. Honoured and excited about his new job, he knew the significance of his place in the succession of *Asia* chefs, including his predecessor, William Gee. U.S. Immigration listed him as five foot six and 145 pounds, but when he turned the galley over to another Gee, William's son, Archie, in 1939, he weighed a sizable 220. For a brief interlude in 1941, Colombo bided time as an enemy alien on the Isle of Man before his release that April.

Chef Pietro Colombo and a delectable selection of his masterpieces.

In addition to the stunning food, the ladies aboard could indulge in a little pampering at the exclusive hair salon. Although hair stylist Phylis Williams was not working that particular voyage in October 1933, her time on the *Asia* in 1932 helped overcome a difficult past, most of which transpired after the family moved to Canada from Britain. In an interview for this book, her son, Bill Attwell, explained that his mother had shared the details of her traumatic adolescence only once, and despite occasional bouts of depression, Williams had managed to remain positive.

Born in Wales and raised in a boarding school from age two to 13, Williams remained with her sister but was separated from her brother. "Absent was her father of whom she had no memory whatsoever except for one short recollection of the silhouette of a tall man in the doorway of the school who identified himself as such," noted Attwell. But young Phylis was tenacious. "She was capable of thriving in difficult circumstances, curious, charming and unconventional, even avant-garde." Like elsewhere, jobs in Vancouver were incredibly scarce, especially for women who wanted independence, but Williams persevered, although her son was unsure if she finished high school or left to find work and contribute to the family income. At 19, after training as a haircutter and hairdresser, Williams received provincial certification starting in 1931. At the time, it was highly unusual for an unmarried woman to not live at home, even if employed, but Williams left and found work at Williams Lake and Penticton. When she returned in July 1932, the family had shockingly moved without leaving an address, although Williams did locate them.

Throughout her youth, the big passenger liners calling at Vancouver always held a fascination. So clutching her training certificates and with a reference from her mother, she applied to Canadian Pacific Steamships, just prior to her 22nd birthday. It proved to be one of her best decisions. The ship's hairdressing salon was located aft on the Promenade deck next to the barber shop — not a huge space — but it was pleasant and offered glimpses into the private lives of passengers. The hours were long, and Williams remained on her feet most of the day, working to keep customers looking like movie stars. Satisfied clients emerged with the latest cluster and swagger bobs, ringlettes, spirals, and as advertised, "indestructible curls."

The *Empress of Asia* hairdresser Phylis Williams.

It was, noted her son, the time of her life, and she did so well that after a layover in Vancouver she transferred to the *Empress of Japan II*. There, promoted to head the hairdressing salon with two assistants, Williams earned respect and received regular invitations to dine at the captain's table. For a young single woman living through the pit of the Depression, she had a dream job. "The *Asia* brought her wonderful affection, warmth, feelings of being special and acceptance," noted Attwell. "It was a good part of her life."

Like other crew, Williams signed Articles of Agreement and was paid the same rate as a barber, 10 cents a month, but she earned additional income from satisfied customers. There are two good photographs of Williams standing on the *Asia*'s Promenade deck. In one entitled "Happy Days," she is dressed in working whites and flanked by two crew members, likely taken in Shanghai or Hong Kong. In the second, she is standing alone at the stern, appearing relaxed with one hand on the rail and her headscarf fluttering in the wind.

Captain Alfred Lovegrove in command, 1932.

At ports of call, Williams would hurry off to purchase salon supplies and often went on longer strolls. One day, while venturing outside Shanghai's International Settlement, she witnessed the shocking beheading of a Chinese man. "She told me he looked at her with sad eyes just as the executioner's blade fell," recalled her son. "Over the years, she sought peacefulness and tranquility in the artifacts she brought home from China. She had a statue of Guanyin [Kuan Yin], the Chinese Goddess of Mercy and Compassion and would gaze upon it and find solace that way or in the paintings of Chinese junks. These were reminders of her best years."[2]

Skippered by Alfred Lovegrove, the *Asia*'s next sailing was August 13 and included a long and diverse passenger list. One particular household name stood out: the Reverend Eric H. Liddell, who was returning to his birthplace at Tientsin, China.[3]

Born in 1902, the son of Scottish missionaries, the shy and modest Liddell was educated at Eltham College, London. While attending the

University of Edinburgh, he excelled at athletics, including rugby, before qualifying as a sprinter for the 1924 Olympic Games in Paris. The devout Christian famously removed himself from his premier event — the 100-metre heats — because he did not want to compete on a Sunday. News of this spread around the world, but despite that, Liddell finished the games with gold in the 400 metres and bronze in the 200. His record-breaking achievement, along with that of fellow British runner Harold Abrahams, was captured years later in the Academy Award–winning film *Chariots of Fire*.

After the Olympics, Liddell returned to China as a missionary for the London Missionary Society, but by 1930, he was back in Great Britain, where on June 22, 1932, he became an ordained minister. He crossed the Atlantic on the *Duchess of York*, reached Vancouver by rail, and while in the city, called on an old school chum, Reverend Roy Stobie, an assistant at First United Church. On the evening of August 12, before boarding the *Asia*, Liddell delivered an inspiring public address at the church known for assisting the poor.

○ ○ ○

Worldwide, the competitive shipping industry was in a severe slump. A high percentage of maritime fleets remained in port, and while not every merchant sailor was unemployed, Walter Oliver noticed how many were suffering. Men who had experienced war now stood in breadlines, hawked fruit for a penny apiece, or took to the road or rails in search of meal money. Alleyways became last stops for the displaced and those without hope. Owners, too, were feeling the pinch. Finding creative ways to overcome reduced passenger and cargo volumes was tantamount to survival and there was growing pressure for Canadian steamship companies to employ white-only crews on subsidized ships. Morgan Berg, president of the Canadian Association of Seaman, stated that while Canadian Pacific had a $900,000 annual subsidy, it employed about 2,450 Chinese in deep-sea and coastal services.[4] Seldom, if ever, was mentioned how proficient those Chinese crews were.

By 1933, the CPR's Empress fleet had four passenger liners on the West Coast — namely, the 20-year-old sisters, *Asia* and *Russia*, the *Empress of*

Canada, and *Empress of Japan II*, completed in June 1930. When the last arrived that August, she set an even higher standard, receiving accolades for design and performance. Oil-fired with a length of 644 feet, compared to the *Asia*'s 570, and a gross tonnage of 26,032, she, too, had come off the slipway at Fairfield but looked too good to be true arriving at such a time.

With or without full passenger lists or cargo, the Empresses maintained their schedules. Mail contracts continued to demand efficient regularity, and people still depended on travel for business or pleasure. Booking agents believed that by keeping a close eye on all classes of passenger volumes, the pulse of the economy could be measured. "Dozens of special tour parties are booked for summer sailings of the C.P. 'Empresses,'" explained J.J. Forster, the company's steamship general passenger agent in Vancouver. "The season looks like one of the biggest in years.... Last year the July voyage of the 'Empress of Asia' was booked so heavily with tour parties it was necessary to use part of the main dining saloon for tourist passengers."[5]

Teacher and tour organizer Annie Stewart of Vancouver caught on and built a franchise around personally conducted tours to East Asia, including

A barge loads tons of West Coast coal onto the *Empress of Asia* in Vancouver during the interwar years.

two on the *Asia* in 1934–35. One, a month-and-a-half-round trip to Hong Kong with stops in Japan, China, and the Philippines, offered tourist-class fares for $375. The all-inclusive package included top-rated hotels, railway transportation in Japan, meals, sightseeing tours, guides, and use of automobiles.[6]

Although Stewart marked her 38th birthday on that same July excursion, the busy tour guide had little time for herself. That May, the *Vancouver Sun* had published a photo of the attractive Stewart with pen in hand, drafting an itinerary. The story explained how travel was in her blood after the first six years of her life was spent in the caboose of a passenger train her parents worked on between Mission and North Bend in British Columbia's Fraser Canyon. Back then, she was "liable to wake up any morning of the year and find herself a hundred miles from the place she went to sleep the night before."[7]

While unrest in China and Japan slowed tourism, it guaranteed travel for diplomats, ministers of trade, medical personnel, military brass, arms dealers, missionaries, and foreign correspondents, most of them eagerly greeted on the waterfront by local press. When G. Ward Price, director of the *London Daily Mail*, disembarked on May 22, 1933, a reporter pumped for an assessment on the troubles in Manchuria following the Japanese invasion and occupation in 1931–32. Price confidently expressed the view that the fighting was about over and the Japanese were doing a good job bringing about orderly government. "Unfortunately," he added, "the Japanese are poor propagandists." Asked if a war between Japan and the United States might develop, Price conveyed what others shared: "Japan is much too busy to look for trouble elsewhere."[8]

The ship also carried passengers bearing fur, feathers, scales, and forked tongues. Embedded in the ship's history is the story of an Airedale terrier named Chow. The dog left Victoria in the fall of 1922 as a lean-looking "stowaway," crossed the Pacific twice, and returned with a bigger belly as the honoured guest of crew and passengers.[9] Among the more recent cats, dogs, fish, prized hens, and racing pigeons was an outbound Alsatian police dog from Winnipeg named "Bill Havon of Southmora," as well as three melodious roller canaries from Britain.[10] Checked on regularly by Chef Colombo in 1938 was a 135-pound Irish wolfhound occupying an unused wireless cabin on the Fiddley deck.

More exotic still were four live Komodo dragons, closely watched as the *Asia* departed Manila in the spring of 1934. Also crated were several blue-green tree snakes and a cobra. Captured on Komodo Island in the Dutch East Indies by archaeologist Lawrence Griswold of Quincy, Massachusetts, and ethnologist William Harkness of New York, the exceedingly rare 10-foot-long giant monitor lizards were bound for the Bronx Zoo and Smithsonian Institution.[11]

The Chinese crew wanted nothing to do with the flesh-eating reptiles when assigned to poop-and-scoop, so the task fell to white crew members, which to his chagrin included Chief Purser William Frederick Taylor, recently of the *Empress of Japan II*. However, as the *Asia* steamed across the

Purser William Taylor proudly displays his autographed rugby ball in 1932. The ball was presented to him by the Canadian rugby team that had returned to Canada on the *Empress of Asia* from a series of games in Japan.

East China Sea from Shanghai to Nagasaki, the smallest of the lizards died, and instead of being committed to a watery grave, the carcass became property of a Japanese taxidermist who advertised the sale of "Dragon Meat" for male virility.[12]

A few days out of Yokohama, one of the remaining animals became seriously ill. The ship's doctor, Hugh Galbraith, prescribed castor oil for constipation — the ailment likely caused by stress and a change in diet. It is not clear if crew members had to draw straws, but a team came together with a rather reluctant Taylor atop the crate pointing a loaded revolver at the dragon's head. The other men had the more onerous task of securing the reptile's legs and wrapping a noose around its powerful jaws.

Galbraith, who had signed on in Vancouver that March, administered the medicine through a long pipe inserted in the dragon's mouth. Unbeknownst to the crew, the dragon's large orifice contained anti-coagulating venom and saliva oozing with bacteria. The procedure worked, and the three dragons arrived alive in Vancouver on May 7 amid much media buzz. However, Taylor later heard the animals had not lived long after reaching their destinations.[13] An October 1934 news report stated the Bronx Zoo specimens had succumbed to colitis while the third was "logey" [*sic*] at the Smithsonian.[14]

Curiously, while Griswold and Harkness accompanied their specimens, there is no mention of any role they filled during the emergency procedure. The *Vancouver Sun*'s Pat Terry interviewed both men, describing Griswold, who sported a Vandyke beard, as "one of the most phlegmatic young men I have ever met."[15]

Capturing flora and fauna and transporting it to zoos, museums, or labs remained a well-funded pursuit that drew interest from a variety of institutions. Killing animals for the sake of trophy hunting and displaying severed heads on a wall was not science, but at the time hardly frowned upon, and certainly achievable with an elephant gun. What Griswold, Harkness, and the crew of the *Asia* discovered was that it took more planning to snare live creatures and transport them by ship around the world. Griswold and Harkness could point to another recent expedition launched in the name of science that also utilized the *Asia*.

After months of specimen gathering throughout the South Sea Islands and East Indies for the Chicago Field Museum, an intrepid party led by explorer-philanthropist Cornelius Crane reached Vancouver. Crane, of Ipswich, Massachusetts, was the grandson of multimillionaire R.T. Crane, whose company manufactured a variety of products, including plumbing and heating supplies. "Short forays from the seashore direct into the jungle" were necessary to collect the specimens, noted the Vancouver press. In attempting to abduct creatures against their will, they "met unfriendly natives, poisonous reptiles, disease-ridden swamps, and ferocious beasts."[16]

Meanwhile, inbound passenger volumes from China were drying up — a result of the ultraexclusive Chinese Immigration Act of July 1, 1923, which froze immigration from China (see Chapter 11).

After the Depression hit, there was not a massive drop in freight volumes, at least not until 1932. That June, the *Asia*, which normally carried approximately 4,000 tons of cargo, called at Vancouver with less than 200 tons.[17] The CPR's overall railway freight traffic had peaked at 40 million tons in 1928, but by 1936 it was 28 million tons.[18] Compared to the 1920s, 1932 and 1933 cargo earnings fell by 50 percent. Exacerbating the problem was the evaporation of the lucrative silk trade, recently displaced by synthetic fibre. In Vancouver, a week before Christmas 1931, at J.W. Taylor & Co. on Granville Street, $20 silk dresses sold for $5.[19]

Shippers also found a cheaper way to transport raw silk to mills. When more than 1,000 tons arrived on the *Asia* at Victoria in October 1931, most of it was loaded onto a Pacific Steamship Company vessel bound for San Francisco where it shipped to New York via the Panama Canal. Only 280 tons crossed Canada.[20]

For the CPR, the Depression forced new thinking, including a redrafting of the company's class-oriented oceanic passenger experience. Steerage became third class, and instead of second-class fares, the company sold "tourist-class" tickets. It also cut fares, adjusted the number of cabins, and in 1931, expanded the *Asia*'s service to Honolulu, which had begun the previous year. This led to marketing efforts for the "Paradise Route" via Hawaii or the "Pacific Speedway Route" via Yokohama.[21] "World Tours," screamed a 1933 newspaper advertisement. "The Opportunity of a lifetime!

Explore those far-off lands of your dreams this year — fares are surprisingly low — the romance of world-wide adventure can actually be realized now! 66 Itineraries — Tickets good for two years."[22]

With the cadence of a midway barker, a 1934 advertisement urged readers to contemplate great expectations while considering no less than 215 itineraries: "Up and over the rim of the World — strange, exotic places — queer outlandish customs — amazing garbs — Babel of conflicting tongues — the old and the barbarous — the new and the sophisticated — meet them all on your World Tour."[23]

On the *Asia*, first-class accommodations dropped from 374 to 346, while the soon-to-be-renamed second class rose from 46 to 74. During the same 10-year period, third-class passage remained at 92, while steerage jumped from 538 to 728.[24]

There were dozens of empty cabins when the *Asia* arrived in Vancouver on November 1, 1930. Steerage occupancy was short more than a third, and while the company still advertised a fast voyage, it took just one seriously ill passenger to force a delay. That summer, the *Asia* was flying her yellow quarantine flag when she arrived for inspection at the William Head Quarantine Station. Thirteen days earlier in Kobe, she had disembarked a steerage passenger with smallpox, but the mandatory quarantine period of 14 days was in effect. Although cleared by station doctors, the *Asia* was still displaying the yellow flag when she reached Victoria, so all Chinese steerage passengers returning to that city had to sail on to Vancouver before doubling back on other vessels to the capital.[25]

Returning on that voyage was Russian-born violinist Efrem Zimbalist, travelling home to New York with concert pianist Harry Kaufman. Earlier that year, the pair collaborated on a recording entitled *Sonata in D Minor*.[26] Now 41, Zimbalist was only nine when he joined his father's orchestra. By 20, he was world-renowned. Waiting at home was Efrem Zimbalist, Jr., who was to become a successful actor, known for roles in the television series *77 Sunset Strip* and *The F.B.I.*

It was hard work keeping the Empress ships in top form, though. In February 1931, all four spent approximately three weeks in dry dock in Hong Kong for annual maintenance and cleaning; however, it was the day-to-day grind at sea and in port that took its toll.[27] A June 25, 1931, logbook entry for the *Asia* while she was alongside in Vancouver states Senior Fifth Engineer H.J. Harper suffered two broken ribs and a possible punctured liver while in the engine room. The 45-year-old, along with his personal effects, landed at St. Paul's Hospital, while promotions from within the ship's engineering ranks filled the vacancy.[28]

Labouring around hot furnaces, inhaling toxic coal dust, completing dry-dock repairs, and exposure to other dangerous or unhealthy conditions remained an occupational hazard for men who could use an overhaul themselves. On May 10, 1934, 46-year-old Sam Smith, the *Asia*'s second engineer, collapsed in the engine room and died before medical aid could reach him. He left a wife but no children.[29]

Other risks presented themselves in the cargo hold or topside near the derricks and open hatches amid high-tension cables and ropes. On Saturday, September 16, 1933, Vancouver doctors were unable to save the life of Chinese sailor Mui Leung, who fell off a ladder and plunged 49 feet into Hatch No. 1. The 45-year-old married man had served on board ship for

Seagulls pursue the *Empress of Asia* as she passes through the First Narrows off Vancouver. The image was captured not long after the Lions Gate Bridge opened in 1938.

four years, and the fall resulted in multiple fractures to his head and lacerations to brain tissue.[30] There was no autopsy, and if an inquest occurred, which would have been unusual, there was no determination of cause other than a fall. The fatality delayed work but not the *Asia*'s departure from Pier B as she sailed on the 23rd with a pontification of Roman Catholic priests bound for missionary work in Japan.[31]

Outbound, everything from automobiles to lumber to zinc was loaded at breakneck speed. On July 6, 1931, ambitious French aviators Louis Cadou

Sanitation engineer John Drummond takes a close look at the ship's massive propellers while in dry dock at Kowloon (Hong Kong) in the 1930s.

and P.H. Petiot met the ship as they supervised the loading of eight large crates containing their precious Dewoitine monoplane. The pair, along with more famous fliers, formed part of a craze aimed at extending the boundaries and speed of global travel by reducing dependency on ocean transport.[32] The CPR's own long-time interest in owning and operating commercial aircraft as Canadian Pacific Air Lines was still several years away. In May 1933, while speaking to members of Vancouver's Junior Chamber of Civic Affairs, aviation expert R. Carter Guest envisioned "flying boxcars" taking hours, not days, to transport silk and other cargos to eastern mills.[33]

Meanwhile, if not burned, scalded, or injured by falls or sharp and heavy objects in tight confines, crew members contracted illnesses ashore or came aboard with pre-existing medical conditions. David Neligan of Victoria was about to quit the transpacific ships and return to work on coastal vessels; however, the steward, who worked in steerage with 25 years' deep-sea experience, died of pneumonia in Kobe in March 1932. He had been with the *Asia* for three years, and when he left Victoria in late February, his health had been good. The Irish immigrant, who came to Canada as a boy, left behind a wife, two daughters, and a son.[34]

On the morning of June 30, 1933, refrigerating engineer Percy Howarth died in his bunk. The cause, noted the ship's surgeon, was coronary thrombosis. Interment followed at Happy Valley Cemetery in Hong Kong.[35] Ten days later, on the same voyage, a ship's messenger entered the forward portside head to find bridge messenger Douglas Barr dead on the toilet. An autopsy in Yokohama revealed he died of thymic asthma, and in late July, his cremated remains were committed to the deep.[36]

By then, the *Asia* and *Russia* had pounded the seas for 20 years, and while factored into a five-ship, $59 million replacement plan announced later in the decade, the cost was prohibitive.[37] Shelved were plans to convert the ships to oil, a cleaner alternative that produced less smoke and was simpler to load than coal bunkering. Oil was more expensive per ton than trimmed coal, but two tons of bunker oil produced roughly the same power as three tons of coal.[38] The use of oil also required less engine crew, and the fuel made it easier to control the fires in the furnaces. Old Herbert James, who had long since left the ship and was ready to retire from the company in 1935, would have applauded any move to save crew from swabbing coal dust off her decks.

o o o

Transporting and containing those involved in criminal behaviour were also part of the *Asia*'s sea-time duties. Handcuffs were a simple solution for the more cordial prisoners; however, those of a more dangerous nature required stiffer measures. Hairstylist Phylis Williams had just left the ship in January 1933 when Vancouver detective inspector James Ellice set off from Victoria in December to take custody of a prisoner accused of attempted murder. When Ellice reached Hong Kong, authorities turned the man over to him, and together they shared a cabin on the return voyage. Newspapers in Vancouver and Victoria followed the story, with the *Victoria Daily Times* gleefully noting that not only were the two cabin mates, but the prisoner had also acted as nurse when the detective was seasick during rough weather. "This morning he [the prisoner] was promenading the deck with other passengers. Attired in a light fawn overcoat, and a dark brown fedora hat, he looked like a wealthy tourist, rather than a man wanted by police."[39]

Harder to spot were more predatory passengers who occasionally slipped out from the shadows. If caught, however, the ship's master was free to respond in the name of passenger safety. The question was where to hold offenders on long voyages. For the record, the "Padded Room," two levels below the Bridge deck on the port side, was for breakable cargo, not people. But if needed, a makeshift brig in the forecastle or some other secure location minimized risk until port authorities hauled the accused into custody.

On the evening of Friday, September 6, 1935, while the *Asia* was in Yokohama, Captain Lovegrove learned a third-class passenger was "annoying a young girl, aged 12, travelling alone in Tourist-Class, by his attentions." The ship's logbook does not indicate what the "attentions" were, but a report from the ship's master-at-arms does: "I saw the girl who told me she had been ashore purchasing various articles and as she arrived at the bottom of the stairs … to go to her cabin, she saw a man standing in the entrance to the third-class quarters. He made some enquiry as to what she had been purchasing and she invited him into her cabin to show him the articles. A few moments later he enquired, 'How old are you?' She replied, 'Twelve.' He then said, 'You don't know much about love then. I'm in love with you, I'll teach you.' As a result of his attitude, she got frightened and ran out of the cabin to the Stewardess. When I saw her she was in a very nervous condition and had been crying."[40]

On September 7, after receiving another complaint involving the same passenger harassing the same girl, Lovegrove placed the passenger in "preventative arrest." In Kobe, the next day when the girl ended her voyage, "full liberties and privileges" were restored to the man. "He was warned against intruding himself and causing fright and embarrassment to other passengers," notes an entry in the ship's log, written by Staff Captain John Bisset Smith. "He was also directed to restrict himself to quarters reserved for 3rd Class."[41] Lovegrove was also skipper when Basil Feodor Shishkin ended his life. The young Russian, described as an "Anti-Piracy Guard from Shanghai," shot himself with his service revolver in March 1936 while the *Asia* was southeast of Shanghai.[42]

When the ship departed Vancouver on January 8, 1938, Canadian surgeon Dr. Norman Bethune was settling into one of the cabins. The weather was typical for January: a woollen grey overcast, showers and fog stirred by a moderate east to southeast wind.[43] The Gravenhurst, Ontario–born doctor was embarking on his historic mission, travelling to China with 26-year-old Canadian nurse Jean Ewen, who had been working in Vancouver and spoke Chinese. The passenger list was light, yet eclectic, due to the Second Sino-Japanese War. It included American surgeon Dr. Charles Parsons, accompanying Bethune and Ewen, and Professor Michael Lindsay, who would visit Bethune in northwestern China. Also on board were B.C. Butler, Canadian trade commissioner to Singapore; two Goodyear Tire and Rubber Company executives, R.M. Astle and C.A. Rich; and R.B. Roxburgh of the Thorneycroft Engine and Armament Company, returning from vacation.

Despite the war and its effects on China, the voyage marked the resumption of visits to Shanghai, postponed since August 1937 when the *Asia* had served as an evacuation ship. In addition to a large fruit shipment, the hold contained four tons of used clothing — collected in Canada by the War Refugee Society — all of it bound for displaced Chinese sheltering in Hong Kong.[44]

Biographer Roderick Stewart obtained a photograph of Bethune taken on departure, showing him in typical men's attire — a double-breasted suit. One published account states the voyage to Yokohama took 21 days. However, such a delay would have made headlines, since the ship was on schedule to arrive in Yokohama on January 20 and Shanghai four days later.[45]

Bethune would have likely devoted much time to thinking, relaxing, and engaging with his travelling companions. In a New Year's card dated the day of his voyage, the surgeon wrote to friends, stating he was China-bound, and if they must know why, then please read Edgar Snow's *Red Star over China*, Agnes Smedley's *China's Red Army Marches*, and James M. Bertram's *First Act in China*. The handwritten note on Canadian Pacific letterhead, which included a photo of the company's Empress Hotel in Victoria, ends with "goodbye & bless you for your kindness & comradeship. With love to you both, Beth."[46]

Bethune had survived the First World War as a stretcher-bearer, beaten tuberculosis in 1926, and invented practical medical instruments. He was again sailing into something dangerous, about to earn heroic status as a battlefield surgeon for Chinese communist forces under Mao Zedong.

Two months after Bethune sailed, the *Asia* arrived in Vancouver on March 12 with her own medical problem — a crew member diagnosed with typhoid fever caused by exposure to *Salmonella typhi* bacteria. The disease is spread through the consumption of food or water contaminated by fecal matter or urine or during close contact. On board were 270 passengers, including many bound for the United States, Great Britain, and Sweden, all of whom disembarked on the 12th. The stricken crew member was diagnosed several days after the ship's arrival, and while undergoing hospitalization, Vancouver's medical officer of health ordered immediate immunization for the remaining crew.

○ ○ ○

As the world continued to unravel through the 1930s, foreign correspondents such as Hallett Abend, based in East Asia, continued to file stories and venture across the Pacific to China where some argue the Second World War began. Well before the Nazis invaded Poland on September 1, 1939, journalists had been reporting on the Sino-Japanese War, including the Battle of Shanghai.

On June 16, Abend was en route to the United States when he disembarked from the *Asia* in Victoria and had breakfast with the parents of Douglas Robertson, the man Abend had placed in charge of the Shanghai office during his absence. The Robertsons, who resided in Victoria, were proud of their son's responsibilities at the prestigious newspaper. It had been Abend who had discovered Robertson's news sense and encouraged him to leave the *Asia*'s purser department and join him in Shanghai. During the bloody Shanghai Incident of 1932, he had more than earned his pay by helping Abend cable or post letters containing stories to New York City.

On August 19, 1939, with the declaration of war in Europe just days away, Abend boarded the *Asia* in Victoria en route to Shanghai where, exactly two years earlier, he had nearly perished touring the battle-torn city (see Chapter 10).

Prior to leaving New York City and Washington, DC, that summer, Abend was dismayed by the lack of interest shown in the United States for unfolding events in China and Japan. By July that year, he, like many, had become alarmed by the rising tension in Europe, and while meeting with the newspaper's publisher, recommended that "in view of what seemed the certainty of war in Europe at a very early date," he should hurry back to East Asia. The journalist believed "that if England and Germany went to war, Japan would certainly take advantage of the world situation and extend her aggressions in the Far East. I told him I believed there was almost a certainty that Japan would attack Hong Kong, the Philippines, and the Netherlands East Indies, and that I was sure she would take over the foreign areas at Shanghai by force."

Aware of how much time his foreign correspondent had spent in China, the publisher suggested it was time for Abend to forget about Asia and Europe for a while and "go into the mountains" of Idaho where he could enjoy a little trout fishing. Abend embraced the idea and was nearly on his way when the situation in Europe worsened, prompting him to book passage on the *Empress of Asia*. "That voyage across the Pacific in 1939 in a ship flying the British flag was one of the strangest journeys I ever made. The passengers included Britons, Canadians, Australians, Americans, Germans, and Japanese. During the first few days of the voyage, with the early outbreak of war in Europe becoming more of a certainty from hour to hour, the Japanese and the Germans herded more and more closely together. Japan had not yet concluded her formal military alliance with the European Axis — that was to come a year later — but the lines of sympathy against the democracies were already clearly drawn."

Abend recalled that when the *Asia*'s radio broke news of the German-Soviet Non-Aggression Pact, "the effect upon the Japanese passengers was startling and comic. They felt, as the government in Tokyo felt, at that

Resplendent in white amid cascading streamers, the *Empress of Asia* prepares to depart Vancouver in 1930.

time, that Hitler had betrayed Japan. From the hour that the news of the … agreement reached the ship, the Japanese not only stopped associating with the German passengers, they even stopped speaking to them when they met face to face on deck."

As the drama unfolded, the *Asia* crossed to Yokohama and was coaling in Nagasaki harbour on September 3 when the European war grew. The next day, the ship reached Shanghai where news from abroad "greatly aggravated the already tense situation." In his memoir *My Life in China 1926–1941*,

From left, three siblings from the Depression era — Kathleen, Ron, and Muriel Bestward — visit the *Asia*'s Fiddley deck while the ship was open to the public alongside Vancouver in 1939.

Abend noted: "Those Idaho trout must be whoppers by now, for I have never yet had the chance to go after them."[47]

o o o

Walter Oliver might not have understood it, but while at sea he had grown accustomed to nature's rhythm of chaos and order. While humans made war and peace, and war again in the span of two decades, the difference was that nature, no matter how powerful, did not make it personal.

10

SEA UPON SKY

I rushed out and found her lying near the door. I dragged her in and went back to look for Leona.

— **Edith Ruth**

There was nothing anyone could do to save Leona Kearns. A wall of frigid water surged over the *Empress of Asia*'s starboard handrail, and in an instant, the 17-year-old semi-professional baseball player from West Union, Illinois, was swept overboard. It was January 21, 1926, and the ship was less than two days out of Yokohama, ploughing through a massive winter storm that spread heavy seas across the North Pacific.

Accompanied by driving rain and snow squalls, the storm punched hard before moderating briefly, creating a false sense of security for the 146 passengers and crew of 520.[1] Unbeknownst to all, the storm would pound the ship all the way to Vancouver Island.

A day earlier, while on the bridge, a grim Captain Lionel Dale Douglas pursed his lips and steadied himself as icy sheets of ocean spray lashed the forward windows, blurring what little he could see off the bow. At times it seemed as though the sea was upon the sky. The ship's barometer had dipped to 95.70 kilopascal, a pressure reading associated with typhoons. As wind velocities approached 70 miles per hour, it keened through the rails and around cables and winches before reaching a screaming pitch through the rigging. Unrelenting large grey-green combers pounded the ship, inundating the forward Promenade deck and flooding the exposed outer decks. Foam filled the air, tumbling and sticking to machinery and the bridge's superstructure before dissolving in the wind. At times during a severe roll, the ship's two outer port and starboard propellers took turns rising out of the water.

For Douglas, it was reminiscent of the November 1924 storm that pummelled the ship a day out of Yokohama, forcing him to reduce speed and heave to for five hours to angle the ship into the wind and waves. On that voyage, the ship fell 12 hours behind schedule, and despite a surplus of nauseous and shaken passengers, Douglas was keenly aware there were 4,750 bales of raw silk in the hold valued at $7 million. He remembered, too, that the *Empress of Russia* had sent a wireless, describing mountainous swells, one of which ripped a lifeboat off its davits, washing it overboard while rendering a second boat useless. On November 10, the soft-spoken Geordie told reporters in Victoria that while "the trip was unusually wild at times," such weather is expected. Losing a lifeboat or other gear is not unusual in heavy seas.[2]

Born to Canadian parents in Newcastle upon Tyne in Great Britain, Douglas was an experienced master who trusted his ship but also understood the awesome power of high seas. In February 1897 after graduating from cadet training at HMS *Conway*, Douglas, then 18, joined the four-masted, iron-hulled barque *Silberhorn* bound for Canada's West Coast. Stowed below were 3,000 tons of general merchandise and stacks of 40-pound boxes containing sheet tin destined for the booming salmon and fruit cannery industries. It was "a rather boisterous voyage," recounted Douglas. South of Ireland, "we ran into fearful weather. The cargo shifted and the sails blew away, and finally we ran to the Bristol Channel [where the ship anchored and the exhausted crew realigned her cargo]."[3]

It took the *Silberhorn* 120 days to reach Victoria, though her arrival was marred by the death of a sailor who slipped off a yardarm, struck the foremast, and disappeared into the water off Cape Flattery.[4] On a more encouraging note, Douglas recalled that his first sense of Canada was not the sight of land but its earthy, vegetative smell from 50 miles out. "Your nostrils picked it up immediately. It was very marked, indeed. It had been a hot summer and I suppose the hot sun on the pine and fir trees — with a light breeze brought this back [to us] and it was delightful to get into Victoria."[5]

In Vancouver in 1905, Douglas joined the schooner-like *Empress of India* as third officer. Eight years later, he became chief officer on the new *Empress of Asia*, and on October 10, 1921, after proving himself on another Canadian Pacific ship, Douglas assumed command of the *Asia*.

In the summer of 1924, the captain and Chief Engineer Robert Henry Shaw demonstrated skill on the North Pacific when the ship averaged 20.2 knots, with the fastest-day average of 20.63. It took just eight days, 14 hours, and 48 minutes for the *Asia* to cross from Yokohama to Race Rocks, and although this remarkable feat did not set a world record for a crossing, it was the quickest voyage recorded by the *Asia* or by her ever-competitive sister, the *Russia*.[6]

Douglas was not the only family member to have survived bad weather. His father, Lieutenant-Colonel Campbell Mellis Douglas, an assistant surgeon with the British Army's 24th Regiment of Foot, was the recipient of the Victoria Cross (VC). Born at Grosse-Île, Quebec, on August 5, 1840, Campbell and four other volunteers earned the VC for rescuing 17 soldiers trapped on low-lying Little Andaman Island in the Bay of Bengal. The 17 had been part of an expedition sent to the island to ascertain the fate of the commander of the *Assam Valley* and her crew.

In fearsome weather, May 7, 1867, five volunteers boarded a rowboat and headed to the island where the soldiers sat stranded between powerful surf and the island's well-armed indigenous Onge, opponents of the expedition. With water filling their boat, the crew made three attempts before rescuing all 17. Citation for the rare, non-combat-related VC states Douglas "stood in the bows of the boat, and worked her in an intrepid and seamanlike manner, cool to a degree, as if what he was then doing was an ordinary act of every-day life."[7]

So, with ever-increasing seas that January 1926, it was vital that Douglas, Shaw, and the rest of the crew maintained a delicate balance between good speed and the safety and comfort of those aboard, often reducing the ship's speed to 12 knots. For this voyage, there were only 17 in first class, 29 in second, 100, mostly Chinese, in steerage, and more than 1,000 bags of mail.

Second-class passenger Kearns, whose nickname was "Slim," pitched for the all-female Philadelphia Bobbies, a name borrowed from the bobbed haircuts the girls wore. The team had proven itself against American men's teams and had been invited to play in Japan against university clubs.

For the Bobbies, participating in the 1925 tournament promised international exposure for women's baseball. The team was a partnership between manager Mary O'Gara and 35-year-old Russian-born Eddie Ainsmith, a

The young and talented Leona Kearns of the Philadelphia Bobbies.

scout and former major leaguer who had played for five different clubs, including the St. Louis Cardinals (1921) and then-New York Giants (1924). Japanese promoters had sealed the deal by covering the girls' one-way, first-class passage to Japan on the SS *President Jefferson* and by promising generous gate receipts to compensate players and staff. The promoters also agreed to cover hotel and meal costs, plus the return voyage. Kearns and teammates were excited, figuring they could each earn $500. "Mr. Ainsmith said that last year he had 28 boys with him and each of them made $830," Kearns wrote to her mother. "Mr. Ainsmith said that we were sure to make good. If I make $500 we will get us a Ford."[8]

But it all landed like a foul ball when the Bobbies lost several games, not just innings. Ballpark attendance tanked, and two promoters disappeared

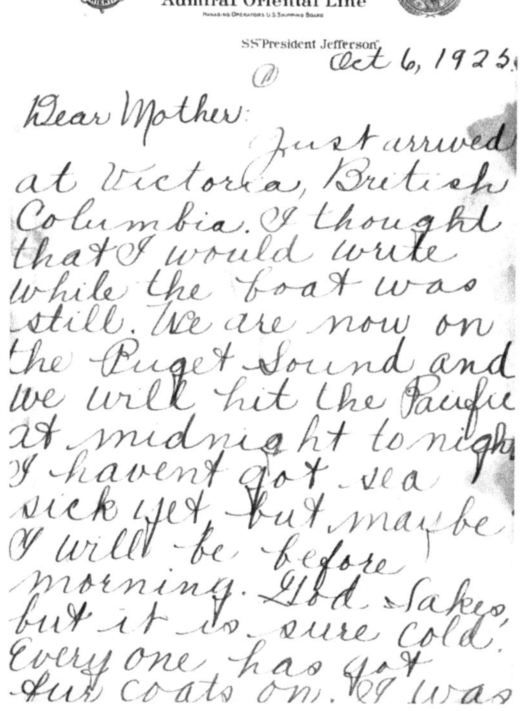

Admiral Oriental Line

SS President Jefferson
Oct 6, 1925.

Dear Mother:
Just arrived at Victoria, British Columbia. I thought that I would write while the boat was still. We are now on the Puget Sound and we will hit the Pacific at midnight tonight. I havent got sea sick yet, but maybe I will be before morning. God Sakes, but it is sure cold. Every one has got fur coats on. I was

An October 1925 letter sent by Leona Kearns to her mother prior to Leona's outbound voyage to Japan.

without paying the bills, while a third hid behind bankruptcy. Arguments led to a split, and by December 6, most of the Bobbies were back in Philadelphia via Vancouver on the *Russia.* Kearns, Edith Ruth, and Nella Shank remained with Ainsmith, who cobbled together a team to compete in Korea. After returning to Japan, the desperate manager attempted to raise additional funds to repatriate his remaining players, but the effort stalled, and instead of staying with the girls, he sailed home after Christmas.

The girls' parents were apoplectic after learning most of the Bobbies had returned home. Leona's father, Claude, who worked as an agent for the New York Central Line, did what any good father would do — he borrowed funds to purchase a second-class ticket for his daughter aboard the *Asia*, departing at 4:12 p.m., January 18, from Kobe.[9] Kearns, along with 21-year-old Ruth and 20-year-old Shank, boarded the Canadian Pacific steamer — for Kearns the voyage was to end with a reunion with her father in Vancouver.[10]

The tall southpaw was swift and agile, but not fast enough for the rogue wave that swept her overboard. Earlier, she and her friends had been waiting for the storm to abate, and for most of the day the sea had seemed less threatening with little to no water on the upper decks. At one point, an order was given to unlock the steel storm doors, including those on the Shelter deck — the second-class promenade.[11]

Like the rest of the ship, the Shelter deck was comfortable during most voyages. There was a spacious indoor dining saloon, and abaft of that, a large women's room to starboard with a public smoking room on the port side. Sheltered open decks allowed plenty of room on fair-weather days to stroll and admire the ocean from behind the railing.

Earlier that afternoon, Kearns, who was likely whistling her favourite tune "Doodly-Doo," bounded up to the Bridge deck to take advantage of the longer wraparound outer promenade. The logbook notes she was "seen running around the deck in a somewhat dangerous manner...." Having been cooped up by the storm, she was anxious for fresh air and exercise and had no doubt weighed that against any danger. Apparently, Kearns was "personally warned by Mr. Vaux, senior assistant purser, to be careful as it was dangerous to run around that way, with the ship rolling heavily."

○ ○ ○

"Mr. Vaux" was 39-year-old Gilbert John Vaux, who happened to know what he was talking about. His office was aft of the second-class cabins on the Bridge deck, and two major experiences in his life had a bearing on the caution he gave Kearns that day. On January 7, 1913, while working as a freight clerk on the 131-foot-long coastal steamer *Cheslakee*, Vaux escaped death when the vessel capsized alongside the wharf at Van Anda on the northeast side of Texada Island, British Columbia. Seven passengers drowned, and a coroner's jury found Vaux responsible, even though he had helped passengers abandon ship. Charged with manslaughter, he went to jail but was exonerated by a marine court of enquiry and released. Speaking for the attorney general, the Crown counsel noted how "Mr. Vaux seems to have done his best in the time at his disposal to warn the passengers."

Vaux's second near-death experience occurred on April 8, 1921, while he was assistant purser on the 6,133-ton SS *Monteagle*, acquired by Canadian Pacific in 1903. A terrific northeaster was hammering ships in the South China Sea, including the reliable *Monteagle* commanded by Geoffrey Hosken's father, Captain A.J. Hosken. Locked in the storm's grip, and with water flooding the engine room, the wooden-hulled French steamer *Hsin Tien* issued a distress signal to the *Monteagle*. Recognizing the peril, Hosken asked for volunteers to row lifeboats to the stricken vessel with the hope of rescuing the crew.

One boat reached the *Hsin Tien* and rescued a man brave enough to jump into the rising and falling craft. The second lifeboat, pushed by high winds and rolling seas, overshot its mark and had to be picked up by the *Monteagle*. Meanwhile, the *Hsin Tien* launched two lifeboats. One capsized and the four men drowned. The second, with 16 on board, got away and reached the *Monteagle*. In the precious minutes that followed, the *Monteagle* repeatedly dispatched lifeboats manned by volunteers to rescue the remaining seamen.

Captain Hosken may not have expected his assistant purser to volunteer, but Vaux was among 10 men to step forward. Risking their lives while battling heavy seas and the hazards of a sinking ship, the *Monteagle*'s crew rescued 66 people. This time, instead of being thrown in jail, Vaux and 10 other crew received gallantry medals from the Liverpool Shipwreck and Humane Society and the French government. Hosken was honoured with the French Médaille d'honneur de Sauvetage de première classe. Attending the Vancouver ceremony was not possible for Vaux, who was fulfilling duties on the *Empress of Asia*.

So, the senior assistant purser was sure of himself when he counselled Kearns on the dangers of a rolling ship, and it appears she discontinued her run and went below. However, at 4:00 p.m., she passed through the storm doors off the women's room and joined Shank, who was on the deck feeling seasick. When the wave rose over the ship's bow two minutes later, Kearns yelled a warning to Shank, who looked up and saw the wall of water. In the next instant, Shank watched as Kearns leaped over a bench and ran toward the watertight door.[12] But despite her quickness, Kearns never reached it. Bowled over by the weight of water, Shank was thrown about like a rag doll but was saved by the railing.

The *Empress of Asia*'s exact location was 40 degrees north, 148 degrees, 28 minutes east — 578 miles as the crow flies, northeast of Yokohama. The logbook states the "accident was due to an exceptional sea coming on board most unexpectedly. No water had been shipped all day, except an occasional spray.... The 2nd class had been perfectly dry all day, and no danger was anticipated at that time." Without other witnesses, the captain concluded Kearns was "either washed or fell over the lee rail." It is possible she was dead before she hit the ocean, given what her body was put through before being washed overboard.

Douglas noted the "ship was immediately slowed down, and turned back on opposite course." The search lasted for an hour but did not include the launching of a rescue boat: "The high sea and frequent snow squalls made visibility very low. Lowering a boat was not possible" without risking further loss of life.

When the ship reached Vancouver on January 29, reporters converged on the pier and pressed Edith Ruth for details. The superstitious young woman told the press that Kearns was the 13th player to join the team before it left the United States for Japan. She had joined the Bobbies at

Seattle, while the rest of the team had met in Philadelphia before travelling to the West Coast. "We three were together most of the time," explained Ruth, still in shock. "But somehow or another all through the tour I remember thinking that Leona was the thirteenth member of the party. Every time we sat down to a meal there were thirteen at the table. It was the same even after we finished the tour, and were coming home on the Empress. Although the three of us left the others [who had travelled to Vancouver on the *Empress of Russia*], there were thirteen at our table still. I did not like that, and neither did the others."[13]

Ruth told reporters that during her fourth day of sailing from Kobe, she and her travelling companions were accustomed to the roll of the ship and were having no trouble keeping food down. "It was about 4 o'clock that the tea gong sounded, and we three went in for a cup of coffee. Miss Shanks [*sic*] stayed on deck for a little while talking with a young man, and I went in to have something to eat with Leona. When she [Kearns] had had all she wanted she went out for a breath of air. Suddenly, on looking up I saw a big wave coming towards the ship. I cried: 'My God! Nora [Nella] is out there.' For my first thoughts were of her. I rushed out and found her lying near the door. I dragged her in and went back to look for Leona. But she had gone overboard in the green water, and there was no sign of her."[14]

Reporters noticed contusions on Shank's neck and arms while "her right eye was a dull purple color and very much swollen." Bruises were also noticeable on other passengers knocked about by the storm.[15]

Some stories recounting the tragedy give January 22 as the date Kearns was lost. The actual date, based on the ship's chronometer and her position relative to the International Date Line, was January 21, and that is how it is recorded in the ship's logbook. The confusion may stem from the date the occurrence entered the logbook.[16]

○ ○ ○

There were other peacetime deaths on the *Asia*, including suicides, but Kearns was the only passenger swept overboard. Her death, however, was not the first disaster to strike the ship on that voyage. Ten days earlier, as the *Asia* was outbound from Shanghai, steaming down the odoriferous and silty Huangpu River (Whangpoo/Whangpu), she collided with the 1,869-ton freighter *Tung Shing*. With 241 passengers and a crew of 520, the *Asia*'s voyage to Vancouver via Japan commenced at 6:20 a.m., January 11, at the China Merchants Steam Navigation Co. Lower Wharf. The more sluggish *Tung Shing* was steaming into Shanghai with a crew of 72 and 19 passengers.

The weather was good, and from *Asia*'s upper decks, passengers had a nice view of the famous Bund, the highly commercialized waterfront where fortunes rose from the shiploads of commerce and tourists flowing in and out of the city. Viewed from the first- and second-class Promenade decks, the city's European skyline left passengers in awe, knowing that up in those boardrooms, hidden by lofty domes and stone facades, Western financiers negotiated high-powered deals and transactions. In a few years, one of them, Sir Ellice Victor Sassoon, opened one of the world's most glamourous hotels — the Cathay — relying on the *Asia* to carry him across the Pacific.

Below, along the dusty, congested boulevard and back streets, women, children, the poor, and the disadvantaged peddled wares or begged near shops displaying art deco, postcards, and other souvenirs. An incessant stream of rickshaws, bicycles, trams, and automobiles paraded by ranks of sweating Chinese labourers — coolies — hauling cargo to and from barnacle-encrusted merchantmen and sampans that had travelled oceans or plied the great Yangtze. Nearby were brothels, opium dens, gambling houses, and a racetrack where foreigners, Asians, and Eurasians bet on Mongolian ponies.[17] Observing or becoming part of this assemblage of Eastern and Western cultures were the tourists and parasol-toting missionaries who stepped off the liners seeking to exchange currency.

After clearing the wharf with the assistance of a river pilot and two tugs, the *Asia* continued along the winding river at full speed on her dual manoeuvring engines. At 6:46 a.m., after passing the Standard Oil wharves, she switched to full speed on main engines, but just before rounding Black Point, six miles from the lower wharf, three freighters were spotted heading toward her. The two leading steamers were in line to pass *Asia*'s port side while the third — the *Tung Shing* — "was a little on our starboard bow," even though the Empress was supposed to be on that side of the channel, noted the captain's logbook.

The *Tung Shing* sounded one blast but did not alter course. "We replied by one blast and helm was put to port, and then hard-a-port. Our ship was sluggish on her helm owing to being near the shoal water on the starboard side of the channel." At 7:00 a.m., anticipating a collision, the *Asia* reduced to half-speed and then a minute later stopped all engines. She then reversed the starboard engine and gave "full speed ahead on her port manoeuvring engines. Our water-tight doors were closed at the same time."[18] But there was no stopping the *Asia*'s downstream momentum.

At 7:02, the ships collided, and *Asia*'s engines were put full astern moments before impact. This likely saved the *Tung Shing* from being sliced in two. "We struck the *Tung-Shing* [*sic*] just abaft the boat deck on his port side with our bow close to the stern: seeing that the *Tung-Shing* was about to founder, the helm was put hard-a-starboard, both engines stopped with the object of keeping our headway and shoving the *Tung-Shing* [*sic*] over to the port side of the Fairway [river channel] before she sank."

The mid-channel collision proved deadly for those on the *Tung Shing.* On the *Asia,* where passengers were in bed or just waking up, witnesses thought the ship had struck a sandbar. "There was a marked absence below deck of any sign of disaster," reported an unnamed passenger. "No whistles were blown; no shouting or roar of escaping steam — just a distant grinding crash and nothing more." Most of the passengers "turned over and went to sleep again. The officers of the *Tungshing* [*sic*] coolly directed the lowering of one of the other boats and the ship's raft was sent away, no easy matter with a heavy list.... The lifeboat crew on the Empress were promptly at their quarters, and the boat lowered and manned."[19]

Within minutes, a flotilla of sampans began plucking passengers and crew out of the murky water. Those who slipped below the surface were undetectable and carried away by the current. "Most of the survivors were in night attire and some who had been immersed in the bitterly cold water were in a pitiful plight, shivering in blankets hurriedly provided for them," added the witness from the *Asia*. "One of the unfortunate European passengers scrambled aboard in nothing but pajamas...."[20]

The Naval Court of Inquiry, headed by Lieutenant-Commander R.D. King-Harman of the Royal Navy, learned the *Tung Shing* had sunk in two minutes in 30 feet of water but acknowledged how the *Asia* had stood by and lowered a boat to save lives. "Ten Chinese passengers are not accounted for, but no bodies have been found...." The report notes some of those thrown into the water might have reached shore, but there appears to be no record of this. Among the missing and presumed drowned were three children. It was one of Shanghai's worst maritime accidents.

The inquiry ruled the master of the *Asia* "navigated his ship in a seamanlike and proper manner, except that, having regard to the circumstances, he should have stopped or reduced speed at the moment he first ported his helm; he failed to do so and the Court for this reason considers that blame attaches to him."

With regard to the *Tung Shing*, "the Court considers that the master was in error of not altering his course to starboard sufficiently at an early enough moment, having regard to the circumstances as shown by the evidence [including the position of the wreck] that he was not holding his starboard side of the channel when he first blew one blast." Of the two vessels, the court attached greater blame on the *Tung Shing*. With damage to her stem (lower bow) and steel plating, the *Asia* delayed four days before crossing the East China Sea to Japan.[21]

The encounter with the *Tung Shing* was not the first or last collision. Before 9:00 p.m., October 2, 1919, while outbound from Hong Kong en route to Shanghai, the *Asia* collided with and destroyed a fishing junk in the South China Sea off Shantou, China. Commanded by Captain A.W. Davison, she was 180 miles into her voyage. Moments before impact, two junks, both showing lights, were spotted, but when the *Asia* tried to clear them a third junk, without lights, was dead ahead. Wind had blown out the junk's hanging paper lantern, and before it could be relit, the *Asia* was "on top of them."[22] A boat was lowered, and the junk's four-man crew were rescued.

After the *Tung Shing* disaster, the *Asia* operated for 11 years without a serious collision. Avoiding monster storms, however, was not as easy, and while the wild weather frightened most passengers, it fascinated others.

During one storm outbound from Vancouver, the large grey-black rollers left young Henrietta Kershaw in awe. "'Don't go outside,' my mother admonished. 'You might get blown away.' But there was an irresistible lure in the danger for me, and when my mother was busily occupied talking to other grown-ups, I would sneak out for a minute or two … clutching the hand rails for dear life and loving the excitement of it and the taste of the salt spray on my lips."[23]

Under the command of Hailey, then 53, the ship was two days out of Yokohama in early February 1928 when at 1:00 a.m. a towering wave crashed across the bow and smashed every window in the fore part of the wheelhouse, 66 feet above the waterline. A torrent of sea water poured into the bridge and came close to sweeping away the helmsman and officer of the watch. "The stout liner successfully absorbed the shock and an emergency watch of sailors rushed to repair the damage."[24] Then, amid the rolling and pitching, the ship struck an unidentified object, which damaged a port propeller blade.

The *Asia* reached Victoria 11 hours late, but at least Hailey could bring her alongside. Not so on January 31, 1921, when a gale prevented the ship with 726 passengers, including 589 in steerage, from reaching the city's outer wharf. The captain elected to stop long enough to discharge the mail at Esquimalt before continuing on to Vancouver.[25]

Storms and emergency distress calls slowed or redirected the *Asia* well into the 1930s. Six-year-old Frances Gillison witnessed the *Asia* pluck a Chinese fisherman out of the South China Sea in 1938. Accompanied by her older sister, Meili, her dad, Dr. Keith Gillison, mom, Kathleen, and three-year-old brother, Walford, Frances dashed to the rail after the ship sounded her horn and began slowing down. It was August 6, and the *Asia* was 10 hours out of Hong Kong when a weather-beaten bamboo raft bobbed into view.

Blown off course by a gale, Lim Cheong Hee, 62, from the village of Wai Lai in Kwantung Province, China, had been adrift four days. "I can see now everyone rushing to the rails and this bit of wood with a small 'mast' and a flapping piece of tattered cloth attached to it," recalled Frances. "When the lifeboat sailors brought him on board, he was cardboard stiff from exposure."

Frances's mother recorded the rescue in her diary, stating the man had "been blown out" of a fishing boat by the storm, and while he was aboard the *Asia* en route to Shanghai, first-class passengers collected $200 for him. Treated by the ship's doctor, the grateful fisherman disembarked, turning to bow toward the waving passengers and crew. Frances's mother noted it was "difficult not to have feelings of resentment" while sailing into Shanghai where buildings flew the Japanese flag.[26]

It had been a remarkable year for the Gillisons. Frances's parents worked for the London Missionary Society at Hankou (Wuhan). "Wuhan had been and was suffering quite serious bombing by the Japanese in 1937 and '38," noted Frances. "My father would not be free to leave the Union Hospital for his furlough until July 1938. My parents decided to send us, their children, to friends in Hong Kong for safety. While my baby brother stayed with my parents, Meili and I travelled to Hong Kong in January 1938 and stayed with friends until Mom arrived in early July, followed by Dad on July 27th."[27]

o o o

In Shanghai, a year earlier, *Asia*'s long-time boilermaker, Thomas McMurtrie, stood topside, leaning against the rail, packing tobacco into his pipe. Looking toward the Bund, the seaman noticed a commotion on the streets; people were yelling and rushing into the banks to nervously swap savings into pounds sterling or American dollars. McMurtrie, whose work included keeping an eye on the steam pressure in the ship's massive boilers, sensed a different kind of pressure that day. It was August 8, 1937, and the *Asia* had disembarked 500 Chinese nationals who had joined the ship at Yokohama and Kobe, fleeing a country about to go to war with China.

In just two hours, 650 Chinese with "piles of trunks and baggage" rushed on board in Yokohama followed by 100 more in Kobe, all desperate to reach Shanghai or Hong Kong. "Little did these poor Chinese know that they were jumping from the frying pan into the fire," McMurtrie later told the Vancouver press.[28]

Arrival in Yokohama a few days earlier was a sad time for the crew. Not only had engineer John Forrest Alexander developed a high fever with severe abdominal cramps, headaches, and vomiting, but Fourth Officer James William Donnelly was also sick, and both had to be stretchered off and left in hospital.[29] Donnelly recovered, but Alexander was not as lucky. His many years of service on the same ship did not quite match McMurtrie's, but his death of typhoid fever 10 days later was a huge loss, and for some, a terrible omen.[30]

On approach to Kobe, the *Asia* encountered a heavy swell, the by-product of a typhoon spinning toward Shanghai, and before departing the "Paris of the East" on August 8, some 900 passengers scrambled on board, seeking protection under the British flag in Hong Kong. "There was pandemonium. The decks were alive with humanity. Some had berths and some had not and there were [*sic*] thousands of pieces of luggage to handle," recalled McMurtrie.[31]

As the *Asia* pulled away from the wharf, the ship's high parabolic stern smashed into a smaller vessel, severing the towline to an assisting tugboat and sideswiping the 6,677-ton cargo liner *Soudan*. It was a frustrating encounter, blamed on the river's tidal current, which had dragged the *Asia* midstream. Damage was minimal, but pride suffered as the "refugee" ship set course for Hong Kong, leaving an anxious city behind.

Within five days of her leaving, the Battle of Shanghai was underway. It began with a Japanese invasion, signalling the first major clash in the Second Sino-Japanese War. Initially, it pitted the powerful Imperial Japanese Army against a hodgepodge of defenders under the Shanghai Volunteer Corps. While resistance was strong, the consequences were disastrous for the city's 3.5 million people. Chinese forces capitulated some three months later after tens of thousands had died during amphibious assaults, house-to-house fighting, and devastating aerial attacks.

On August 14, amid gale-force winds and storm-warning flags notifying residents of an approaching typhoon, American-made bombers, with inexperienced Chinese crews, miscalculated an attack on the 9,500-ton Japanese flagship *Idzumo*. Two bombs fell into the Whangpoo, nearly striking the British cruiser HMS *Cumberland*; two more fell in the densely packed International Settlement. Others caused significant damage as well as casualties just outside Sassoon's Cathay and in the Palace Hotel across the street. Like Shanghai's French Concession, the enclave was an autonomous district, inhabited by British and American families and business interests. But on that day, the settlement included hundreds of thousands of displaced Chinese who had fled neighbourhoods to the north.

Scattered in the burning wreckage were broken, dismembered, and incinerated bodies. Wounded children, minus parents, wandered or sat crying on debris-clogged streets. Those who could not find refuge fled beneath a pall of dense, oily smoke. Many, including the wounded, found passage on small vessels strafed by enemy aircraft. At night a crimson glow hung over the city, seen by ships far out to sea.

Westerners were also desperate to flee, and by the 18th, the *Asia* returned as an evacuation ship for foreign nationals. On her chains in the Yangtze, six miles from Woosung at Blockhouse Buoy, she arrived with 700 members of the 1st Battalion, Royal Ulster Rifles, from Hong Kong. These reinforcements, in support of the British garrison in the International Settlement, were transferred to naval ships for the 12-mile journey upriver. Heading in the opposite direction were three British destroyers crammed with evacuees anxious to board the *Asia* on the 19th.

Captain Goold told the Vancouver press he had been in his cabin aboard ship at Hong Kong when he received instructions from the commodore's office to transport the British troops to Shanghai.[32] He had also received a telephone call from the CPR's Montreal office advising him to place his ship at the disposal of British military authorities to help evacuate British women and children from Shanghai.[33] "It was an anxious trip," recalled the captain. "As we steamed down the [Yangtze] river (with the evacuees) … we could see Japanese warships bombarding the city, while airplanes circled overhead dropping bombs. From a distance of twenty miles, the burning buildings looked like huge bonfires."

One evacuee was Canadian Margaret H. Brown, a nurse and missionary from Tiverton, Ontario. In a letter to the *Vancouver Daily Province*, written while on board the *Asia*, she noted half a dozen Canadians among the "fifteen hundred women and children evacuated from Shanghai by" the Vancouver ship. "We would like to tell Canadians everywhere how grateful we are to the British authorities, the British navy, and the Canadian Pacific officers and crew of this ship." Brown, who worked as editor for the Society of Christian Literature, was far less impressed with the Canadian government's response. "So far as we are aware our own Canadian Government have [*sic*] shown no concern about our fate."[34]

Recounting the August 14 errant bombing, Brown stated she was in her office in the International Settlement when the second bomb missed the Japanese warship by 50 feet and it sounded like "the crack of doom."[35] The explosions and fire killed hundreds, including some Westerners who had been watching from rooftops, keeping score as if the aerial attack was a cricket match. However, the bloodiest carnage caused by errant bombs occurred at a congested intersection near the city's amusement centre.

Like Brown, Liza Benjamin had made it on board a smaller vessel before getting on the *Asia*. The big ship had been a welcome sight, looming large and white on the muddy Yangtze. Noticeable were the large Union Jacks painted on the ship's hull and deck to identify her as non-combatant. While catching her breath in a deck chair with her 21-month-old son, David, the no-nonsense Jewish refugee was grateful to be among 1,368 evacuees on board. She was several months pregnant and missing her husband, Benjy, left behind in Shanghai. Liza wondered if she would ever see him again or if he would ever get to hold their unborn child.

Her daughter, Ester, born three months after the *Asia* reached Hong Kong, recalls her mother as a courageous and resourceful woman — a former high-diving champion and Girl Guide leader who never backed down from a challenge, someone who always took charge whenever there was a crisis. There would be many more such instances after they returned to Shanghai and became prisoners of the Japanese during the Second World War. "She was a powerhouse in an under-five-foot package," recalled Ester.[36]

Goold told the press the ship steamed on "with every possible flag flying" and "though Japanese warplanes roared overhead for nearly an hour after we left, no bombs fell near the liner."[37] Before extinguishing his pipe and returning below, McMurtrie took in the "gruesome evidences of warfare" with bloated bodies of Japanese and Chinese soldiers, some dismembered, floating down river. "The continuous muffled boom of the guns told us we were quite close to the fighting.... We were thankful for the privilege of taking … the women and children out of that blazing inferno and transferring them to safe territory."[38]

On August 21, as the *Asia* neared Hong Kong, another pregnant woman, much farther along than Liza Benjamin, gave birth. Her name was Zoya Raphaelovna, born in 1909 in Nikolayevsk-on-Amur in eastern Siberia near Sakhalin Island. She was married to a British national and possibly evacuated from the International Settlement. Goold became the child's godfather, and the boy was aptly christened Michael Asia Rowland.[39] He lived to be 70 in 2007, coincidentally dying on the same day and month the *Empress of Asia* was lost.

○ ○ ○

Young Frances Gillison and her family did make it home to Britain in 1938 after visiting relatives in Canada. While Frances and Meili divided time between a boarding school and "much-loved guardians," their parents and little Walford returned to Hankou, which was by then overrun with Japanese forces. "As it turned out we would not see our parents for six years, and they were, with all other foreigners, interned by the Japanese for nearly three years. It was a strange world then and sadly it doesn't seem to lose any of its strangeness now!"

11

SHIP TO SHORE

In a lifetime spanning most of the 20th century, she was able to see all her grandchildren through college and spread throughout North America.

— **Patricia Wong**

Low Shau Wah did not know what to expect when she emerged from the *Empress of Asia*'s steerage at seven o'clock on the evening of Monday, January 31, 1921.[1] The 22-year-old was exhausted and in need of fresh air, and while grateful to set foot on land in Vancouver, she was entering a country that viewed Chinese immigrants as a threat.

Fresh in her mind was the recent appalling weather, described in the press as a "fifty-mile gale," forcing Captain Hailey to off-load Victoria-bound mail in Esquimalt where hard-faced customs and immigration officials came aboard. The voyage from Hong Kong had been relatively smooth until it approached British Columbia's rugged coastline. "As fine a trip as we ever had across the Pacific," Hailey told the press.[2]

Their hats and overcoats slick with rain, the immigration officers made their rounds while the *Asia* steamed through the southern Gulf Islands to calmer waters in Vancouver. Working deck by deck, and aided by the chief deck officer, all documents and faces were scrutinized. Much of their focus was directed at the Chinese passengers in steerage, although they also kept an eye out for illegal activity, including signs of drug or people smuggling, as well as stowaways. Hailey would have been familiar with the latter throughout his career. Keen to go to sea at age 12 in 1886, the lad from Bradford, Yorkshire, might even have stolen aboard himself had he not signed on as a "boy seaman."[3]

While arrangements were made to transport Low Shau Wah and others aboard coastal vessels back to Victoria, stevedores went to work transferring enough Chinese eggs from the *Asia*'s holds to fill 38 railway cars, much of it destined for eastern cities. It took less time to unload the ship's bales of silk — swiftly removed and secured in 17 special freight cars on a "silk train" leaving at midnight.

When Low Shau Wah left the village of Sack Kee, Chungshan (Zhongshan) County, China, she was without family or friends. Located within the fertile Pearl River Delta in South China, her village was 56 miles northwest of Hong Kong where her life crossed paths with the *Asia* on January 13. While January is typically the coldest month in Hong Kong, the weather was pleasant with daily temperatures between 12 and 18 degrees Celsius.

Descending the cavernous ship toward "Oriental Steerage," Low Shau Wah entered an alien world of steel. She had never been on such a vessel, let alone crossed an ocean. Steerage was noticeably humid and smelled of oil and hot metal. Primarily intended for single women and families, the accommodation consisted of four bunks, but no toilet or natural light. For fresh air, she would need to visit the small third-class promenades at the bow and stern. Single or married, these women were a rarity for a variety of reasons, but mainly due to the highly restrictive immigration laws, which made it difficult for all Chinese, but especially women, to settle abroad. Another factor was the long-standing Chinese patriarchal and polygamous tradition that permitted primary and secondary wives or concubines. Often, when a husband moved away to establish a new life — say, in North America — his primary wife would remain behind to raise the children while his concubine travelled with him.[4]

Low Shau Wah was neither. At the time of her arrival, there were 3,431 Chinese residing in Victoria's Chinatown, but only 503 were female.[5] Like others, she had to pay the $500 head tax to enter Canada, an amount equivalent to $7,892 today.[6] Originally set at $50 as part of the Canadian government's Chinese Immigration Act of 1885, its main purpose was to restrict specific immigrants. This came after many Chinese died building a difficult portion of the CPR through British Columbia.

The head tax increased to $100 in 1901 but jumped dramatically to $500 in 1903. Steamship companies on the transpacific routes had to be mindful of the act, since they were permitted to transport one Chinese passenger for every 50 tons of ship tonnage. This meant the *Asia* could legally transport a maximum of 338 Chinese immigrants per voyage to Canada. The restriction did not apply to those returning to Vancouver or in transit to plantations in the West Indies or the United States; nevertheless, it meant the CPSS had to distribute immigrants among its fleet for legalistic reasons.

In 1902, the Royal Commission on Chinese and Japanese Immigration concluded that Chinese and Japanese "are unfit for full citizenship ... and are so nearly allied to a servile class that they are obnoxious to a free community and dangerous to the state."[7] Despite this, Canada's Chinese population continued to rise, and by the time Low Shau Wah arrived, it numbered 39,587.[8]

As proof of paying the head tax, Low Shau Wah was issued with the required C.I.5. certificate from the Dominion of Canada Immigration Branch, Department of the Interior. In the black-and-white head-and-shoulders photograph, she looks healthy, calm, and intrigued by the camera. Although she was 22, her age on the certificate states 23. This is because Chinese often add a year to their ages, based on the assumption they are one year old when born.

The certificate confirms she "arrived or landed at Victoria" at the end of January; there is a declaration date of February 2 and signatures of the local controller of Chinese immigration and Malcolm Reid, Dominion immigration inspector for British Columbia. What happened next might have stemmed from arrangements made in South China, but there is no evidence of such arrangements. On February 2 — the same day her head

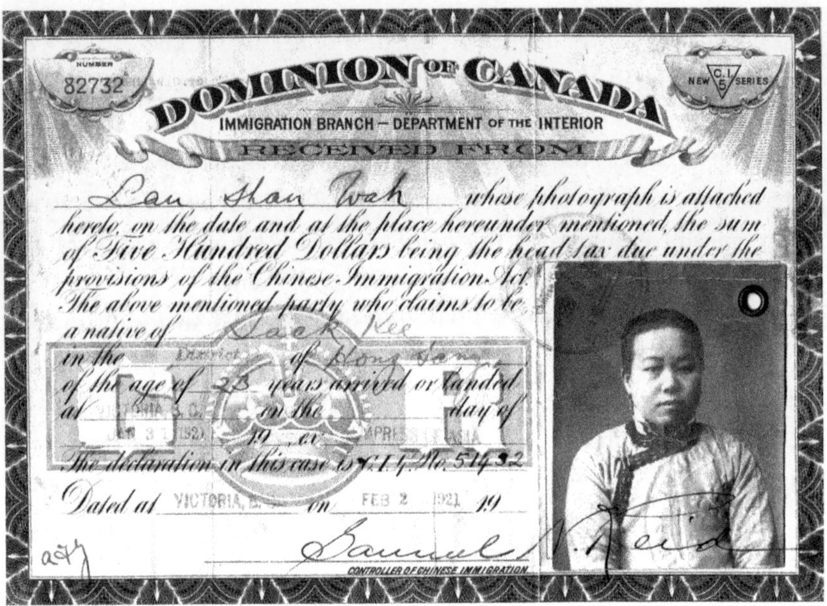

NUMBER 82732

DOMINION OF CANADA

NEW C.I. 5 SERIES

IMMIGRATION BRANCH — DEPARTMENT OF THE INTERIOR

RECEIVED FROM

Lau Shau Wah whose photograph is attached hereto, on the date and at the place hereunder mentioned, the sum of Five Hundred Dollars being the head tax due under the provisions of the Chinese Immigration Act. The above mentioned party who claims to be a native of Sack Kee in the District of [illegible] of the age of [illegible] years arrived or landed at VICTORIA, B.C. on the JAN 31 1921 day of 19 ex EMPRESS OF ASIA

The declaration in this case is C.I. 4 No. 51432

Dated at VICTORIA, B.C. on FEB 2 1921 19

Samuel N. Reid

CONTROLLER OF CHINESE IMMIGRATION

Low Shau Wah's 1921 Dominion of Canada immigration certificate.

tax certificate was stamped — a marriage took place between Low Shau Wah and 28-year-old Sue Lee, a Victoria farmer. The licence was signed by a Presbyterian minister and witnessed by Lew Yew of Victoria and Alice Cronkhite of the Women's Missionary Society of the local Presbyterian church. Like Low Shau Wah, Sue Lee could not read, write, or speak English.[9] The ceremony unfolded without glitter or celebration in a damp, cold immigration shed along the waterfront. There, she became "Mrs. Lee," but there is no proof she actually resided with Sue Lee, so it was likely a marriage of convenience (see "paper marriage" below).

Low Shau Wah soon returned to Vancouver and secured lodging with a Chinese family in Chinatown. That December, she entered into a union with another "Mr. Lee" — Lee Sing — whose ancestral village was not far from hers. While there is no provincial registration of marriage, it is possible a traditional Buddhist-based marriage ceremony occurred.

If the young woman had not arrived on the *Asia* when she did, she might never have reached Canada, because by 1921, pressure was even more extreme to keep Canada white and Christian. On Dominion Day 1923, the federal government passed the Chinese Immigration Act, legislation not repealed until 1947, but even after that, restrictions remained in place for another 20 years. While the 1923 act dropped the head tax, it excluded all Chinese immigration, including those of British nationality — the exceptions being diplomats, select foreign students, and merchants.

The terms "paper marriage," "paper son," and "paper daughter" are not well known, but there were numerous cases of ethnic Chinese relying on deception to enter Canada and the United States. In discreet Chinese communities, for good reason, the subject was not spoken of above a whisper. However, more recently, grandchildren have been researching and sharing family histories, including the paper gymnastics used to vault past immigration barriers.

For example, a resident in Canada would relocate to China and transfer his or her immigration documents to someone planning to board a ship to Canada. Or residents of Canada would claim to be parents of children in China, requesting that "their children" join them in Canada. Before boarding a steamer in East Asia, the supposed child would receive documents stating who their "parents" were.

The *Asia* was one of many transpacific ships bringing "paper sons and daughters" to North America, a practice so widespread that in 1960 the Canadian government announced a Chinese Adjustment Statement Program, granting amnesty for paper sons or daughters if they confessed to the government. Thousands came forward up until 1973 when the program ended.[10] It is also clear that the names of the individual ships they arrived on remain an important part of family histories, often cited as the starting point.

Although Low Shau Wah's arrival was tenuous, it was the first thread woven into the fabric of a modest but successful life with Lee Sing, who had resided in Vancouver for more than a decade. As a houseboy for Canadian families of British descent, he taught himself to read, write, and speak English. Quietly ambitious, he made friends and developed important contacts. By 1923, he had enough money to establish a growing Chinese importing business that moved three times within two years to locations along

East Pender Street. Ten years later, Yuen Fat Wah Jung Co. Ltd. was a fixture in Chinatown, occupying the ground floor of the red-brick May Wah Hotel. Behind the window displays, the store's imports drew orders from restaurants and stores across Canada. On the retail side, it sold everything from local vegetables to imported herbal remedies to household goods, including Chinese lanterns, tablecloths, and tableware.

While Mr. Lee managed his inventory and volunteered on the Chinese School Board, Low Shau Wah raised four children, all born in the small apartments behind each successive location. Lily was the eldest, followed by three boys — George, Harry, and Alfred — named after English kings by Low Shau Wah's midwives. In 1943, the Lees moved out of Chinatown and purchased a doctor's house in Vancouver's South Cambie neighbourhood. By then, Low Shau Wah was in her midforties and likely would have heard about the wartime loss of the ship that transported her to Canada.

Back in June 1921, nearly five months after Shau Wah's arrival, 13-year-old Yau Shek Ying disembarked from the *Asia* in Vancouver. Although born in 1908, 1910 was the year shared with Canadian authorities to make him appear younger. His uncle, Yau Shiu Wun, who had been quietly working with Yau Shek Ying's mother in China, greeted him at the pier and paid the $500 head tax, claiming his nephew as his son. The boy's name on the *Asia*'s passenger list had been spelled "Yau Shack Ying," but when he passed through the cavernous immigration building, the name had been anglicized as "Hugh," and he remained Albert Hugh for the rest of his life.[11]

Albert's journey had begun on foot from Nam Wai, a fishing village set among semi-mountainous terrain four miles northeast of Hong Kong on the Kowloon side of the harbour. In addition to his mother, there was an adopted brother. His father, a fisherman and seaman, had died in 1913, and so his mother ran the tiny household with little money, barely managing to save enough for the boy to reach Hong Kong and buy a ticket to board the *Asia* on May 26.

While crossing from Yokohama, the ship, commanded by Hailey, set a new record by cutting two and a half hours from the former record, set on her previous voyage. Passengers disembarked at Victoria, but there were still 732

Yau Shek Ying (Albert Hugh) arrived on the *Empress of Asia* in 1921.

aboard when the ship reached Vancouver. Yau Shek Ying was among 429 in steerage. "He boarded the ship not knowing anybody," said his son Richard Hugh, one of Albert Hugh's four children, the others being Dennis, Lorne, and Mildred. "It must have taken a tremendous amount of courage."[12]

"I have nothing but admiration for my grandfather's journey to a new life in a new country: a journey I don't think I'd be brave enough to make — and that's why I'm fascinated with the *Empress of Asia*," added his granddaughter, Shana Hugh. "That ship is the beginning of the story for my family's existence; my existence; my daughter's existence. What would that have been like, to reinvent your life in a country where you don't speak the language? Where you suddenly have a brand new everything — including your name?"[13]

Albert Hugh settled in Chinatown with his uncle, learned English, found work on the Alberta oil fields at Swan Lake, and later in a Vancouver restaurant as a short-order cook. He was married more than once, helped run a coffee shop, was a founding member of the Hakka Society, and helped raise his children. "He was proud of his heritage," explained Richard Hugh. "He would instill in us the importance of always 'holding your chest up high while pulling your shoulders back, so when people looked at you they are impressed with the way you are walking.'"[14]

Albert Hugh's big break came at the Tahsis sawmill on the west coast of Vancouver Island where he was cook for the main chef. His son Dennis, who also worked at the mill, recalled how his "five-foot-nothing" father peacefully ended a dispute between two large angry workers — one a member of the green chain and the other a longshoreman — both intent on battle. He also recalled the rejection his father faced when he volunteered to serve as a wireless operator at the start of the Second World War.

○ ○ ○

For other Chinese and Japanese, Canada was often an easier entry point to the United States. Word had spread about the exhaustive interrogations and lengthy detentions Asian immigrants faced at the Angel Island Immigration Station in San Francisco Bay. European immigrants screened at New York's Ellis Island faced relatively far fewer barriers.

Asia steerage passenger Gim Suey Chong (Zhang Jin Rui) was eight when he reached Vancouver on April 11, 1932, although the paperwork for entry into the United States had begun in earnest when he was five. From his great-great-great-grandfather, Cheun Saan Jeung, in 1849 to his father, Moi Chung, the family had tied their dreams to North America.[15]

As a "paper son," Gim Suey depended on his "paper father" uncle, Hung Quock Chong (Zhang Guo Xiang). The 44-year-old Hung Quock was a naturalized American working as a cook in a Boston chop suey house. His last name "Chung" had been misspelled as "Chong" by immigration authorities. The boy's real father, who also resided in Boston, was not a naturalized American and so could not legally claim his son.

In 1929, quoting a provision that allowed his "son" entry into the United States, Hung Quock swore on an affidavit that he was Gim Suey's father. The affidavit lists the boy as older than he was and states he was an American

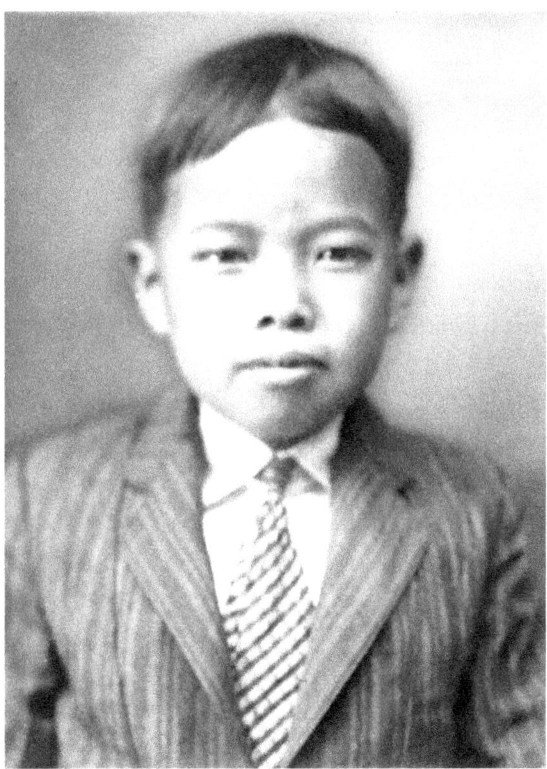

Gim Suey Chong faced a world of adventure when he left China bound for Canada and the United States in 1932.

born abroad. Family records provide two possible birthdates: March 3, 1924, and November 5, 1923. In completing the affidavit, Hung Quock affixed separate photographs of himself and the boy and stated he had retained a Boston lawyer to represent their interests.

Although Gim Suey embarked on a 19-day voyage to Canada's West Coast, his journey began in a pair of slippers. In March 1932, he had said goodbye to his mother, Cun Chuen Wong, and half-sister in the tiny grey brick house tucked into an alley at Yung Lew Gong (Yang Lu Gang). It was Gim Suey's first trip beyond the village in southern China, and as he walked along the dusty road, he took one last look at the old stone *diaolu* built to protect the village from bandits. He bade farewell to neighbours, and beneath the Mount of the Eight Immortals, passed others working in the rice fields, banana groves, and vegetable plots.

Accompanied by his uncle and a "family friend," Mee Fong Gee, Gim Suey hiked four miles to the town of Baisha in the Taishan district of Guangdong Province. There, they boarded a train for a 62-mile journey to Jiangmen — alternately romanized in Cantonese as Kongmoon — and Pakkai (a placename that seems to have disappeared) on the muddy Si Kiang River. Gliding downriver in a ferry, the youngster already missed his mother; ahead lay a journey full of intrigue, one that would alter his life forever.

After reaching the South China Sea, the ferry completed its 106-mile voyage by heading east to the Portuguese colony of Macao and then to the Crown colony of Hong Kong. On March 25, following several restless days in Kowloon, the three stood in line to board the *Empress of Asia*.[16]

Gim Suey was half an inch shy of four feet. His eyes followed the thick mooring lines from the wharf to a height he could not measure. Pointing skyward were the ship's golden-yellow funnels and a steel mast resembling a cross. The ship was larger than anything he had seen, save for the sacred mountain near his village.

In steerage, the boy and his companions kept to themselves. The monstrous ship seemed too heavy to float, yet somehow did. There were creaks, vibrations, and strange smells, and just before it began to move, one long blast and three short blasts from a whistle. The big ship was reversing, moving away from the pier.

From Hong Kong, the *Asia*, skippered by Alfred Lovegrove, called at Shanghai, Nagasaki, and Kobe before commencing her 10-day voyage from Yokohama to Vancouver. Far below, the boy and his uncle quietly rehearsed their roles as paper son and father. It is probably just as well the *Asia*'s cargo holds were out of bounds to passengers, for on this voyage there were large crates of gold specie, a consignment of raw silk valued at $1.5 million, and more intriguingly, strings of lemons containing considerable amounts of opium.[17]

The passenger list on this occasion was light, only 15 first class, 37 tourist, and 129 steerage. Among first-class notables were two arms manufacturing representatives returning from Japan — Captain E.G. Paget from Hadfields Limited in Sheffield, England, and Frederick Vaatz of the German industrial giant Krupp. The former was a steel manufacturing and engineering giant that produced cast-steel wheels and axles, as well as war *matériel* such as armour-piercing artillery shells and armoured structures for warships. The latter was one of Europe's largest armaments manufacturers.[18]

Paget and Vaatz paid no heed to those in steerage and likely remained politely evasive around other first-class passengers if asked about their business. It was easy to retreat to the lounge, smoking room, or cabin.

Dense morning fog in the Strait of Juan de Fuca delayed the *Asia*'s Victoria arrival by three hours. The murk was so thick, according to the *Victoria Daily Colonist*, that it caused the "navigating staff some anxiety until the ship reached Race Rocks."[19] Captain Lovegrove, who had taken over from Douglas and was marking his first return voyage as her commander, recalled "a man-sized fog.... I could not see land all the way up the straits and at times I could hardly see the bow ... from the bridge." According to passengers, the captain was ever watchful from the night before. "The fog horn on Race Rocks [Lighthouse] was booming out its warning as the ship came around there this morning, and it was not until she was at the [William Head] quarantine station that the fog lifted and passengers could see [Race Rocks]."[20]

By 11:00 a.m., the ship was alongside Rithet's Piers in Victoria where clamouring newshounds pressed Paget for his views on the conflict between China and Japan. They wanted to know if it was good for the arms industry. In his opinion, the troubles between the two countries were probably not serious enough to warrant the purchase of arms on a large scale by either side. In the next breath, he noted any business relating to the purchase of armaments was confidential and then emphasized that in no way could recent fighting between China and Japan be called a war. Another exaggeration, he said, was to suggest the combatants were purchasing small arms and ammunition from European countries.[21]

In Vancouver, Gim Suey and his chaperones nervously entered the immigration building where they hoped the carefully fabricated cover of false paperwork would conceal their true story. It did, and they boarded a CPR passenger train for Montreal's Windsor Station. By the time he and his uncle arrived at the seaside town of Yarmouth in southwestern Nova Scotia, Gim Suey was still wearing the slippers he had put on when he left home. On

April 19, he and his paper father boarded a steamer for overnight passage to Boston without "family friend" Mee Fong Gee, who had remained in Montreal. Beyond earshot, while crossing the Gulf of Maine, the two held one final rehearsal before U.S. Immigration and Naturalization Service detained the pair on April 21 for interrogation. It was intense and stretched into the following day, but it was expected. On day one, there were 10 questions, including:

> **Q:** How much money did you have on your arrival?
> **A:** Fifty cents, American money.
> **Q:** Did anyone accompany you on your journey to the United States?
> **A:** Yes, my father.

On day two, the boy identified his uncle as his father and answered 108 questions, most seeking names and ages of relatives and details about his village:

> **Q:** How many houses and rows are there in your home village?
> **A:** Eight dwelling houses and one school-house; three rows.
> **Q:** Who lives in the house in front of yours, the 2nd house, 3rd row?
> **A:** Chong Thick Teung lives there with his wife, son, and daughter.
> **Q:** Is there a wall around your village?
> **A:** There is a wall on the west side and on the north. There is also bamboo on the east.
> **Q:** Are there any gates to the village?
> **A:** There is an opening on the west side but not on the east side. There is no gate on the east side but if you push the branches of the trees apart you can enter the village.

When Hung Quock's answers compared favourably to the boy's, the examining officer, Inspector J.E. Fitzgerald, ruled the relationship between the two as reasonably established, and Gim Suey entered the United States.[22] Within hours he met his real father, Moi Chung, manager of the popular Imperial Restaurant in Cambridge, Massachusetts. Moi had first arrived in the United States in 1912, via Angel Island. By 1917, he was managing a New England restaurant where his father was silent partner. Like many restaurants during the Depression, the Imperial struggled, so the boy and his father collected their few possessions and moved to Los Angeles, where like many others, they fell into poverty in Little Tokyo.

○ ○ ○

Gim Suey, Albert Hugh, and Low Shau Wah were just some of many immigrants who arrived throughout the late 1800s and early 1900s. The harsh conditions they left behind were myriad, but they were not, by definition, refugees — people forced to flee their homeland "to escape war, persecution, or natural disaster." They were economic migrants hoping for a prosperous future.

On September 22, 1923, the *Asia* reached Vancouver with 31 refugees, including an infant and child, all survivors of one of the world's worst natural disasters.[23] The Tokyo-Yokohama earthquake, or Great Kantō earthquake, killed approximately 140,000 people. Thomas Maitland was among those refugees. His wife, Jess, and eight-year-old son, Robert, were killed when their apartment building collapsed. The September 1 temblor created a massive tsunami, spawned fire whirls that immolated thousands, and triggered untold ethnic violence. Across Yokohama and Tokyo, buildings pancaked and infrastructure was razed, including bridges, roads, power lines, railways, and prisons. The resultant tsunami was 39 feet high when it swept ships and other debris inland along the Honshu coast.

Three CPR Empress ships were involved in rescue and relief work during and after the 7.9-magnitude quake. At 590 feet long, the 21,861-ton *Empress of Australia* was alongside at Yokohama, ready to sail for Vancouver — her mooring lines singled up, and two tugboats with lines made fast to her bow and stern.[24] On her starboard side, passengers tossed confetti and streamers to those blowing kisses and waving goodbye on the wharf. Then, less than two minutes before noon, the huge ship began to shake "all over in a most terrifying fashion," noted Captain Samuel Robinson, "and also rocked very quickly and violently...."[25] For many attempting to escape the crumbling pier and waterfront, the ship became a lifesaver.

What Robinson had missed in the way of action on the *Asia* during the First World War, he experienced now on the *Australia* — in a much different form. Standing on the bridge, he fully expected the first shockwave to "carry away" the mast and funnels. In horror, he watched large sections of wharf fold and crumble beneath the feet of hundreds of shocked well-wishers. Huge fissures opened and closed, swallowing and crushing several people. "In one case a motor car and its occupants ... disappeared entirely," he noted. Gazing toward shore, Robinson witnessed "the land rolling in waves ... 6 to 8 feet high like a succession of fast moving ocean swells...."[26]

In that moment, Thomas Maitland, who managed the Japanese office of the Manufacturers Life Insurance Company, was on a commuter train five miles from home in Yokohama. He was convinced the train had collided with another at full speed as the carriage rocked and leaped like a freakish carnival ride, while the railway corridor's six sets of tracks writhed like "snakes twisting and squirming." Buildings fell amid a roar of smoke, mud, and plaster, choking out sunlight and leaving an eerie orange glow. Maitland stumbled from the twisted remains of the coach and fought to maintain balance as the ground shook and cracked.

Covered in dust and ash, he reached Yokohama harbour the following morning and boarded the *Australia*, desperately searching for his wife and child. Unsuccessful, he tried returning ashore, but the situation was too dangerous. The previous evening, he had wretchedly kept hope alive, even after staring at the rubble where his apartment had once stood.[27]

Maitland had relocated to Canada from Scotland in 1909, working in both Ontario and British Columbia before arriving in Japan in 1922. As a Great Kantō refugee, he was among thousands who lost loved ones.

Throughout, Robinson's foremost priority was to protect the *Australia* by getting her away from the wharf. "Every fire-hose possible was connected, special water pressure put on, and the decks and upper works [of the ship] kept flooded."[28] Gale-force winds had whipped the harbour into a foul, jostling mess. Chunks of burning debris fell toward the ship and freight shed. When the tugboats let go, the wind on her port side made it impossible to clear the wharf. She could not move ahead more than a few yards because of shoal water, and immediately astern was the 5,719-ton cargo ship *Steel Navigator.*[29] Compounding the situation were the ships, once at anchor, now loose and creating a domino effect. A large Japanese steamer struck the *Australia*'s port quarter, damaging her steel plating, bulwarks, and rails before bouncing off and ramming her again. The second blow obliterated a cargo vessel caught in the middle.

With the freight shed ablaze, Robinson ordered reversed engines, and his ship pushed back against the *Navigator*, which paid out her two anchor chains. Adding to the chaos, lighters loaded with lumber were crushed between the two ships and the wharf. Moving slowly astern, the *Australia* screeched along the *Navigator*'s port side until her port propeller snagged the other ship's port anchor chain, dragged out of position by another ship.[30]

By 4:00 p.m., the *Australia* was no longer in imminent danger and several relatively secure lines held fast to posts along the damaged wharf. Turning their attention to shore, the captain, crew, and several passengers formed rescue parties and embarked on lifeboats with food, water, and medical supplies. The situation grew more precarious when fuel from ruptured oil and kerosene storage tanks formed massive slicks that burned and swirled across the harbour. There were countless additional hazards, but by the evening of September 2, the *Australia* was at anchor beyond the breakwater, and around 7:00 p.m., the ship's purser informed the captain he had 1,901 earthquake refugees — 1,000 fewer than the previous evening because many had transferred to other ships.[31]

Robinson's old ship, the *Asia*, skippered by Douglas, was in Hong Kong when news arrived, but within hours an emergency relief committee was formed, comprised of British, Chinese, and Japanese residents. With 500 tons of rice stowed, the *Asia* steamed full speed toward the earthquake refugee camps in Kobe, approximately 1,553 miles away.[32]

Off Yokohama, the anchor chain torn from the *Navigator* wrapped three times around *Australia*'s propeller, requiring a Japanese diver to remove it, but within four days, most of the remaining passengers who found safety on the *Australia*, including Maitland, transferred to vessels bound for Kobe. The *Empress of Canada*, with Hailey in command, arrived on September 3 from Vancouver and resupplied the ship via lifeboats full of provisions. She took on 509 refugees, many of them injured, and reached Kobe the next morning.[33] The *Australia* remained at anchor for a week, dispatching relief parties and taking on more injured; however, before setting course for Kobe with 684 refugees, Robinson went ashore with a small party and beneath the ruins of the CPSS office discovered the charred bodies of two employees — J. Reed and 67-year-old Frederick Wevill, a married man with two grown children.[34]

Dozens of refugees left Kobe aboard the *Australia*, reaching Vancouver a mere day after Maitland arrived on the *Asia*. In addition to carrying refugees, *Asia* delivered a letter from the American Relief Committee, addressed to Robinson. "You, Sir, must be proud of every member of your staff and crew, who served you as one man and who never failed in any emergency, and the kindly, courteous and efficient manner in which you all cared for the refugees, regardless of condition, nationality or colour...."[35]

The largest wharf-side greeting was for Robinson and his ship. Among numerous international honours and accolades, the captain received the Commander of the Order of the British Empire.[36] Captain Hailey's contribution earned him the Spanish Cross, First Class, Civil Order of Beneficence, and a silver bowl inscribed by the emperor of Japan.[37] No honours were bestowed on the *Asia*'s officers and crew.

○ ○ ○

Two years after the earthquake, and seven years before Gim Suey's voyage, the *Empress of Asia* transported 59 Russian refugees. It was July 1925, and the men, women, and children were in steerage, destined for farming communities in northern Alberta. Most knew little to nothing about farming, but that was unimportant; they were relieved to be fleeing Russia.

All had endured dangerous journeys before reaching Harbin in northeastern China. During the Russo-Japanese War (1904–05), Harbin was a Russian military base for operations in Manchuria. However, when Russia lost the war, control of the town fell to Chinese and Japanese forces. For several years, it remained a large construction and maintenance depot for the Chinese Eastern Railway, linking the Trans-Siberian Railway through Manchuria to Vladivostok on the Sea of Japan, but its Russian population had been steadily rising.

The First World War, Great October Socialist Revolution, and Russian Civil War, occurring between 1914 and 1923, led to an exodus of Russians. More than 100,000 White Russians and White Russian Guards fled to Harbin, the largest enclave outside the newly formed Soviet Union. While many refugees elected to settle, others used it as a transit point to the south, where ships left for North and Central America. This exodus coincided with a period of nationalist agitation in China that led to a general strike, boycotts, and violence. It began when police in Shanghai fired on Chinese demonstrators calling for an anti-British strike. After several demonstrators died, anger and violence spread swiftly, and when the *Asia* reached Hong Kong at 6:01 a.m., June 21, 1925, all but 150 of her Chinese crew of 450 deserted.

Its ship stranded through insufficient personnel, the CPSS recruited a multinational crew and tasked the *Empress of Canada* with transporting a separate party of White Russian refugees from Shanghai to help fill the void in the *Asia*'s stokehold. Although the Russians arrived in Hong Kong five days late due to a shipping delay, the *Asia* sailed on June 30, commanded by Douglas. The ship kept excellent time, averaging 18 knots with her greenhorn crew that included the 60 Russians, 70 Filipinos, 40 Portuguese, 20 British, 15 Americans, and several Japanese — in addition to her remaining Chinese firemen. Newspapers were eager to tell readers about the *Asia*'s unusual crew, including a report in the *Victoria Daily Times* noting that female missionaries were working as stewards after escaping the disturbances in China.[38]

Eleven-year-old George Miasnikoff and his parents were among the Russians who boarded at Kobe. He, his mother, Anastasis, and father, Theodore, were from Samara (Kuybyshev), 435 miles north of the Caspian Sea. Forced to abandon his hardware store, Theodore had little choice but to leave with his wife and son, and a wicker basketful of belongings. Their eastward journey on a Trans-Siberian Railway train marked the start of an epic story as they were jammed onto the creaking, swaying carriages along with endless numbers of malnourished and disease-ridden men, women, and children. The stench from festering sores, vomit, and bodily waste was overpowering. George remembered seeing dead people tossed from the moving train; at other times, the train sat idle for weeks.

Finally, by 1921, the Miasnikoffs reached Chita in southeastern Siberia, 2,610 miles as the crow flies from Samara, and three years later they were in Harbin where Theodore chose Canada from among other possibilities. The family eventually travelled south and crossed the Sea of Japan to Osaka, near Kobe.

Their voyage on the *Empress of Asia* was another deplorable experience, even with the White Russian crew complement. "What a journey that was! There was a strike of the crew on board the ship. The crew seemed to feel their only duty was to make miserable the lives of the steerage passengers." The meals were so "infrequent and inedible that if it had not been for the food we had brought along from Osaka, we would have been half starved by the time we reached our destination." And the accommodation, recalled George, "consisted of large wire mesh compartments with about 40 persons in each.... Privacy was completely lacking. In each compartment were a series of bunk beds, tables, and a few benches."[39]

After passing through the immigration building on July 17, George and his parents boarded the train for Alberta where they eventually purchased a farm near Wetaskiwin. By 1926, the Miasnikoffs had acquired 28 acres of wheat, later ruined by rain and snow. Hail also destroyed the following year's crop before the Depression sent commodity prices tumbling.

Shortly after the Miasnikoffs' arrival, the *Asia* commenced another outbound voyage. Ongoing unrest in China meant her passenger list was once again light, but before she cleared Victoria there was a gracious farewell ceremony. Several of the Russian refugees who helped stoke the ship on her inbound voyage were high-ranking Russian officers returning to Japan. In the Far East Russian region of Priamurye, they had served under General Spiridon Merkuloff, the deposed leader of the anti-Bolshevik provisional government. Merkuloff was an ardent supporter of Russian refugees and immigrants, having escaped to British Columbia with his wife. He did not want to miss this opportunity to wish his men safe passage and honour their service. Mustered on deck in their old uniforms, the returning Russians doffed their hats, bowed, and kissed the extended hand of Madame Merkuloff. "It was one of the most picturesque scenes witnessed at the docks for some time," noted the *Victoria Daily Colonist*.[40]

○ ○ ○

Like many, Low Shau Wah, Albert Hugh, Gim Suey Chong, and the Miasnikoffs persisted throughout the Depression. Federal exclusion laws had diminished the population in most Chinese communities, yet Shau Wah and the Lees held fast and prospered. Mr. Lee's status as merchant under the Exclusion Act, and his success as a businessman, made it possible for his family to travel to China twice — once in third class, and again much later, in first class. On VJ day, the Lees, who had sponsored newspaper ads welcoming the 1939 Royal Tour, demonstrated patriotism by flying the Canadian Red Ensign, even though they were still not naturalized Canadians.

Twenty-six years after she stepped off the *Asia*, paid the head tax, and added her name to a marriage licence, Shau Wah finally obtained the right to apply for Canadian citizenship, receiving it two years later in 1949. While two of her children settled in Vancouver, the other two moved away. Meanwhile, Mrs. Lee continued to exhibit the same adventurous spirit that had first brought her to North America in 1921 by regularly travelling across the continent alone, now by air, to visit her children, grandchildren, and great-grandchildren. Although she could read and write fluently in Chinese, Mrs. Lee never demonstrated any interest in speaking English until after Mr. Lee's death in 1965 when she moved into an apartment and learned to communicate with her English-speaking neighbours.

When she died on April 25, 1994, Low Shau Wah was 95. "In a lifetime spanning most of the 20th century, she was able to see all her grandchildren through college and spread throughout North America," explained her granddaughter, Patricia Wong. Like the *Asia*, her life embodied a tumultuous period of history.

Thirty-five years after Albert Hugh arrived in Canada, he became a Canadian citizen. Lamentably, diabetes put him in hospital in December 1992 and he passed away from organ failure. Granddaughter Shana Hugh explained:

> Even though I am only one quarter Chinese, it is the biggest cultural part of me. I have an eleven-year old and we celebrate Lunar New Year traditions like giving lucky money in red envelopes and eating Chinese-food dishes that I remember Grandpa used to order for our family dinners. I feel like he gave me something really important in my identity. I like to tell the story of my grandfather coming to Canada on the *Asia* and an immigration officer hearing something and writing down the anglicized version and wondering what my life would have been like if I were a Yau instead of a Hugh. He left us with something very important about who we are, even though he came to a place that wasn't very welcoming to Chinese.[41]

Still missing his mother, Gim Suey started school in Little Tokyo where the majority of students were of Japanese ancestry. In high school, he earned average grades and met new friends from Chinatown. After school and on weekends, he swept streets and picked up trash with a stick. He also worked as a busboy while his father waited tables — the two of them sleeping in the café's storage room. By April 23, 1943, however, after studying at the Curtiss-Wright Technical Institute of Aeronautics, he became a certified master aviation mechanic and was hired by Pan American Airways. He and his all-Chinese ground crew worked at Treasure Island Station in San Francisco Bay, not far from Angel Island. The young man helped maintain the transpacific flying boats, including the *China Clipper*, and in 1944, he worked at Pan Am's Honolulu Station in Pearl Harbor.

Postwar, Gim Suey married and started a family. Son Raymond Chong remembers his father as the "Quiet Man," a "parent who said few words and showed few touches of affection. But he showed sacrificial love by his discreet gifts." Raymond believes his father's dark secret as a paper son left a lingering fear. "He never shared stories about his life in China and America and never returned to his home village." Yet, added Raymond, "he believed in the American dream."

On December 2, 1979, in the Year of the Ram, 47 years after he arrived on the *Asia*, Gim Suey died, predeceased by his mother and his father, who became naturalized American citizens in 1953, and his paper father.[42]

For the Miasnikoffs, it remained a hardscrabble existence, but they persevered. Theodore died in 1941; his wife, who continued to farm, succumbed in 1972. George married and had a son, George Jr., but polio claimed his wife in 1954.

12

SMUGGLERS AND STOWAWAYS

Stop in the King's name!

— **Special Customs Officer Norman DeGraves**

Customs officer Frederick Quinlan had no idea he was part of an ill-fated sting operation on the night of October 10, 1921. He was on surveillance duty near the corner of Richards and Hastings Streets. Another block north and he would have an excellent view of Vancouver Harbour and the CPR station. Fog blanketed the harbour, and the damp air reverberated with the low sound of the Brockton Point bell ringing one stroke every five seconds.[1]

Around midnight, Quinlan observed a man sidle from the misty shadows. The figure approached and offered Quinlan $200 to "make himself scarce" by 2:30 a.m. Recognizing the fellow as a former customs guard, Quinlan accepted the bribe and then reported the shady encounter to his boss, the inspector of customs. Along the waterfront, gleaming under the wharf lights at Pier A, sat the *Empress of Asia*, due to sail on October 13 with a large passenger list and much cargo, including the last-minute rush of mail. She had been there nearly a week, having arrived October 3 with 744 passengers, including 591 in steerage.

At 2:00 a.m. on the 11th, a small watercraft paddled by two men in dark overcoats emerged from shore at Coal Harbour and began gliding silently through the fog toward the *Asia*. Unbeknownst to its occupants, a Vancouver police boat, the *William McRae*, was lying in wait behind a barge. On board were several officers, including Captain Alton S. Hann with one hand on the wheel, the other on the throttle. Crouched behind was 30-year-old customs officer Norman J. DeGraves with a hand on his revolver.

The clandestine transfer of opium from the *Asia* was about to begin when the small craft suddenly veered from the passenger liner. What happened next took Hann and DeGraves by surprise. Instead of rowing or paddling away, the men "roared" off. "Instantly the engine was driving the frail craft … at top speed away from the circle of the wharf lights toward North Vancouver."[2]

In hot pursuit, DeGraves, a former acting sergeant-major from the First World War, stood and fired several shots at the phantom vessel without result. The patrol boat, even with her experienced captain and skillful engineer, could not keep up.

Witnesses described the getaway vessel as a dark green canoe powered by a "high-speed automobile engine without a clutch."[3] The colourful image captured the public's attention through the early days of a commission of inquiry into Royal Canadian Mounted Police (RCMP) drug operations in British Columbia. Evidence to the contrary surfaced during the stunning testimony of RCMP Special Agent 23, Francis William Eccles, who had arranged the sting operation.

Eccles told the inquiry a rowboat, not a canoe, approached the *Asia* to collect $10,000 to $30,000 worth of refined opium from a smuggler on the ship. The plan, he said, was to transport the narcotic to shore, rendezvous with the dealer, and make an arrest. However, the operation imploded because essential customs and law-enforcement officials were unaware. No doubt, Eccles felt the operation could spring a leak if too many people were looped in.[4]

Agent 23 identified the two men in the rowboat as individuals working for him undercover. One was a "confidential informer" named Ray who formed a connection with a disgruntled former customs guard — the man who handed Quinlan the cash. Asked if a canoe was deployed to reach the *Asia* on the night in question, Eccles said a canoe had been purchased by his informant but was at the time "high and dry on a barge," needing repair. He said plans to resurrect the operation the next night failed because the waterfront was crawling with police searching for a rowboat or something similar.[5]

Whether the getaway boat was a motorized canoe or some other narrow-beamed vessel, it left the *William McRae* in its wake on October 11. More significantly, the episode sparked a commission of inquiry lasting two years.

With dozens of witnesses, contradictory evidence, and undercover informers to protect, the inquiry had more twists than one of the *Asia*'s mooring lines. Among them, Eccles faced narcotics and perjury charges, concluding with a federal pardon on the former and acquittal on the rest.[6] In addition the press, quoting from testimony, often identified the ship as the *Empress of Russia* even though the *Russia* was not in port at the time. As for the disposition of the drugs, one scenario is they may have remained hidden on board when the *Asia* sailed.

○ ○ ○

The *Asia* and green-canoe caper is just one example of how opium, cocaine, and morphine were smuggled into Victoria and Vancouver on passenger steamships during the early part of the 20th century. The frequency of such activity came as no surprise to Walter Oliver, given how easy it was to conceal illegal cargo and people entering port. His memoir did not shed light on the subject, but he was aware of how readily items could be slipped on board, especially during the Depression or whenever ships were due to call at U.S. ports during Prohibition. It was easy to recall the numerous news stories and yarns spun by fellow sailors in the smoky clubs along East Hastings Street. Dark recesses behind a ship's bulkheads could become perfect hiding places, but as Oliver discovered — after he concealed four tins of tobacco aboard the *Anglo Colombian* for resale purposes — anything hidden could easily slip out of reach if the ship rolled in bad weather.[7]

Following a 1907 visit to Vancouver by Deputy Minister of Labour William Lyon Mackenzie King, the House of Commons passed the Opium Act of 1908, making it an offence to import, manufacture, or sell opium for non-medicinal purposes. The act did not prohibit possession, but the Opium and Drug Act of 1911 did. It also included the prohibition of possessing morphine and cocaine.

The narcotic, extracted from poppies and shipped dry or raw for further processing — boiled into a dark, brown paste — was unregulated until 1865 when sellers were required to purchase a licence. That regulation applied to what was then the Colony of Vancouver Island. Opium was imported and used legally, and governments benefited from the duty flowing from its importation, revenue that reached its peak in the late 19th century.[8]

King had been briefed on the existence of several opium factories in Victoria's Chinatown and knew there was widespread use of the drug in Chinese and white populations. It is useful to acknowledge that while there are various opinions behind what motivated the act, there are both race- and class-based theories. Proponents of the former believe those agitating against the drug trade were opposed to any breakdown in the isolation of the Chinese community that could lead to more contact between whites and Chinese. The latter theory suggests the inspiration behind the agitation came from middle-class Christians, a group that saw the working class, whether white or Chinese, as particularly vulnerable to addiction and social degradation.

Soon after the 1908 legislation passed, the Vancouver press reported detectives had unearthed an opium smuggling ring involving an Empress liner, a member of her crew, and wealthy suppliers in China.[9] This seizure occurred more than three years before the completion of the two new Empress ships, the *Asia* and *Russia*.

Vancouver police arrested and charged a Spanish man, Thomaso Hernandez, for attempting to sell drugs he had obtained from the *Empress of India*. At the time, the Chinese Reform League was co-operating with police to counter the city's drug trade. Law enforcement also understood that a large portion of narcotics entering North America via Vancouver and Victoria went south to customers and dealers in the United States.[10] This pipeline had existed for many years, especially after 1880 when Chinese opium dealers in the United States could not legally import the drug directly from China.

In 1910, a royal commission investigating fraud and opium smuggling on the West Coast adduced that "opium enters Canada freely as it ever did, in fact more freely, as there is no longer duty to pay." The narcotic was stored in half-pound or one-pound tins — nothing of great bulk but in any shape convenient to the smugglers.

One convicted smuggler, caught with a belt full of tins, claimed he had made numerous trips to shore without any trouble. He also testified that while the Empresses were being coaled, some of those assisting with the process at Vancouver would place packages of opium from the ships onto the coal hulk. Once retrieved, distribution commenced.[11]

It did not take long for smugglers to utilize the *Asia*, starting with her maiden voyage. Ah Sing of Vancouver was charged soon after the ship, skippered by Robinson, called at Vancouver in 1913 (see Chapter 2). On September 10, the day the liner was outbound from Vancouver, the courts convicted a man of trafficking a large quantity of opium obtained from a Chinese sailor on the *Asia*. Police knew there was a supplier on board but were either too busy or caught flat-footed because the ship sailed early, needing to work against an 11-foot tide. Several passengers stranded on the wharf had to board coastal vessels to make their connection with the ship in Victoria.[12]

On Tuesday, April 27, 1920, Health Minister Newton Wesley Rowell rose in the House of Commons to report how the importation of narcotics had risen in recent years. Rowell was a lawyer, politician, and churchman, raised in an evangelical Wesleyan theology that was driven by a desire to obtain moral perfection. He was also an early advocate of conscription in the First World War. In this instance, he was speaking to a bill aimed at amending legislation that would establish more control over opium, morphine, heroin, and cocaine trafficking.

During the fiscal year ending March 31, 1914, Canada imported 4,598 pounds of crude opium. The fiscal year ending March 31, 1918, that imported quantity had risen to 12,471 pounds — an increase of 171 percent.[13]

For Rowell, that consumption of drugs in Canada did not explain the increase in imports, and his department believed the more stringent regulations in the United States caused drug traffickers there to continue importing supplies through Canada. He told MPs this meant American traffickers were in possession of quantities of drugs not registered with authorities and therefore were free from regulations requiring the reporting of each ounce distributed to users.

Rowell reminded MPs that Canada was party to the International Opium Convention signed in The Hague in 1912, reinforced with a clause in the Treaty of Versailles. These moves stemmed from the first international conference on narcotic drugs, convened in Shanghai in 1909. Up until then, despite an opium epidemic in China and widespread use elsewhere, there was little interest in controlling distribution, let alone examining addiction. The drug proved very profitable among suppliers, port distributors, shipping interests — including the monopolizing British East India Company — bankers, insurance agencies, and governments. "Many national economies were as dependent on opium as the addicts themselves," notes a 2008 United Nations report on international drug control measures. "Even a country the size of British India derived 14 percent of state income from its opium monopoly in 1880."[14]

From Shanghai and The Hague grew international attempts to control production and trafficking. To give an idea of how prevalent opium use was, consider that in 1906–07 the world's population was 1.7 billion.[15] Twenty-five million people, or 1.5 percent of the population, used opium, with millions of addicts in China alone. Total production for the same period was 41,000 tons.[16]

Between 1915 and 1927, federal customs officers seized $254,000 worth from vessels in just Vancouver's harbour. The figure does not include drugs confiscated by local police and other harbour officials. "From these figures it may be conjectured that, in spite of best efforts of the authorities, not much less than $1,000,000 worth of narcotics has found its way to this continent through the port of Vancouver in the last 10 years," stated an April 1927 article in the *Times* of London.[17]

That article, written by its Vancouver correspondent, led with the observation that opium valued at $10,000 was seized from the *Empress of Asia* during her last visit to Vancouver. "Such a seizure is not rare, but the amount of the drug seized was larger than usual."[18]

Clever detective work, including use of confidential informers and sometimes just plain luck, helped locate drugs on ships in their myriad hiding places. "Across the ceiling of a lower deck, above one of the boilers, ran a wide girder," the *Times* reported. "To absorb the heat [from the boiler], cotton packing was stuffed around the girder, and the whole was encased in an iron covering about twelve inches wide."

The investigator noticed a nut missing from a ceiling manhole cover that when removed gave access to the stuffing. A dab of black grease had camouflaged the missing nut, but in removing the manhole cover, the officer found nine pounds of cocaine and morphine and thirty-eight, five-tael (227 grams) tins of smoking-variety opium. "The chief obstacle in the way of checking this traffic is the great profit to be made.... Cocaine, which comes chiefly from Germany, can be purchased wholesale through legitimate channels for $12.75 an ounce." The newspaper correspondent noted that in Vancouver, illegitimate or underground dealers sell the drug for $50 to $100 per ounce. "Morphine [made from crude opium], also imported chiefly from Germany, can be purchased, legitimately, wholesale for $7.75 per ounce."[19]

Dropping a cache of narcotics from a passenger liner's outer deck or through one of its many portholes to an awaiting boat worked, too, but it was risky. DeGraves and other experienced customs and drug squad members knew what to look for on ships and while patrolling the waters. There were also eyes along the waterfront near the CPR station where security was tight given the million-dollar silk shipments.

Australian-born customs officer Norman DeGraves was a no-nonsense workaholic, known for 15-hour days. He had joined the Canadian customs department in 1911 and returned to the job following wartime service in the Canadian Expeditionary Force during which he overworked himself into severe depression. In the course of his duties, he frequently swore warrants, participated in raids, and helped prosecute criminal cases. In the course of drug raids, he was known to yell "Stop in the King's name!" His work often involved the use of patrol boats, but on at least two occasions counted on seaplanes to chase and apprehend smugglers.

Success, however, depended on co-operation between various enforcement agencies. At 2:30 a.m. on Tuesday, November 18, 1924, H. Brady, a Canadian Pacific Police constable, spotted three individuals acting suspiciously in the fog alongside the *Asia* at Pier A. The ship, skippered by L.D. Douglas, had arrived on November 10, several hours behind schedule on the same uncomfortable voyage that saw her heave to for five hours shortly after departing Yokohama (see Chapter 10).

Constable Brady yelled for the men to halt and then witnessed them toss three bundles into the murky water next to the *Asia*. Brady drew his revolver and fired, but the men escaped. Two special customs officers, including DeGraves, along with Captain Hann on the *William McRae*, directed the investigation and supervised the daytime dragging near the pier. By nightfall, the lines hooked two sacks containing 105 tins of opium.[20]

In January 1925, the *Daily Colonist* reported the outcome of a trial involving a "former quartermaster" on the *Asia* who was sentenced to a year in prison and a $200 fine for violations under the drug act. The accused instructed his mess boy to deliver 12 tins of opium to a customer who was an undercover customs officer.[21]

After a particularly stormy crossing with strong westerly gales and a snowstorm off the Aleutian Islands, customs and RCMP officers made one of their largest drug seizures. It occurred on March 24, 1926, and DeGraves was again in on the action that located $10,000 worth of cocaine, morphine, opium, and a small quantity of whisky. The cache, discovered in the ship's lower engine room, was described in the press as the "biggest haul … made in many months."[22]

Under the command of Douglas, the *Asia* had reached port with many high-powered business executives and bankers completely oblivious to the "haul." Among them was newspaperman Keith Murdoch, who in the months ahead led efforts to acquire several Australian newspapers.[23] Handsome and well dressed, the conservative Murdoch soon married an attractive and much younger Elisabeth Greene, a union that produced four children, including future media tycoon Rupert Murdoch.[24]

Four months later, RCMP officers had probable cause to visit the *Asia* while she was alongside Pier D. After making a careful search, officers found an ingenious apparatus, complete with instructions, used for submerging tins of opium in deep water. The discovery made headlines, less for the drugs than for the method.

A diagram and letters, written in Chinese, were hidden above the bunk of Tchung Mee, the ship's number one pantryman. During the trial that fall, Corporal John Healey described the use of the apparatus while the ship was at sea. Someone on board, known as a "runner," started moving the second he received secret signals, often just the flash of a white handkerchief from another vessel or from shore. Given the cumbersome nature of the device, it is presumed more than one individual was involved.

Attached to a sinker were two ropes, each measuring 250 feet. At the opposite ends of each rope were wooden floats with identifying marks, and tied to the iron sinker was a heavier rope of 120 feet connected to a 50-pound iron weight or anchor. The smugglers tied opium packed in watertight copper tins to the lower rope before the apparatus went overboard at a rendezvous point off the William Head Quarantine Station or Vancouver Harbour. "The pieces of wood floated, and a confederate from shore would detect these, through a secret mark on one of the floats."[25] He or she would haul up the drug like a catch of fish, load it into the boat, and make haste for a secluded inlet.

In keeping with the *Times* report, Healey told the court how hundreds of tins of opium had been fished out of Vancouver Harbour by RCMP officers with the assistance of a rowboat and grappling irons. It was far more difficult, however, to snag the traffickers, let alone register a conviction. The case of pantryman Tchung Mee ended with acquittal. Most of the time it was difficult to prove who was in possession of the drugs, and furthermore the courts, even then, were reluctant to give up on the concept of reasonable doubt. Cases based on the testimony of informants and cohorts often appeared self-serving, contradictory, and unreliable, giving defence attorneys a significant advantage.

The discovery of the apparatus on the *Asia* came after a group of women out for an evening stroll in South Vancouver phoned police to report a wild car chase that ended with the abduction of three Chinese men. The women believed white "gangsters" had overtaken and forced the lead car with the

Chinese to pull over whereupon a mysterious package flew out the car window. It was, of course, a drug takedown involving three arrests and the seizure of morphine and cocaine with a wholesale value of more than $2,000. The arrests prompted a search of the *Asia* with Healey and DeGraves among the men described in the chase as "gangsters."[26]

Customs officers believed aircraft surveillance from Vancouver's Jericho Beach Station prevented smugglers in March 1927 from transferring $10,000 worth of opium from the *Asia* while steaming into Vancouver. Thirty-eight tins of opium and nine pounds of morphine in the ceiling above the boiler room were due to be dropped into a small boat in the gulf.[27]

If not stashed in ceilings, behind bulkheads, or inside wardrobes or trunks, illicit cargo was often hidden in plain sight. The string of lemons innocently hanging to dry during young Gim Suey's voyage in April 1932 had absolutely nothing to do with the lad or his travelling companions, but they contained cores of opium (see Chapter 11). "The customs men seized the drugged fruit and some paraphernalia for smoking, but made no arrests. The drugs and implements were assumed to be the property of members of the ship's crew."[28]

Shipping companies risked serious fines if narcotics showed up on their vessels at American ports. In July 1933, the *Empress of Canada* faced a $40,500 fine after the seizure of $50,000 worth of opium in Honolulu. The Chinese steward on board told authorities the shipment was bound for Vancouver, but American authorities claimed the steward had the name of a Honolulu contact. When the ship reached Canada's West Coast on August 3, suspicious Canadian customs officers paid a visit and seized another $10,500 worth of opium.[29]

During the 1920s and 1930s, stopping drug shipments was like trying to prevent waves crashing on shore. The best authorities could hope for was a reduction in flow through raids, seizures, and arrests, and perhaps a conviction or two.

DeGraves, who sometimes worked 48 hours without rest, was clearly devoted to the job. In addition to knowing the port and ships inside and out, the special customs officer carefully cultivated a network of sources. Membership in various clubs, including the prestigious Vancouver Club, the

Vancouver Yacht Club, and the Vancouver Board of Trade helped, as did his public speaking engagements that drew attention to the war on smuggling. On the evening of November 13, 1937, as DeGraves stood up to deliver an address to his beloved wartime artillery unit, he collapsed and died. "He was a tireless worker, possessed of much energy, vision, and a passion for detail, which he had always at his fingertips, and ready for the service of whoever might need it," noted Vancouver port manager Kenneth Jardine Burns.[30] Wise smugglers paid close attention to men such as DeGraves, since their nefarious activities involved outbound ships, as well.

After a gale forced her to ride her anchors for two hours on the evening of June 5, 1924, the *Asia* came alongside in Victoria. Minutes before she reached the pier, two large trunks arrived on the wharf, tagged for the liner. After prying open the mysterious cargo, customs officers found guns and ammunition bound for accomplices in China where political and paramilitary groups were active prior to the disturbances of 1925–26. With the shipment seized, the *Asia* cleared Victoria with 3,000 tons of legal freight, including 500 tons of box shooks (components for the assembly of boxes), a consignment of new automobiles, asbestos, and lead, plus 1,600 bags of mail, much of it from Britain.[31]

○ ○ ○

The same opportunities existed for stowaways, and the *Asia* had her fair share, starting with her first return voyage to Vancouver on October 25, 1913, under Samuel Robinson's command. The ship endured a rough crossing with a light passenger list when two young men landed in custody after stealing on board at Yokohama. Immigration Superintendent Malcolm Reid displayed no compassion for the two — in his mind, stowing away had become "too prevalent" and needed to be "dealt with in a most stringent fashion.…" Reid disclosed that in addition to the two in custody, there were two others awaiting transportation to the United States and as many as 40 men in Yokohama awaiting the opportunity to stow away on Canada-bound steamships.[32]

It is unknown whether the no-nonsense Reid viewed the next stowaways with zero compassion. Two Japanese boys, classified as stowaways, were accidental passengers after the ship cleared Yokohama. Exhausted, both young stevedores fell asleep in the cargo hold and woke up after the *Asia* put to sea. "Without delay, they set about to get out of their prison, and hammered with energy on the hatches," noted the *Vancouver Sun*. "This attracted the attention of the sailors who released them."[33] Soon after the ship reached Vancouver on December 21, 1913, the boys left for Japan, and presumably remained awake for most of the voyage.

One of the *Asia*'s youngest stowaways was 15-year-old William Henry Jackson, whose failed attempt in August 1918 made for a brief entry in one of Captain Davison's logbooks during the First World War (see Chapter 5). The fair-haired boy was a Dickensian character, straight from a seaman's orphanage. He was five foot two and weighed 134 pounds, and very likely the son of a merchant sailor lost at sea. Listed as next-of-kin was Mrs. L.G. Jackson of Liverpool, but she may have been an aunt, not his mother.[34] Before *Asia* left Liverpool, the elusive Jackson stole aboard and hid beneath a pile of lifebelts in the ship's baggage room, which is where the crew found him on the crossing to New York City.

It seems Davison may have taken pity on the orphan and instead of discharging him to port authorities allowed Jackson to sign on with the deck crew. When the ship embarked American troops for overseas service in France, Jackson remained on board for the return voyage to his home port. He was still part of the *Asia*'s deck crew on November 10 when the troopship embarked more "doughboys" in New York City and called at Brest, France, but is thought to have signed off when the *Asia* returned to Liverpool later that month.[35]

Shortly before departure from New York's Pier 97, at least four of the ship's Chinese firemen were severely beaten by customs officials searching for contraband. The violence occurred on November 5 in the firemen's forecastle in the belly of the ship. All four suffered appalling lacerations and bruises to their heads, likely caused by batons or bully sticks. Injury descriptions from the captain's logbook reflect the anger Davison felt after the attack on his crew. One man, Fong Yuet, required extensive treatment at Bellevue Hospital and could not rejoin the *Asia* before she left for Brest.[36]

In the fall of 1926, the discovery of 15 Chinese and two Russian stowaways underscored the plight of those seeking a better life and the workload experienced by immigration officers. Discovery of the Russians occurred two days after the ship cleared Yokohama for Vancouver; however, the Chinese stowaways escaped detection across the Pacific. The men were "concealed in various parts of the ship, the majority being in the laundry," stated the *Vancouver Daily Province*. "It is suspected they were stowed away with the connivance of members of the Chinese crew, as it would be impossible for such a party to remain concealed from all on board during the entire trip." After checking travel documents belonging to Chinese steerage passengers about to disembark in Vancouver on November 1, immigration authorities posted guards before searching the *Asia* top to bottom. The 15 landed in the immigration building's detention wards.[37]

Charged under the Immigration Act with entering Canada illegally, one of the accused, Lee Bah, appeared on the court docket the following week, serving as the trial's test case. Bah's lawyer entered a plea of not guilty, claiming the group had boarded the ship at Hong Kong believing they were en route to Singapore.[38] In his ruling, Magistrate J.A. Findlay stated the men were undoubtedly stowaways, but there had been no evidence of their attempt to enter Canada illegally. The men were deported and sent home on the next available ship.[39] The two Russians, age 18 and 25, were also deported and escorted onto the *Asia*, which left November 12. In addition to having one of the largest passenger lists in years — 967 — the ship cleared with a heavy cargo of 1,000 bars of silver in her strongroom.[40]

The *Asia*'s logbooks further recorded the discovery of stowaways on voyages between the Philippines, Hong Kong, Shanghai, and Kobe. In one instance, crew found British able seaman William Morgan as the ship crossed Japan's inland sea toward Nagasaki on the morning of December 22, 1921. Morgan, formerly of the cargo ship *Gaelic Prince*, completed his journey under escort to the British consulate in Shanghai.[41] Eight years later, British siblings Jack and Leslie MacDermott, ages 21 and 20, failed to produce tickets after departing Manila and were met by authorities at the end of the gangway in Hong Kong.[42] There is also the story of 18-year-old Vladislav

Kerbutt, who embarked in Shanghai and remained hidden in a lifeboat until the evening of March 9, 1930, when the *Asia* made her regular coaling stop at Nagasaki. Hailey's logbook notes the port's "water police" took the Polish man into custody.[43]

More endearing perhaps was an episode that unfolded just days prior to the start of the Second World War when a Chinese boy boarded without a ticket in Shanghai. Rather than conceal himself, he casually mingled with other passengers until ship's officers intervened, ending his socializing prior to Hong Kong. It was just one more event in a voyage that saw the ship, skippered by Captain Maurice Mayall, ride out two typhoons.[44]

○ ○ ○

Of course, not everyone intent on entering Canada illegally was a stowaway. Before he decided to jump ship at Victoria, Ho Lang Took was living his dream as a sea cook. He had signed onto the *Empress of Asia* in Hong Kong, but the moment he laid eyes on Vancouver, he was smitten. It was during Took's second return voyage that a less-than-foolproof plan was set in motion. On July 2, 1920, while the passenger liner was alongside Rithet's Piers, Took consumed a quantity of drugs — cocaine or opium — believing it would keep his body warm. Next, he pulled out a money belt filled with gold coins and secured it around his narrow waist. To counter the weight of the gold, Took added a rubber belt filled with air.

He entered the cold, oily water either by slipping over the side or through a porthole and swam until he was beneath the pier where, according to the press, he "clung to a pile like a barnacle" until the ship cleared the wharf. "When the coast seemed clear, and the *Asia* was but a memory on the skyline, he started to swim toward shore." However, "he had stayed too long in the cold water, in spite of heating drugs and warm clothing. Exhaustion came, and he was about to drown."[45]

Lingering on the wharf that night was the assistant manager for the steamship line who, according to the *Vancouver Daily Province*, "personally sees to it that every vessel sails in apple-pie order and that the wharves are clear after departure...." After spotting something black bobbing in the water, the assistant manager hollered for assistance, and within minutes, he and a few others pulled Took in on the end of a rope, "more dead than alive."[46]

Took was treated, detained, and then escorted onto the next China-bound ship.

PART THREE

THE SECOND WORLD WAR

13

THE "ACCIDENTAL" BOMBING

The fifth bomb made a direct hit on the Boat Deck, tore its way through the steel plates ... smashed through ... and onwards down to the galley.

— Sanitation engineer John Drummond

Passengers in the dining saloon did not panic when three cooks from the *Empress of Asia*'s galley burst into the room with "blood streaming from faces and arms."[1] Most of the diners stopped chewing and remained seated as they contemplated what horrible accident had occurred in the galley.

The injuries to cooks Lau Wing Fai, Kwok Shing, and Mak Kwong were not the result of exploding cookware, although there was a loud bang. Their wounds resulted from an aerial bomb — one of seven or eight — dropped on and near the passenger ship by a country that had not yet entered war on the side of Nazi Germany and Fascist Italy.

Incredibly, the Vancouver-bound *Asia*, commanded by Captain George Goold, was 50 miles from completing her regular 353-nautical-mile passage from Kobe to Yokohama — and sailing in broad daylight — when the bomb pierced the Promenade deck near the second funnel. With 193 passengers and a European and Chinese crew of more than 500, she was just north of Oshima Island, southeast of Tokyo Bay, with four other merchant ships nearby.

The *Asia* did not have cannons, but she was grey, the standard defensive colour that made her less visible to enemy forces. On board were four naval ratings serving as DEMS, paid 10 cents a month by the CPR, in addition to their Royal Canadian Navy pay. DEMS stood for defensively equipped merchant ship; however, Leonard Morris, John Mead, Andrew Bing, and LeVern Michel Petrek had no access to heavy firearms. Their job was to provide basic security and guard against sabotage.

While Imperial Japan remained busy pursuing an expansionist agenda in China, the bombing of a Canadian passenger liner on September 14, 1940, was alarming. It occurred 13 days before Japan signed the Tripartite Pact with Germany and Italy and 15 months before it fully entered the war with attacks on neutral United States at Pearl Harbor, the British colony of Hong Kong, and other Pacific and Southeast Asia targets. However, for those who had been carefully monitoring tensions in Asia and the Pacific, Japan's pact with the Axis powers was less shocking.

A September 27, 1940, memorandum to William Lyon Mackenzie King, Canada's prime minister, from Oscar Skelton, the long-time head of the Department of External Affairs, focused on the "new Japanese-German Agreement." In recapping his telephone conversation with senior assistant Loring Christie that day, 62-year-old Skelton noted that American diplomat Sumner Welles, the major foreign policy adviser to President Franklin D. Roosevelt, had shown Christie the text of the agreement. "Apparently Japan recognizes Germany's leadership in Europe, and Germany recognizes Japan's leadership in the Far East," Skelton told the prime minister. "There is a provision for economic and financial and military assistance by each to the other if attacked by a party not now at war. This, I gather, is clearly aimed directly at the United States."

In the memo, Christie indicated that Welles described the pact as "a healthy sign that they [Japan and Germany] had now come out into the open. There was nothing more in the document than had been implicit in the activities of both countries in the past two or even four years." In reply, King scribbled a note stating how he and the Liberal Party caucus had been making the same observations in regard to Germany and Japan for three years.[2] As an example of the friction, an incident at Yokohama in April 1938, which also involved the *Asia*, is noteworthy.

Twenty-seven-year-old American pilot Elwyn Gibbon was returning home with his wife when Japanese police boarded the ship and forcibly removed him for interrogation. Gibbon had served in the Chinese air force as a member of the Foreign Volunteer Squadron amid hostilities with Japan. By then he had resigned, but Japanese authorities wanted to know about any involvement he might have had in Chinese air attacks on Japanese positions. The airman was released six days later, but not before the *Asia* — with his wife aboard — had continued on her way to Vancouver. The incident prompted a flurry of communication between high-ranking Canadian government officials, the CPR, and Gibbon's legal representatives. Skelton agreed with the British consul general, who stated there was little to protest and no representation could be properly made to the Japanese government because the ship was in Japanese territorial waters and Japanese police were within their right to interrogate Gibbon.[3] Still, it reinforced the belief that tensions between Japan and the West had increased.

On the West Coast of Canada and the United States, anxiety over a war with Japan was palpable. Decades of public distrust and racism directed at Japanese immigrants in Canada and the United States had been well documented, culminating in the rounding up and incarceration of thousands of innocent people. By September 1940, it was not a question of whether there would be war with Japan but how soon. A month earlier, in reply to a cable from Canada's prime minster, Thomas Dufferin Pattullo, British Columbia's premier, expressed his concerns about the West Coast. "No mention is made in your communication as to the stationing of a unit at Prince Rupert. There is, however, a unit there at the present time and I assume it will be maintained. My understanding is that the Queen Charlottes and [Prince] Rupert might be especially vulnerable."[4]

Although war with Japan seemed inevitable, passengers and crew on the *Empress of Asia* figured they were relatively safe. In fact, the September bombing came as such a surprise that some of the 193 passengers assumed the aircraft had to be German, not one from the country they were two hours away from visiting.[5]

At 1:30 p.m., passenger Alan Aitchison Barnes, a Scottish-born Dutch East Indies rubber planter with self-professed knowledge of civil and military aviation, was standing next to the rail on the foredeck when the bomb whistled in on an angle and struck the starboard side amidships.

Fortunately for Herbert Connell and another Marconi operator, the bomb missed the wireless room on the Fiddley deck by just under 10 feet before tearing through the falls on two lifeboats and punching a jagged, two-and-a-half-foot hole through the Promenade deck. It was not the first time Connell had come close to death. The First World War veteran from Newcastle, New Brunswick, became a wireless operator in 1921, joining the *Asia* 10 years later. After moving to British Columbia, he worked as a bank clerk in Chilliwack before enlisting with the Canadian Expeditionary Force in March 1915. The scars to his arm and leg were the result of shrapnel wounds suffered on the Somme on September 8, 1916, which led to a "medically unfit" discharge and repatriation in 1918.[6] Marconi hired him in 1921, and two years later in July, he married his sweetheart, Agnes Candlish. Around midnight on Christmas Eve 1930, Connell was off the West Coast on the Union Steamship *Cardena* when Agnes and his sister-in-law were fatally struck by a hit-and-run driver — a loss that still haunted him nine years later, especially near Christmas.[7]

The Japanese bomb also narrowly missed the centrally located first-class children's playroom and dollhouse on the top deck, a nearby writing room on the starboard side of the Promenade deck, and first-class cabins 122 and 124 immediately below. "It was darn good shooting," said Barnes as he regaled reporters in Vancouver. "He was trying to hit the *Asia* — there's no doubt of that."[8]

Passenger W.R. Heath of Vancouver was aft on the Shelter deck in the tourist-class smoking room "killing a beer" while waiting for a friend. It had been a tough year for the young man. Three weeks earlier, he had badly injured his hand in Manila while loading cargo onto a freighter and had been anticipating a smooth crossing with more time to heal. "Everybody was curious, but not very frightened," he told the press. "Everybody rushed on deck to see what was the matter but there was only a bit of smoke and a small hole in the decks where the bomb had hit."[9]

Thirty minutes earlier, American passengers John Clark and Arthur Slettedahl, who both embarked at Manila, were jogging around the Promenade deck when they witnessed a bomb plunge into the sea astern, producing a high column of water. Deep thuds from subsequent bombs straddling the *Asia* convinced the pair this was an attempt to find range in order to sink the ship. Both told the press visibility was good and "specks of the planes could be seen overhead, indicating several bombers were operating in manoeuvres."[10]

"We went below, not knowing what to expect and were surprised to find there was no undue commotion," added Clark. He, too, had been to sea before, employed by the General Steamship Company with experience on the SS *Miraflores*, later sunk during the war.[11]

Speculation mounted on responsibility and types of bombs used. "They might have been time-action bombs," offered Clark, while his friend from Enumclaw, Washington, suggested 50-pound bombs, excepting the larger one dropped astern, which appeared to be 500 pounds.[12] Barnes believed the one that struck the *Asia* was "a small one — about twenty pounds with about an eighth charge."[13]

What these passengers could not see was damage below. In a letter not published until now, the ship's long-time and trusted sanitation engineer John Drummond recalled seven bombs, although he noted some witnesses counted eight. The first one fell at 12:52 p.m. "right alongside the ship and splashed into the water with a loud crash. The fifth bomb made a direct hit on the Boat Deck [Promenade deck], tore its way through the steel plates of the deck and smashed through A Deck [Bridge deck] and onwards down to the galley where it exploded...."[14]

Vegetable cook Mak Kwong was peeling potatoes for the evening meal when the projectile ripped through the ceiling and slammed into the cement floor just outside the door to the vegetable storage room. The impact blew the steel door off its hinges and sprayed the galley with splinters of metal and wood.[15] Some 20 chefs and cooks, including Vancouver's Archie Gee, were at "ground zero" along with an assortment of stewards under the supervision of venerable chief steward Francis Wright, whose employment with Canadian Pacific Steamships began in 1917 following stints as barkeep and cellar man at Hotel Vancouver. The unmarried Wright was used to crowds and working with limited space. One of 13 children, he grew up at Lowestoft, in East Suffolk, the most easterly settlement in the United Kingdom. While issuing orders to the stewards who shuttled trays of dirty dishes and utensils from dining saloon to the scullery, Wright had been keeping a watchful eye on the galley staff prepping food or sweeping up spills — all perfectly chaotic until the blast.

One diner, who declined to give his name to the Vancouver press because he had to return to Japan, recalled hearing a bang and what sounded like crockery "scattering all over the place." There ensued a "scramble for the deck" where passengers lined up for lifebelts. "Nobody got the wind up, though, and in two shakes we were ready for anything."[16]

In the galley, shrapnel had lacerated Kwok Shing's forehead, pierced and broken one of Mak Kwong's fingers, and sliced through flesh and muscle on Lau Wing Fai's left forearm, fracturing bone. "There was a terrible bang as it landed," declared one chef. "I had just entered the galley from the dining saloon when it struck."[17]

On the bridge, Goold had calmly ordered his helmsman to put the 27-year-old ship onto a zigzag manoeuvre while Chief Officer Donald Smith dispatched the fire crew to investigate and extinguish any flames along with a messenger to alert the surgeon and nurse, whose offices were near the stern below the Promenade deck. The bridge also advised passengers to muster on deck, don lifebelts, and be prepared to board lifeboats.

As surgeon Robert Swan and nurse Julie Harwood dashed toward the galley with bandages and surgical tools, they both wondered how bad it would be. Up to that moment on the return voyage, they had only dealt with mild dysentery, one flu case, and the usual bumps, bruises, heartburn, and seasickness. Cleaning, suturing, and dressing shrapnel wounds were considerably farther down the list of expectations, so Swan immediately enlisted the assistance of passenger Dr. S.L. Chiu and Frederick Gillespie, the ship's chief electrician.[18]

Swan had grown up in Winnipeg South and graduated from the University of Manitoba, Class of '37. Although only recently signed onto the *Asia*, he had worked at the Calgary Associate Clinic before becoming an acting surgeon-lieutenant in the Royal Canadian Naval Reserve at the *Naden* in early 1940. Juliette de Lotbiniere-Harwood, who had taken over from Vancouver nurse Hilda Mary Smith, had connections with the de Lotbinieres, a prominent family involved with the governance of New France. She had graduated from the St. Paul's School of Nursing in Vancouver, Class of '39, and acquired further nursing experience aboard the SS *Letitia* and SS *Athenia* before joining the *Asia*.

Lau Wing Fai's and Kwok Shing's injuries were so serious they required immediate surgery with a general anaesthetic and additional hospitalization in Vancouver. Born in China, Lau had signed on in Hong Kong on July 20, 1939, with three years' sea time and familiarity with the ship, having served on her in 1936 and 1937. Mak, who had also signed on in Hong Kong, was fortunate his recent promotion had taken him out of the scullery and immediate vicinity of the blast.

Working through the confusion, Drummond and the ship's engineers scrambled up and down the metal stairs between decks assessing damage and collecting evidence. "The heavy tiles embedded in cement on the galley floor doubtless prevented the bomb from tearing its way right through to C Deck [Upper deck]," wrote Drummond. "But the repercussion did severe damage to the [tourist-class] cabin immediately below the galley, bringing down woodwork, chips of white paint & splinters, both in the cabin and Alleyway."

Soaked to the skin through his white coveralls, he made swift running repairs. "Water was flying all over the place from burst pipes and the noise from the crash of bombs still falling was terrifying, and the awful feeling of helplessness was one that never could be forgotten."[19]

The "great shining light was the morale and courage of all passengers and crew, both European and Chinese," noted Drummond. "The marvellous behaviour of the Chinese was from the beginning of the 'incident' more than exemplary. They carried out orders instantly, quietly, and with absolutely no sign of alarm, though fear was in their eyes. Passengers donned lifebelts and proceeded quickly and in a most praiseworthy and orderly manner to their

Muster Stations. There was never at any minute any sign whatever of panic, even the children choked back their tears and were brave and helpful." He also witnessed crew conducting themselves with "precision and no outward sign of undue haste or excitement."[20]

For courageously remaining at their vulnerable posts, the wireless operators earned high praise. "The radio men were the heroes," said Barnes, who in less than two years landed in the same Japanese internment camp in Sumatra as six other *Asia* crew members. "The wireless operators got in touch with the Japanese immediately, and in a few minutes a couple of [Japanese] pursuit planes shooed the bomber off."[21]

Drummond's recollection closely matched Barnes's story. "Japanese planes came in answer to the call for help, and chased away the attacking plane which was too high to be seen without glasses [binoculars]." The appearance of Japanese naval aircraft chasing away one of their own was a peculiar but welcomed sight, one that brought relief and opened minds to the possibility it was an accidental bombing, although many found that difficult to accept. Ernie Higgs was one. Recently promoted from boy seaman to ordinary seaman, he was earning $30 per month — double his previous salary — when the bombs fell. Years later, whenever people asked him about the "accidental" bombing, there was an unmistakable glow of amusement followed with "Accidental bombing? Ha! Ha! Ha!"[22]

There were other crew who held the same view or remained skeptical months, even years, after the Japanese issued their official statement on September 15. At least 17 other crew members, including Drummond, Chef Archie Gee, and Boy Seaman Owen Gillett, were aboard the *Asia* when Japanese bombers finally destroyed her in 1942.

Texas mechanic Rosser Douglas, who embarked in Hong Kong, also shared his take with the press. He called the bombing "meddlesome mischief," speculating it was engineered to spark another international incident.[23] Harsher words of indignation spewed from New Yorker John H. Kerr, an importer-exporter from Shanghai who wrote to a friend in Australia. "A mistake! A mistake! Forty minutes trying, in 100% visibility, beautiful clear weather, NINE bombs — and a MISTAKE. They wanted to create a first-class 'incident' no doubt...."[24]

But the Imperial Japanese Navy was clear. It was an accident, and when the *Asia* reached Yokohama at 5:00 p.m., officials rushed on board to apologize and offer medical assistance and carpenters. However, Goold held on to an important piece of evidence: the bomb's fuse cap with clear Japanese markings.

The Japanese naval office stated its air force, "which was carrying out practice bombing operations on the sea near Oshima, mistook the mark and scored a direct hit on the *Empress of Asia* with a small-sized bomb of the kind used in manoeuvres." The official statement described damage as "extremely light," noting four crew members "sustained extremely slight injuries."[25]

When Tokyo reporters sought Goold's reaction, the captain favoured brevity and diplomacy. According to the *Japan Advertiser*, "he could say nothing except that such incidents are likely to happen under present world conditions with wars going on in Europe and China."[26]

Later that afternoon, E. D'Arcy McGreer, Canada's chargé d'affaires at the Canadian legation in Tokyo, sent an urgent cable to Ottawa: "I have been trying to telephone you concerning the bombing of the 'Empress of Asia' this afternoon off Yokohama by Japanese aeroplane. Details follow."[27]

At 10:00 p.m., following repairs to patch the damage, the *Asia* steamed out of Tokyo Bay bound for Canada. As she crossed the Pacific, naval attaché D.N.C. Tufnell and air attaché W.E.G. Bryant of the British embassy in Tokyo questioned Captain Ichimiya, Japan's minister of marine. The attachés were chiefly concerned about the target, the aircrew's experience, visibility, the ship's behaviour, and punishment of the aircrew. After completing their verbal interview with Ichimiya, Tufnell and Bryant filed their report on September 18 and shared it with McGreer, who forwarded it to Ottawa along with enclosures, including the bomb's fuse cap — with instructions to return the device to Goold.

The Tufnell-Bryant report documented what the Japanese told them: that the target for the bombing practice was a ship, but details of this ship "could not be given, as it was a 'secret target.'" The attachés were told "glass bombs," were used, "and it was one of these glass bombs," according to Ichimiya, that struck the *Asia*.

With regard to visibility, the report noted it "is agreed that at sea-level, horizontal visibility was very good, but at the great height from which the bombing practice was being undertaken, there was a haze, and it was not clear for bomb aiming." The bomb aimer and pilot were, explained Ichimiya, "students only. For such students, it is difficult to distinguish a particular ship at a great height, and practically impossible to distinguish whether a ship has three funnels or not. Often destroyers have been mistaken for cruisers, and vice-versa."

Tufnell and Bryant learned that the target area was east of Oshima and away from the passenger traffic route, but according to the Ministry of Marine, the *Asia* "was north of Oshima" and on the same course the aircrew was expecting to find the target ship. "As soon as they saw the Empress of Asia they thought it must be the target ship," noted the report. "Although it [the *Asia*] was not in the bombing area."

Furthermore, the ministry stated the target ship was to carry out avoiding tactics and commence zigzagging as soon as the bombing practice began. "When the Empress of Asia commenced zig-zagging the aircraft crew saw nothing unusual, whereas had she [the *Asia*] continued on a straight course they [bomb aimer and pilot] might have realized their mistake."

The attachés also noted the "Japanese Navy gave us assurance that in future such an accident will not occur" and that the "persons responsible will be punished severely under Naval Law."

Tufnell and Bryant queried a number of points and noted that "against a ship, glass bombs would be highly desirable, so they state that the bomb used was a glass one. The A.A. [Bryant] without seeing the entire bomb cannot be absolutely certain, but he is convinced that no glass bomb would penetrate three decks, and cause damage such as he saw." When the attachés asked the Japanese captain whether it was possible for the aircraft to have put on a dummy metal bomb in error, he stated he thought it highly improbable.

Tufnell and Bryant also observed the following: "Only in the final stages in bombing training would a target ship, if used, be 'attacked' on a zig-zag course, and in such a final stage of training, the pilot and bomb-aimer should quite easily be able to tell the difference between the 'Empress of Asia' and the target ship." And furthermore: "if unable to tell the difference due to height, come to a lower level, and make sure," and finally, "be able to tell the difference between NORTH and EAST of OSHIMA."[28]

Meanwhile, a front-page story in the *Vancouver Sun* on September 19 quoted an official statement made three days earlier by the Japanese government expressing "profound regret" and that it was wished the "error of sight target" would not affect the friendly relations between Canada and Japan.[29]

Four days after the Tufnell-Bryant report, McGreer of the Canadian legation in Tokyo wrote to Yosuke Matsuoka, the Japanese minister of foreign affairs, that the "Dominion of Canada is prepared to accept the explanation of the circumstances of the bombing and the expressions of regret...." However, McGreer went on to state that Canadian authorities "find it difficult to understand how the accident could have occurred under the conditions existing at the time of the bombing if due care had been exercised, and they expect that all possible steps will be taken to ensure the safety of Canadian vessels from attacks of a similar nature...."[30]

Cook Lau Wing Fai entered hospital in Vancouver on September 23, but when he sailed home to Hong Kong on the *Empress of Russia*, he still did not have full use of his left hand. In determining the amount of compensation, the CPR's "Oriental Manager" quoted a medical report citing a six-inch incision scar on the back of Lau's left hand with "definite loss of power of the left hand." This was a hard blow for a man who made his living with his hands. The proposed compensation called for $200 for Lau and $100 each for Mak Kwong and Kwok Shing, paid in Hong Kong dollars.[31]

While the investigation's initial overtures were between Japan and the British embassy, the incident illustrated Canada's growing independence or sensitivity in regard to international affairs. A passage in British ambassador Sir Robert Craigie's report on the bombing, sent to the British Foreign Office, noted the ship was of Canadian registry but "was under the control of the British Admiralty for the period of the war."[32]

In response, Canada's Department of External Affairs clarified that "during the war merchant ships registered in Canada have been instructed to accept orders from Canadian or British Naval Officers when the local situation makes naval intervention necessary. This however is not what would be suggested by the passage quoted. The vessel in question is of Canadian registry and under contract with the Department of Trade and Commerce."

The ambassador replied that he hoped the Canadian government would not attach any particular significance to the use of the word *control*.[33]

For the crew, however, the main concern was how easy it had been for a small bomb to penetrate deep into an unarmed passenger ship. The *Asia* still had two more round trips before she completed her final, 307th Pacific crossing and be formally requisitioned for war on January 13, 1941.

o o o

Even prior to the bombing there had been signs of worse things to come for the old ship that creaked, groaned, and belched black smoke, but those contemplating her deployment as a troopship knew she could still get good steam with a reliable crew, and best of all, she had plenty of hammock space. In August 1939, the *Asia* had steamed east across the Pacific with 68 non-commissioned British officers, all reassigned from Far East postings to serve as instructors in Britain's new conscript army. A skirl of camaraderie, compliments of the 16th Canadian Scottish Regiment, welcomed the men to Victoria on August 7, but within three hours the contingent was en route to Vancouver for a Quebec City–bound train.[34]

Threat of war affected the travelling public, too, especially those whose country of origin was the enemy. For example, possessing Austrian passports did not help the *Asia* passengers Dr. G. Halpern and his wife, who only discovered this predicament upon reaching Vancouver from Shanghai on the same day the British troops arrived. Pulled aside by immigration officials, the couple landed in detention until their stories checked out. "That country no longer exists as an entity, and German passports are now required by traveling Austrians," noted the *Vancouver Sun*.[35]

While the rattle of war discouraged North American travel to Europe, it did the reverse for thousands heading west to Hawaii, the Philippines, even Japan and China.[36] Still, along the waterfronts at Vancouver and Seattle, the mood was much dampened by mid-August 1939. "Port of Vancouver shipping felt the first definite effects of the tense European situation today," wrote the *Sun*'s Don Mason, reporting on the cancelled visit of a big Hamburg-America liner. "Capt. Franz Harder, master of the

'Portland,' has been ordered to tie up at Everett, Wash., until further notice." He continued by noting that the *Asia* and *Russia* must now adhere to an order that "placed all Canadian-registered vessels under control of the Dominion Government."[37]

Two other Empress ships, the *Canada* and *Japan*, commenced troopship duty in November 1939, and over the next several months, encountered difficulties that likely affected the *Asia*'s use of Chinese crew members. As the *Canada* steamed for Wellington, New Zealand, the *Japan* called at Sydney, Australia. By the second week of January 1940, both were part of a large convoy, escorted by warships bound for the Suez Canal. Less than four months later, the *Canada*, *Japan*, and the *Empress of Britain* were back off Melbourne with an impressive convoy that included the *Queen Mary* — all bound for Colombo until the threat of Italy entering the war sent it to Cape Town.[38] Upon arrival, Chinese crews aboard the *Canada* and *Japan* refused to sail through an active war zone to Britain, so left with no alternative, the *Japan*, skippered by Douglas, was obliged to transport Chinese crews from both ships back to Hong Kong, forcing the *Canada* to cobble together a scratch crew to reach Britain.

○ ○ ○

Now tagged as a troop carrier, the *Asia* discharged her Chinese crew, which had signed on under Hong Kong Articles — a significant loss since these men had vast experience working all over the ship from engine room to galley and upper decks. Only four, Chan Kam, Chan Lam, Chan Tin Yau, and Wu Chiu remained with the ship. Years later, Geoff Tozer recalled that the four were kept on to ease the transition to a white crew. While two of those men had been reassigned from bosun and carpenter to storekeeper and assistant storekeeper, Tozer felt they "knew the deck operation of the *Asia* inside and out, whereas their white counterparts [serving as bosun and carpenter] were so used to the Chinese doing all the work they were somewhat out of touch."[39]

In preparation for her February 13, 1941, requisition voyage to Greenock and Liverpool via Panama, almost an entire engine room crew of engineers, firemen, greasers, and trimmers boarded trains to Vancouver from the

Great Lakes shipping industry. Unfortunately, few personal records seem to remain from the list of 53 names belonging to men who gave Collingwood or Owen Sound, Ontario, as their home addresses, and two others who had served on the SS *Keewatin*. Although they would have had a choice of staying or leaving once the *Empress of Asia* reached the United Kingdom for refit, many would have preferred employment on the lakes because they also worked as farmers, bush workers, and trappers during winter.[40]

The only Great Lakes man to sign on and remain with the *Asia* until her loss was 20-year-old fireman Franklin Wesley Glover. Born in Collingwood, he had served on the 142-foot Great Lakes steam tugboat *Sulphite*.[41] It is likely Glover's decision to stick with the troop carrier had to do with steady work and world travel.

Owen Sound native Barrett Lumley, who was 18 when he signed onto the *Empress of Asia* as a fireman on February 10, 1941, had served on the Great Lakes with the *City of Hamilton*. He had tried joining the Royal Canadian Air Force but faced rejection on medical grounds. Undeterred, he pursued his dream after reaching Liverpool on the *Asia* and went on to serve in the Royal Air Force and RCAF.[42]

The loss of Chinese crew opened opportunities for West Coast men who were unemployed or looking for a switch. Without a pension and wounded in the First World War, veteran Patrick Conlin offered his "humble services but was found medically unfit" to join the regular service. However, on the *Asia*, during her requisition voyage, Conlin earned his pay washing pots, pans, and dishes in the scullery.[43]

New to the ship in February were 16-year-old mess boy Frank Davis, born in Springfield, Massachusetts, and 18-year-old Clarence Mitchell of Summerland, British Columbia. The youngest of three boys, Frank's father's occupation was listed as salesman, but he was engaged in bridge construction before moving the family from Quebec to Springfield and then Toronto. The couple divorced in 1928, and Frank and his brothers remained with their father until mother Kathleen purchased a rail ticket for Frank to join her in Vancouver where she had married a British Indian Army officer.

So it was that at age four, small suitcase in hand, Davis boarded the train to North Vancouver by himself. Not long after, his mother enrolled him in

Frank Davis was a toddler when he crossed Canada by rail to be with his mother in Vancouver, British Columbia.

a private school and then announced she was temporarily moving to India with her new husband. After leaving school, Davis earned a bit of money picking fruit until his mother became ill with stomach cancer and died in Vancouver in 1940. Davis then found work in a logging camp, and it was his brother, William, concerned about Frank's welfare, who arranged a meeting with the Merchant Marine.

Clarence Mitchell had also learned to be resilient. He weighed only two pounds at birth, and his parents, James and Minnie, welcomed him into the world by placing him in a shoebox lined with cotton batting. "They kept my father in the opened top drawer of their dresser," noted his son, Dale Mitchell. Minnie died in 1929 when Clarence was five. His father, gassed during the First World War, was still suffering from "shell shock" and it was not a pleasant home life for Clarence or his brother, Murray. The boys moved from one place to another with regular visits from their father. At 13, Clarence quit school and earned board and $5 per month on a farm. During the Depression, he rode the rails looking for odd jobs and his next meal while trying to stay ahead of the "railroad bulls" who evicted the homeless from the trains. "Sometimes his only sustenance was a crust of bread," remembered his son. "But on one occasion, a Chinese farmer left lower-grade produce in a special horse trough, just for the 'boes.'" The life of a hobo led Clarence to Vancouver where he signed onto the *Asia* as a trimmer.[44]

Making her return to the ship in February 1941 was Almonte, Ontario, native Hilda Mary Smith, who had signed up for the requisition voyage. At 36, she was the only female among 279 crew members after taking the place of nurse Julie Harwood. Smith, who never married, had a twin sister, Blanche, both having trained as nurses at Ottawa Civic General Hospital.

Nurse Hilda Mary Smith was the only female crew member during the *Empress of Asia*'s Second World War requisition voyage.

During the Depression, the big ocean liners offered opportunities that went beyond the ward, and Smith, who reported to the ship's surgeon, benefited from this. Like Harwood, her experience on long voyages taught her what to expect in terms of seasickness, contagions, burns, and lacerations, but there were surprising events such as a storm that tore a large ventilator off the foredeck, or the ever-present and worrying threat lurking beneath the Atlantic.[45]

Earning $40 per month as a mess boy, Frank Davis usually served food and cleared dishes, but on occasion likely had cause to visit nurse Hilda. "The merchant navy had no problem taking in young guys and for me it was something to do — a good job and it paid well," he recalled. "They fed you and I had a good time on board, going down through the Panama Canal. We stopped at Colón and Cristóbal and went into a tavern. It was pretty hot, and when I got back on ship I developed quite a rash around my stomach and everybody was joshing me, telling me I picked something up during shore leave. I was just a snot-nosed kid and so didn't know any difference. They had me pretty worried for a while.... Overall, the crew was pretty good to me. I was never abused or disrespected, but I didn't make long-time friends either."[46]

Time in Panama helped the ever-resourceful Mitchell, who purchased 200 green bananas for 50 cents, hoping to sell them for a shilling apiece in Liverpool.

In Jamaica, the *Asia* coaled and loaded up with nitrate before heading northeast, on her own, to Bermuda and then across the treacherous Atlantic. "I remember they just shovelled coal to her and we crossed as quickly as we could," recalled Davis. "There were a lot of U-boats out there — the Battle of the Atlantic was in full swing; a lot of apprehension being a single ship like that, but we were able to slip through."[47]

○ ○ ○

Reassured by a half pack of unfiltered Player's and a few ounces of Lamb's Navy Rum, Walter Oliver punched away at his Smith-Corona about how he joined the ship on Monday, January 14, 1941, a month less a day before her departure. He recalled Cristóbal's nightlife where sailors mingled in the cocktail bars, longing for female companionship. But mostly it was the North Atlantic he remembered, particularly when the ship passed through the Mona Passage, zigzagging across the cold grey expanse loaded with nitrate. "They believed U-boat positions were received at regular intervals, and one such position coincided exactly with the ship's position. We learned there had been twelve U-boats operating in the North Atlantic during February, sinking thirty-eight ships. This averaged to more than one ship a day with no submarine losses reported."[48]

It was no less dangerous when the *Asia* reached the Firth of Clyde in late March 1941 before heading south to Liverpool where Clarence Mitchell got a shilling apiece for his ripened bananas. Shortly after Frank Davis settled into the attic of the local YMCA, the Luftwaffe returned and released their payloads. One bomb slammed through the roof. "It [the concussion] pushed me down into the bed springs and I buggered up my face and lost some teeth, and ah, later on, as a result of that, I lost all my teeth. At the time I wasn't hospitalized, just First Aid. There were a lot of people worse off, so I considered myself lucky."

While the *Asia* was fitted out for war duty and busy signing local replacements for her engine room and stokehold, Davis hung around and met a girl named Monica Hunt before returning to Canada on the *Britannic*, the same ship boarded by Mitchell. Eighty years later, Davis said it was "funny" how he could remember Monica's full name and street address. "At the time, of course, my mother was dead and my brothers were both in the services. I was only seventeen, just living day to day. People don't understand the condition I was in. I was a homeless boy with no adult direction, nobody to talk to … just on my own."[49]

Mitchell became homeless again, riding the rails before rejoining the Merchant Navy, which took him back to Liverpool where he sheltered in a doorway during another bombing raid.

Davis's daughter, also named Kathleen, explained that her father found work with the United States Coast Guard. Asked in May 2021 how he had survived to the age of 97, Davis offered: "I was still warm. I was alive. That's how I got through it. I guess they don't make them like they used to."[50]

As one 16-year-old left the *Asia*, another signed on. Cadet Maurice Atkins had completed his training at the *Conway* and came aboard in April with his sea chest, the wooden box containing his clothing, toiletries, shoes, hats, and uniform. "Everything you owned was in that box and there were two of us to a cabin big enough for one."[51]

Cadet Maurice Atkins survived two wartime ship sinkings.

14

ARK OF WAR

For it's Tommy this, an' Tommy that, an' "Chuck him out, the brute!" But it's "Saviour of 'is country" when the guns begin to shoot …

— **Rudyard Kipling, "Tommy"**

Walter Oliver loved Kipling's poem about the "red-coat" entering a "public 'ouse to get a pint o' beer." While in uniform during the war, the *Asia*'s fourth officer felt respected and rarely had to reach for his wallet to buy a drink. For him, Kipling brought to mind the contrast between the treatment many First World War veterans faced in the Depression and the sudden need for those same men to return to war.

After returning to Liverpool's Gladstone Dock, the farthest from the main thoroughfare, Oliver saw "scars of war concentrated around the docks." During one raid, German bombs pummelled an ammunition ship. "That was the end of the dock and ship; also, the end of a car with four occupants that passed the dock at the time of the explosion — car and occupants never seen again."

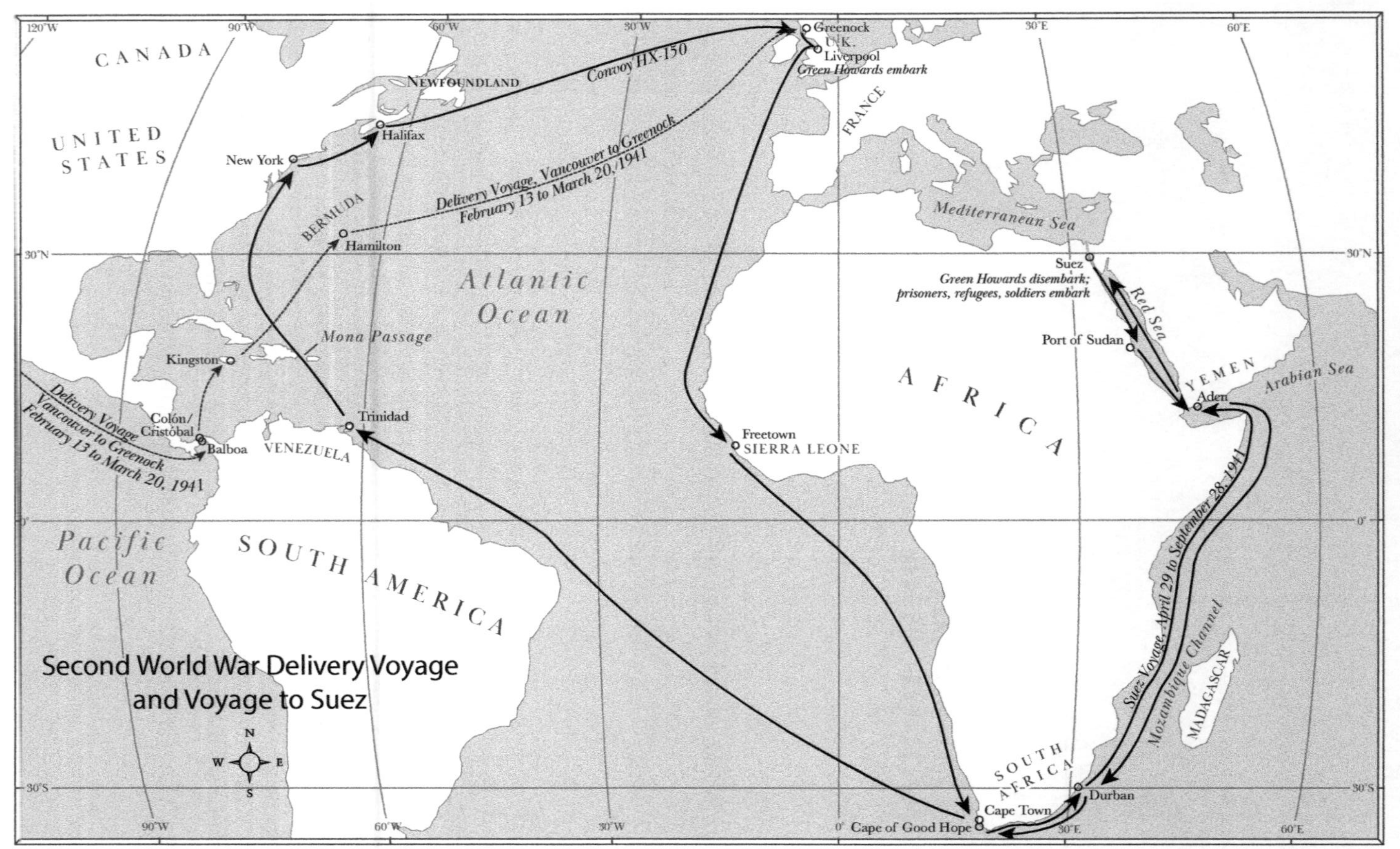
Second World War Delivery Voyage
and Voyage to Suez
Delivery Voyage, Vancouver to Greenock
February 13 to March 20, 1941
Convoy HX-150
Suez Voyage, April 29 to September 28, 1941
Green Howards disembark;
prisoners, refugees, soldiers embark
Green Howards embark
CANADA
UNITED STATES
NEWFOUNDLAND
SOUTH AMERICA
AFRICA
SOUTH AFRICA
FRANCE
U.K.
SIERRA LEONE
VENEZUELA
BERMUDA
YEMEN
MADAGASCAR
Atlantic Ocean
Pacific Ocean
Mediterranean Sea
Red Sea
Arabian Sea
Mozambique Channel
Mona Passage
Greenock
Liverpool
Halifax
New York
Hamilton
Kingston
Colón/
Cristóbal
Balboa
Trinidad
Freetown
Cape of Good Hope
Cape Town
Durban
Suez
Port of Sudan
Aden

The *Asia* was 28 years old, but of all the ships he had sailed on, Oliver was partial to the *Empress of Asia* — she was the one he remembered most. Still, it had been nonsensical to turn her into a troopship without converting her to oil. "She is too old, too slow and when cleaning fires throws enough smoke that may be seen for miles," he wrote from his apartment as though he were still back in Liverpool. "She is a menace to herself and any convoy she joins. In her present state, she might make a good hospital or depot ship, but useless otherwise. Still, the powers-that-be let her remain a troopship, but alas, 'ours is not to reason why.'"[1]

Oliver's feeling was shared among many of the long-time crew even after the DEMS firepower was installed — two rapid-fire Bofors anti-aircraft guns; a 12-pounder (three-inch gun) fitted aft; a six-inch-gun, also aft; six large-calibre Oerlikon guns; and 10 Hotchkiss light machine guns, six of them secured within nests above the bridge.

The conversion at Gladstone Dock also included a rocket system for propelling two parachutes with wires running between, presumably to tangle the props of low-flying aircraft. Oliver remembered a kite installed for the same purpose, kept aloft by the ship's speed. The refit called for cement casing around the wheelhouse, extra ventilation fans to expel fumes and body odour, alterations to the galley and mess, plus work below the afterdeck to accommodate thousands of troops, many of whom had never been to sea.

Her skipper was another Scottish-Canadian, John Bisset Smith, who replaced Goold in Vancouver. Remembered by some as a "quiet, calm man," Captain Smith was a veteran of the transpacific service with decades at sea, which commenced with a sailing apprenticeship from Glasgow. He was 48, and although not the oldest, everyone called him the "old man," but not to his face. At five foot nine, Smith was of average height and build with a lean face, aquiline nose, and keen eyes.[2]

One rung below him was another Smitty, Chief Officer Donald Smith, 55, also from Vancouver, though born in the tiny fishing village of Barvas on the Isle of Lewis in Scotland's Outer Hebrides. Fluent in Scottish Gaelic, Smith was the son of a fisherman and had known the sea from an early age. After a few years in the Royal Navy reserves, Smith, at age 22, commenced homesteading in Canada in the Mud Lake District near Macleod (later Fort

Macleod), Alberta. Three brothers joined him on the farm, and during the First World War while he was on the Atlantic in the Merchant Navy, his brother, Alexander, suffered wounds in France and died.

Back in Canada, Donald Smith joined the CPSS and served on five Empress ships, including a long stint on the *Asia* before signing on as her chief officer in June 1939. A patient, soft-spoken man of six feet with a square jaw and a determined look, he provided the main liaison between the ship and the officer commanding the troops.[3]

While renovations continued into April, shore leave was granted and most of the Canadian crew, including Crofts, Ewart, Higgs, Hosken, and Tozer, boarded trains for Old London Town and the blacked-out nightlife. "One does not look for war damage," noted Oliver after his visit. "It is almost everywhere, particularly the area behind St. Paul's Cathedral."[4]

Favoured by the *Asia*'s thirsty, pent-up crew was Café Anglais where taps remained open, unlike the once-popular Café de Paris on Coventry Street. The building above the nightclub had taken a direct hit; customers and employees did not stand a chance. Some, noted Oliver, had oxygen sucked from their lungs, leaving them "sitting at tables as in real life."[5]

London's gloomy nightlife was not part of cadet Maurice Atkins's travel plans when he graduated at HMS *Conway*, moored off Rock Ferry, Birkenhead. Before graduation, his father had lobbied Canadian Pacific to hire his son directly from *Conway* where discipline meant a great deal and where the lad was prepared for a life chosen by his father and circumstance. "There was a set of rules and one had to read these very carefully … and follow them because this was back in the good old days when if you stepped out of line you bent over, touched your toes, and somebody beat the dust out of your pants."

The academy's motto, "Quit Ye Like Men, Be Strong," stuck with Atkins, who found himself whispering those words to himself when faced with any difficult situation. The motto, he recalled, "was good for a lot of *Conway* boys."

Clutching papers to report aboard, Atkins pulled his sea chest from under his bunk and set off for the ferry to Gladstone Dock. He had no idea where the *Asia* was berthed until a constable pointed the way. "When I got there I didn't know what to do. I was just standing with my face hanging out and then the master-at-arms [William Cottingham of Vancouver] came along — a prince of a man — who treated me like someone human, and escorted me up the gangway and stairs to where the cadets lived, and that was it. I was aboard."[6]

In addition to attacks on the Cammell Laird shipyards, repeated bombings reduced areas of Liverpool and other Merseyside communities to rubble. In late August 1940, some 160 bombers wreaked havoc for three nights, one of dozens of raids that killed or maimed thousands, and destroyed tens of thousands of homes.[7] "The [possibility of] air raids each night are with us," wrote Oliver. "When the sirens sound out the alert, the officer of the watch, along with three regular gunners, stands by the anti-aircraft gun. To date, only one bomb fell close to the ship and that turned out to be a dud. Otherwise, life in Liverpool flows along without too much interruption."[8]

The *Asia*'s newly signed-on barkeeper Bernard Bree knew about the carnage all too well. His last ship, the *Empress of Canada*, had arrived in Liverpool five days after a German bomber levelled his house and flattened the family car. Fortunately, his wife, Nell, had rushed herself and the children — one of whom had suffered a broken collarbone — to safety in the small corrugated steel Anderson Shelter dug into the garden. The Brees were very proud of their tidy brick semi-detached house at Wallasey on the Wirral Peninsula where the dunes rolled up from the Irish Sea. Always a hardworking fellow, Bernard had fallen in love with Nell, marrying her in 1916. She had been a ship's nurse but gave that up because the couple were intent on raising a family. Nell found work in a Liverpool hospital, looking forward to seeing Bernard between sailings. By 1928, they had two girls and a boy born with Down syndrome. Relatives recall that they only heard Bernard swear once with the expletive aimed at Hitler.[9]

Led to his shared cabin, Atkins unpacked, stowed his chest, and did something he regretted. "There were three other cadets on board … and they asked me if I wanted to go ashore and go skating. Well, I don't like to skate … [but] they had given me this kind invitation and I was stupid enough to say, 'No thanks, I'm busy.' Well, they just looked at me and left and I realized I should have said, yeah, I'll go with you, but I didn't."[10]

Bernard and Nell Bree around the time of their marriage in 1916.

One of the cadets, 17-year-old Art Le Patourel, quickly let it pass, and as senior bridge cadet, "A La P," as he was called, had his own cabin kitty-corner to First Officer Leonard Johnston's. Born in Shanghai, Le Patourel was six when his parents shipped him off to an English boarding school until the Depression forced a transfer to a less friendly school in Hong Kong. At age 10, he entered the Canadian Academy in Kobe, Japan, and was comfortably speaking Japanese six years later when his father helped sign him onto the *Asia* in Nagasaki.

Atkins and A La P became good pals, and during off hours, enjoyed listening to music on a portable record player owned by Le Patourel. "Sugar Blues," written in 1919 by Clarence Williams, with lyrics by Lucy Fletcher and recorded by Ella Fitzgerald and William Henry "Chick" Webb in 1939, was a favourite played repeatedly and loudly:

Cadet Art Le Patourel became good friends with fellow cadet Maurice Atkins. Born in China, Le Patourel had been a crew member during the ship's 1941 requisition voyage — signing on at age 17 — and remaining on board for the final voyages.

> I could lay me down and die;
> You can say what you choose,
> But I'm all confused;
> I've got those sweet, sweet sugar blues …

The activity sparked at least one "tongue lashing" from First Officer Johnston, who stormed in from across the hallway, ordering the lads to pipe down or face another kind of music.

With little to occupy himself prior to sailing, Atkins "boned up" on the ship. The upper bridge where he stood watch contained the wheelhouse, while the lower bridge had the chart table and clinometer used to gauge the roll and pitch of the ship. "Cadets were junior officers and so I didn't play around [work] on deck and didn't speak to the [deck] crew or anything. Most of the time I simply kept my eye on the clinometer and did what the officer above me ordered, reporting directly to Second Officer Cecil Crofts, an excellent fellow in every sense of the word."[11]

Several decks below in the engine room and stokehold, the Great Lakes crew, with the exception of Wesley Glover, signed off. Most opted to return home where they continued to work in support of wartime shipping. Although too few in numbers, young and lacking deep-sea experience, the delivery crew had performed well, which likely resulted in a false sense among the ship's senior officers looking to replace them. If the Great Lakes crew could steam her, men raised in Liverpool could surely fill the void — besides, the *Asia* had some of her old "black gang," including trimmer James Jackson, who became somewhat of a father figure — and peacemaker — when disputes arose.[12]

The replacements were local men, Liverpool Irish, some with ocean experience and from seafaring families, some not. To understand how the engine room problems that plagued the final voyages developed, it is necessary to be aware of the Irish experience in Liverpool when the men signed on. It is also helpful to know that in terms of sheer numbers, the incoming Irish engine room crew was considerably smaller than the Hong Kong–sourced Chinese crew that existed prior to the ship's wartime requisition.

For example, in 1935 there were 81 firemen, 63 trimmers, and 24 greasers. When the *Asia* struggled into Cape Town after leaving Liverpool in the spring of 1941, she had 65 firemen. What was more, the Chinese crews had been older, more experienced, and familiar with the complicated and sometimes temperamental workings of the furnaces, boilers, and turbines. There was, however, something else at work, setting the stage for rebellious behaviour.

Liverpool had a substantial Irish community, dating back well before the 1930s. Its Irish enclave predates Ireland's Great Hunger, or Potato Famine, of the 1840s. By the late 19th century, those of Irish ancestry in Liverpool numbered between 150,000 and 200,000. These migrants arrived seeking a better life and many had dreams of emigrating to the United States or Canada. Liverpool's Scotland Road neighbourhood, between the city centre and waterfront, became home to many who ended up on the streets or in overcrowded, shabby tenement housing. The community, however, shared a cohesive identity, and during the 1920s and 1930s, there was an ideological shift at Scotland Road.

Previously, the neighbourhood had been politically insular and inward-looking, electing an Irish Nationalist to Parliament. However, during the 1930s, the neighbourhood became a bastion of Labour Party support, placing it within a mainstream political society. Residents became exposed to Labour Party and trade union beliefs on how workers could achieve success, if necessary, by forcing concessions through work stoppages or disruptions. It is more than likely that some of this transferred to the *Asia* (and other ships) where engine-room crews assessed their situation with a trade-union approach, meaning grievances could be resolved through work stoppages and agitation.

During this same pre-war period, the Roman Catholic population of the Scotland Road area was roughly 95,000, most of Irish descent.[13] The harsh feelings among Liverpool's non-Irish were omnipresent, although they fluctuated. Many feared the existence of the Irish, and these racialized attitudes led to ongoing conflict and violence. The city's Protestant community, with its active Orange Order and ultraevangelical Anglican diocese, rejected Roman Catholicism, pigeonholing it as a foreign theology that encouraged ungodly beliefs and behaviours. Following Irish independence in 1922, racialization of the Irish became more public and incendiary, especially in the press and among political leaders — characterizations that regularly pegged the "Paddies" as inferior, violent, drunken, and dirty.[14]

In 1932, during the Depression, Liverpool's unemployment rate soared to 32 percent, but this did not slow Irish migration to Merseyside. The result was more anti-Irish agitation, much of it based on fear that the Irish were stealing jobs and taking advantage of social welfare. Serious racialized attempts to restrict the movement of Irish labour followed, and in April 1937, the evangelical Anglican bishop of Liverpool, Albert David, spoke of how the Irish would soon gain control of local government.

Therefore, when the Liverpool Irish came aboard to stoke the *Asia* in 1941, they had already been condemned for ethnicity and understood that their jobs placed them at the bottom of the ship's hierarchy. The men worked, ate, and slept with limited outside contact and only reported to the chief engineer, who then reported to the captain. Overall, the situation did not auger well for the ship's operation when she left for Suez on April 24.

Eighteen-year-old Patrick James Brown was among those below. He had grown up on the rough backstreets of Liverpool and had engaged in heavy manual labour at an upholstery shop. At five foot nine, with blue eyes, fair hair, and a scar on his left temple, Brown signed on as a trimmer — considered lowest of the low — on April 17.[15] For the next several months, the husky lad sweated rivers in the dark, dust-filled bunkers, hour after hour shovelling coal onto a rickety wheelbarrow, which was then shunted into the undermanned boiler room.

The Browns' tenement on Burlington Street was noisy and crowded, tucked behind a characterless building with a rattling refuse shoot hanging off the facade and laundry on the railings. The apartment lacked privacy and variously housed nine children, including Patrick, his mother, Mary Ann, and her brother, who was employed as a ship's sealer for Harland & Wolff. "It was a tough time to bring up a family, especially as my grandfather [James] jumped ship at Melbourne, Australia, and my grandmother [Mary Ann] had to bring up a family on her own," explained his son Al Brown of nearby Southport. "My father had to go to school in his bare feet as they couldn't afford shoes."[16]

Patrick was fifth oldest, born on September 23, 1922, and there is a photograph, circa 1934–35, showing him holding his infant brother, Joey, who died a short time later. Patrick, with hair ruffled, seems like the responsible older sibling. In front to the right stands an angry-looking boy, arms crossed in defiance, and a less robust lad with a bandaged head. Brown's discharge book lists his character as "Very Good," suggesting he kept his head down and worked hard.[17]

Fireman James Cranny, 32, was the son of a dock labourer and in 1921 had two older brothers already at sea, one as a trimmer. Twenty months before Cranny signed on, his name appeared in the local paper under the headline "Police Used Field Glasses to Watch Gambling on Foreshore." Caught along with 13 others for tossing copper pennies on the Seaforth shore, Cranny stood before the magistrate while the police explained how the men were spotted from Gladstone Dock and had given police "endless trouble."[18]

At a landing upriver, the ship had been towed into place by tugs before lowering three gangways to receive members of the 5th Battalion Yorkshire Regiment of the British Army, also known as the Green Howards; the 72nd Field Regiment Royal Artillery; and the 150th Field Ambulance. Eighteen-year-old *Asia* sailor Ted Evans had noticed the ship had room for 2,000 soldiers, "but they pushed another 500 on us."[19]

Lance Bombardier Oliver Perks, 21, of Bristol was among the Suez-bound troops. Like many, the outbreak of war had interrupted his education. From a posting at Weston-super-Mare, he, along with other members of the 72nd, had journeyed north by train. The sound of kit clattering against railings and boots on metal stairs echoed and reverberated as Perks descended into the troop quarters — until space opened up near the waterline.[20]

From the Mersey, the *Asia* steamed northwest across the Irish Sea, skirting the southern tip of the Isle of Man on her journey north to Clyde. Ted Evans had grown up at Graigadwywynt, a hamlet tucked into the green folds of North Wales, where his parents had a cottage. However, this was not his first voyage. In the fall of 1937, at age 15, he crossed the Atlantic on the rusty 5,127-ton steamer *Cragpool* as cabin boy, earning 10 shillings a week, plus food and a "bucket to wash in." En route from Galveston, Texas, to Rotterdam with 6,000 tons of wheat, the *Cragpool* slammed into a hurricane and went adrift after suffering a near-fatal malfunction.

The *Asia* also gave Evans an early scare: "Very soon we suffered a fire and as most of us were new to the ship we had problems finding fire equipment, smoke helmets, hoses, and axes to get at the seat of the fire. We eventually put it out, but by then there was damage to wiring and to the officers' quarters. By the time we reached the Clyde it was doubtful we could sail, but after inspection at Loch Long [northwest of Greenock] and, considering no other ship had room for our troops, we were ordered to join the convoy and carry out repairs at sea."[21]

Convoy WS.8A, composed of troopships, fast freighters, and escorts, travelled north around Ireland then south. Oliver estimated the number of troops at 15,000, but it was likely around 25,000.[22] While several vessels

Fifteen-year-old Ted Evans (left) and mate Joe Bates. Evans, who survived the war, joined the Merchant Navy while Bates, who did not survive, served in the Royal Navy.

detached at Gibraltar for Malta, the rest, including the *Asia*, continued south past Madeira, the Canaries, and Cape Verde. Unlike the convoy's oil-burning ships, coal-burners faced separate challenges. Funnel smoke made them highly visible but also drew attention to the entire convoy if suddenly ordered to reduce speed. By slowing down too quickly from a full head of steam, the safety valve could sometimes blow off steam, creating a piercing noise heard for miles.[23] Coal-burners also routinely dumped furnace ash, but this became prohibited at night because smouldering embers could draw unwanted attention, so significant ash buildup was a problem.[24] And with convoy speed of 14 knots, the *Asia* struggled to hold station, even before reaching Freetown, Sierra Leone.

Perks recalled how the *Asia* lagged behind, mostly in the morning, and it was "a miracle" she was not torpedoed.[25] Evans remembered this deeply annoyed the escort ships. "It was meant to be a fast convoy ... but the *Asia* was old. Relations were, to be honest, very strained from the start. It became clear to us the stokers feeding the boilers with coal were reluctant to work harder to keep steam pressure up."[26]

Evans also remembered the "food was not good and the men were not working as hard as was needed, blaming the quality of coal. Eventually, we fell behind the convoy and had to have our own battleship escort and our crew had to keep extra watches on deck for enemy activity while some of the deck hands and troops had to go into the boiler rooms to help bring coal from the bunkers to the stokers.... This led to complaints from the deck crew and two stokers were arrested and eventually sent home for insurrection."[27]

After bunkering at Freetown, the ship was to head to Durban but made an unscheduled call at Cape Town. "There was definitely a mess with our crew, and so we spent two days in Cape Town before coming around to Durban," recalled Atkins. "There had been a Chinese crew in the engine room and they were good. Unfortunately, these men left and I never did find out why, but there was a kerfuffle and we ended up with a crew of — well I shouldn't say this — white men — Liverpool Irish. They were a hard bunch … they could steam ships, but they could not steam her. We were always behind the convoy or dragging our feet. It was very embarrassing, but we couldn't keep up steam and it would drive the old man mad."[28]

The situation placed the ship at tremendous risk, but it is important to note that the responsibility to ensure an efficient and orderly engine room rested with the chief engineer, and his superior, the captain. It is also worth noting that when the *Asia* had a Chinese engine crew, there was an organizational structure for the stokehold which included three senior ratings of Fireman No. 1, Fireman No. 2, and Fireman No. 3. These positions were analogous to the Chinese bosuns of the deck crew, who were responsible for transmitting orders to the lower ratings. When the *Asia* transferred from a Chinese to a European engine crew, these senior positions were not replicated, and this may have contributed to an organizational breakdown that made the successful distribution of orders and the enforcement of discipline more difficult.

○ ○ ○

Lack of steam pressure was not the only problem as both Oliver and Perks recalled. "Anchoring or trying to anchor with a strong wind blowing outside the breakwater of Capetown [*sic*] the ship lost the anchor due to poor handling," observed Oliver, allowing himself — in a rare moment years later — to point a critical finger. "It is expecting too much trying to anchor a ship with so much freeboard, broadside to the wind."

On deck, Perks was "surprised to see the anchor chain run right out, including the last link of the chain." Minus an anchor, the ship, noted Oliver, waddled around in rough seas until a harbour pilot directed her to anchorage where the second anchor was "paid out." On the upside, the unexpected visit impressed locals who cheered the Green Howards as they marched through the streets.

After coaling in Durban, the convoy followed the Mozambique Channel across the equator, up the east coast, and around the Horn of Africa into the Gulf of Aden and Red Sea. At the time, the East Africa campaign was underway, pitting British Empire forces against Italian forces. Enemy aerial attacks were expected, but none materialized.

Able Seaman Evans was thirsty and sticky with sweat as the ship entered the Gulf of Suez to disembark troops at Port Tewfik (Suez Port) near the entrance to the famous canal. It was late June and the *Asia* received orders to turn her Bofors over to the British for the North Africa campaign. This made sense, although in the midst of an air threat the crew, including Oliver, had hoped to keep all the anti-aircraft guns.

With preparations underway to utilize the ship as a kind of ark of war, Evans looked up and cringed when ordered to paint the main mast, including its top-most section, 38 metres above the deck. Far below, in the ship's secured after-holds, hundreds of Italian prisoners of war stood toe-to-heel in the scorching heat. Bound for the dusty prison camp at Durban, many of these men looked awful, and a few would die on route. Overall, "they were a sorry sight, dirty and poorly clad," Evans noted.[29]

The "ark" also embarked a few hundred army officers and ratings, and surprisingly, at least 70 Norwegians. Mixed in were a number of British, Free French, Belgian, and Dutch officers and soldiers, as well as some merchant sailors. The Belgians and Dutch had, recalled Oliver, experienced a very "rough time" in Crete and Greece, but like the Norwegians and Free French, their hope was to resume the fight. Most noticeable were civilian evacuees, mainly women and children from Yugoslavia, Syria, Greece, and Crete. Hungry and travelling with worn-out clothes, they looked thin and exhausted.

The Norwegian refugees had reached Port Said earlier that month with many stories to tell. They had fled Norway in February after the German invasion in 1940. To reach Egypt, they took an overland route from southeastern Norway, across Sweden and the Gulf of Bothnia to Finland, from there through Russia, Ukraine, Romania, Bulgaria, Turkey, and eventually Syria.

Per Hvidsten was the son of a dyed-in-the-wool newspaperman who ran a left-leaning paper in southeastern Norway at Sarpsborg. Hvidsten began working for his father at 17, wiping down the ink rollers and setting typography before graduating to reporting. He had met Norway's compulsory military service standard and found employment on a Swedish farm when the Nazis arrived. According to his son, Peter, "time has clouded details as to what happened to my father, but he most likely joined one of the many groups hiding in the forests to continue the battle against the invading Germans."

With few dirt roads and paths, the hinterland between Norway and Sweden was mostly inaccessible. Anyone caught without a resident card risked arrest. On the plus side, the forest was difficult to patrol, and working against the Germans were scattered groups of Norwegian resistance assisting those wishing to escape. Hvidsten suspects his father crossed near Halden, on February 3, marking the start of a 3,728-mile odyssey to Egypt.

Joining the *Asia* at this point was a tall, dashing young midshipman who was to become the longest-serving British royal consort — Prince Philip of Greece and Denmark. Recently turned 20 in June, Philip had joined the Royal Navy in 1939, the same year he began writing to 13-year-old Princess Elizabeth, and while on the *Asia*, ably performed his duties until he disembarked at Durban. Well before reaching that coaling port, Philip became acquainted with Chief Officer Donald Smith, resulting in a lasting friendship that included social calls in Canada and the United Kingdom. From an early age, Philip, born in Greece, had become familiar with navy ships. As an 18-month-old in December 1922, he arrived on board a Royal

Per Hvidsten in his Norwegian Air Force uniform in 1945.

Navy ship in an orange crate during his family's evacuation from the Greek island of Corfu.[30]

As some predicted, two Italian prisoners of war died while the ship crossed the Red Sea to the Gulf of Aden, their remains committed to the deep. A third prisoner, recalled Ted Evans, perished off Africa. "Despite their priest asking for the dead to be kept on board until the next port, we had no cooler to store bodies." As heat and body odours rose to inhumane levels in the prisoners' hold, the men were permitted access to the afterdeck during daylight hours only. This went well until some tried to avoid going below by hiding in lifeboats or storage lockers. Walter Oliver noted that "one prisoner hid in a refrigerator and died the next day from pneumonia."[31]

In a particularly nasty incident, some off-duty stokers on the Shelter deck used ropes to lower empty buckets to accomplices in the stokehold — a tug on the rope assured the men up top that the buckets were full of hot

A photograph showing a young Prince Philip (left) posing with Leonard Johnston, the *Empress of Asia*'s first officer. The prince also became close friends with Donald Smith, the ship's chief officer.

water from the boilers. The firemen then waited until a prisoner stuck his head out of a porthole for fresh air and emptied the contents. Although the water had mostly cooled by the time it slapped the man on the back of the head, the shock was enough to cause alarm. Oliver confiscated the ropes and buckets but could do nothing to remove the stokers' deep-seated hatred for the enemy.

However, there was also love in the air to reinforce the idiom "all's fair in love and war." Second Lieutenant Douglas May of the Royal Engineers, whose wife and mother had perished in Hertfordshire during the Blitz, had joined the ship at Port Sudan in the Red Sea. On board, he met Rosalie Pryce, an adroit Auxiliary Territorial Service volunteer who had driven a three-ton truck in Cairo. Although May was returning home on compassionate leave, both he and Pryce understood the fleeting and difficult circumstances of war. While the ship steamed from Durban, the couple announced their engagement at a dinner party.

Unfortunately, the working relationship between the Admiralty and the constantly dawdling ship had worn thin by the time the *Asia* had returned around the Cape of Good Hope. "Admiralty had become sick and tired of us and so we crossed the Atlantic to Trinidad alone," recalled cadet Atkins.[32]

o o o

Pausing in front of his typewriter, Walter Oliver reflected on the route to the West Indies and recalled an attractive Greek woman who stopped to chat while he was having a smoke. Bound for Trinidad, she had been having quite a time keeping the men at bay. "All the men want to luff me," she told him as they stared out to sea.[33]

In New York City, the *Asia* was greeted by cheers and the whirl of newsreels that September as she slid past the Statue of Liberty looking more like a

Rosalie Pryce of the Auxiliary Territorial Service and her truck in Cairo, Egypt.

homeless tramp, with scarred grey paint, a bulging plate on her starboard side, and barnacles above the waterline.[34] The press however, rushed aboard to interview some of the 393 weary passengers and ship's crew. It was understandable because this was the first transport to arrive in the city with refugees from the Mideast. "Needless to say," Oliver observed, "we made the headlines of every New York newspaper. The caption of one read: 'The Ark of War.'"[35]

Associated Press/British Movietone and Reuters/British Pathé newsreels captured passengers' expressions as the ship entered the harbour. There were interviews with young Norwegians, two of them explaining how they narrowly escaped by skiing into Sweden. Several relieved women with young children and four Australian sailors — one with a bandaged arm — recalled the loss of their warship in the Mediterranean.

On the darker side, reporters asked about problems in the *Asia*'s stokehold and quoted a sailor who stated 20 naval ratings had to go below to tend to fires

British Second Lieutenant Douglas May.

Looking worse for wear, the *Empress of Asia* arrives at Pier 61 in New York City, September 1941.

after the civilian firemen "mutinied."[36] In answer, Captain Smith took a deep breath and referred all questions to the British Ministry of War Transport. "Don't believe anything you hear on this ship" was his parting quote.[37]

Grateful to set foot on land, Per Hvidsten headed to Toronto and prepared to enter pilot training at Little Norway, a camp that attracted displaced Norwegians anxious to fight the Nazis. It later moved north to Muskoka where Hvidsten trained and became an instructor, and by February 1945, he had kept his word and returned to Europe.[38]

While the dilapidated *Asia* entered dry dock, Pryce and May tied the knot at Chelsea Presbyterian on September 11. Fully anticipating a dangerous North Atlantic crossing, Oliver recalled the newlyweds boarded the ship and spent "their honeymoon sleeping in separate bunks and cabins."[39]

Now with fewer passengers, the *Asia* headed to Halifax where she offered her deck crew, including Boy Seaman Geoff Tozer, "a real Canadian

Evacuated from Palestine, 22-month-old Katrine Anne McQueen was one of the younger passengers on board when the *Empress of Asia* reached New York City in September 1941. Her mother, Alice Elizabeth McQueen, smiles as she hugs the ship's railing. The girl's father, Angus, who had served as a policeman in Palestine, was also on board.

feeling in the air" with a perfect view of "yellowish cliffs and deep-green fir trees in the clear, morning sun." Writing to his nervous parents in British Columbia, Tozer explained there "really was no reason to worry" because "the Germans wouldn't waste a torpedo on the ship." He also mentioned the problem reported in the New York City press. "For some reason the Englishmen are very jealous of all the Canadians left on board so there is quite a lot of trouble between the two elements. There have been strikes and nearly mutinies, while not long ago, the firemen climbed over the coal and got into the beer in No. 2 hatch and proceeded to get roaringly drunk so our speed dropped to nothing. But now things seem to be a bit better."[40]

In the harbour, ships of all sizes and condition sat upon rusty chains, awaiting troops and cargo arriving by rail. The *Asia*, which would embark approximately 2,000 officers and enlisted men, received instructions to join HX.150, the war's first North Atlantic convoy escorted by U.S. warships.

Several sailors, including *Empress of Asia* crew members (back row, second from left) James McAinsh, Terrence Walding, John Hall, William Pritchard, Arthur Russell, and (front row, right) Donald Piercy, gather for a photograph shortly after the ship reached New York City in September 1941.

Assigned to centre column, second row, station 52, the *Asia* was the largest of nearly 50 merchant ships forming a grid composed of nine columns and six rows with 600 yards between columns. Among the men saying prayers or nervously awaiting departure were 21-year-old Willard "Dusty" Perrin of Kingston, Ontario; newly minted Australian pilot Colin Albert Alt; and Second Lieutenant Herbert D. Roistacher, a radar trainee from New York.

Although the United States had yet to declare war, its navy had been conducting neutrality patrols off the North and South American coasts since September 1939. At Ogdensburg, New York, in August 1940, President Franklin Delano Roosevelt and Prime Minister William Lyon Mackenzie King had authorized the Permanent Joint Board on Defence, and later that year, a destroyers-for-bases deal led to U.S. military bases

U.S. Second Lieutenant Herbert Roistacher joined the *Empress of Asia* in Halifax, bound for Liverpool, England, as part of Convoy HX.150.

in Newfoundland and the British West Indies. Other significant developments followed, not least of which the signing of the Atlantic Charter between Churchill and Roosevelt.[41]

Roistacher, along with Perrin and Alt, hoped the voyage would be safer because of the escorts and the Admiralty's efforts to plot U-boat locations. "My feelings were rather mixed, mostly sad of course," commented the 24-year-old Roistacher the day after he left New York's Grand Central for Halifax via Montreal. "Sad, that I was leaving home without opportunity to return for goodness knows how long. Yet there was also a tingling feeling of excitement at having the unknown future."

With orders to report to London for duty as an observer with the Electronics Training Group, Roistacher boarded the *Asia* on September 15. Meanwhile, the American escorts moved into position 348 miles off Nova Scotia as they anticipated the arrival of HX.150, courtesy of a Canadian escort group. "I have a lower berth in a former drawing room with five other brother officers," wrote Roistacher. "On board are Australian, British, Canadian, New Zealand and African flyers and Canadian troops as well as several Canadian and British officers. The boat is well armed, but some airmen refused to get on board."[42]

The air force personnel were graduates of the British Commonwealth Air Training Plan (BCATP). "The Army came aboard, but the CO of the air force contingent we were supposed to take refused to put his men aboard," recalled Geoffrey Tozer. "He said the accommodation was dirty and not suitable for air force types. Actually, he was quite right. It was not suitable for anyone, but we did say a few harsh things about the 'Brylcreem Boys' after that."[43]

For Albert Alt, a BCATP graduate, it was "apparent not enough cleaning had been allowed" and "conditions ... were quite primitive." He recalled how "disgruntled Italians, turned prisoners by their inept offensives ... had ... left faecal calling cards in cabins and oozing sacks of rotten potatoes in the hold." The "authorities persuaded us to go on board after a long harangue, promising us better conditions, but after getting on board [we] found we had been fooled and could not get off again.... I slept on the bare floor, but it served as a reminder we were joining the war zone."[44]

The complaints reached Ottawa, but years passed before the Canadian government released classified War Cabinet Committee reports on the matter. An investigation showed the ship had left the Suez Canal without fumigation or proper cleaning. "She had undoubtedly been in very bad shape, in fact verminous, at Halifax, when the RCAF personnel went aboard." In response, the Admiralty cited the extreme shipping shortage, which meant no time to properly clean and fumigate transports, and that "crowded conditions were inevitable."[45]

It did not end there. With the War Cabinet Committee's approval, Canada's minister of national defence for air sent a strong cable to his British counterpart urging him to have the Admiralty take all necessary steps to prevent a recurrence.[46]

Former tinsmith Dusty Perrin of the Royal Canadian Electrical and Mechanical Engineers was aware of the complaints but remembered the *Asia* as a "nice ship." His focus was on staying out of the water. "Dad was a non-swimmer and was petrified of the ship sinking and being lost at sea," noted daughter Barb Paquette of Gananoque, Ontario. "He had enlisted in 1939 and recently gotten married on Valentine's Day. He was very much in love."

The fear Perrin felt was well justified when Convoy HX.150 left Halifax on September 16. Less than a week earlier, off the southeast coast of Greenland, U-boats sank 16 merchant ships in Convoy SC.42. In addition to troops, HX.150 transported steel, flour, lumber, refrigerated goods, and flammables such as benzene, oil, gasoline, and paraffin.

Writing to his mother, Roistacher stated the "10-knot convoy" started out on water "mild as a millpond" and "proceeded along smoothly" while troops were organized and briefed on eventualities. "At sundown … there is a blackout … portholes are covered and windows barred so no light appears. No smoking or lighting matches allowed on deck. It can be so dark that you can hardly see your hand in front of you."

While engine trouble forced some ships back to Halifax, 44 merchantmen continued northeast, and on the afternoon of September 17, Roistacher witnessed the historic handover from Canadian escorts to a "goodly number of American warships." Now, like never before, it seemed the U.S. Navy was in the war, defensively tasked with protecting HX.150 to the mid-ocean meeting point south of Iceland for rendezvous with British escorts.

The route took it from beneath air cover, and as the Americans practised newly acquired naval tactics, the merchant ships slid into fog over placid, deep seas. There were a few scares and false sightings, but the enemy did not show, although new intelligence regarding U-boat positions caused course adjustments on the 18th and 22nd.

Perrin began feeling his darkest fear when the wind picked up, but riding a storm on the *Asia* was better than enduring it on one of the smaller vessels that dropped alarmingly into troughs. Roistacher described 10-foot waves: "Our boat really rode high and took the waves well. I didn't feel even that queer dizzying feeling. I sure have water legs."[47]

The storm lasted into the night of the 24th when a "smoulder" broke out in the bunker of the 5,993-ton *Nigaristan*, a general cargo ship off the *Asia*'s port bow. It occurred near a place called Torpedo Junction and forced the

Nigaristan to fall behind. She "quickly went to the rear and it looked like she just missed two other ships," noted Roistacher. "She disappeared over the horizon in a veil of smoke."

By the time USS *Eberle* reached the burning vessel, the fire was out of control. Her crew piled into lifeboats, and amid rolling, scattering seas, it took the *Eberle* three and a half hours to rescue everyone, including a man who had slipped overboard.

On the 25th, British escorts loomed over the horizon and took control, then later that afternoon, Roistacher noted the *Asia* had "pulled out from the convoy with two escorts and made a break for England. Our speed was increased by about six or seven knots and we expect we will probably get to port a day or two sooner." On the evening of the 26th, one of the escorts "dropped several depth charges and calcium flares. Rumours came back that a submarine had been forced to the surface and sunk."[48]

Two days later, 13 days after the *Asia* left Halifax, the American woke with Ireland on one side and Scotland on the other. The ship was plying the Irish Sea, passing the Isle of Man toward Liverpool.[49]

○ ○ ○

From his apartment, Walter Oliver again pictured himself back at Gladstone Dock leaning against the ship's rusty, paint-peeled railing. "Liverpool is one port that never seems to change," he wrote. "Air raids nightly as usual. They never, however, interfere with the working of the port too much — bombs or no bombs."[50]

15

LAST IN LINE, FIRST IN TROUBLE

This convoy appears to be heading into the most dangerous area in the Far East.

— **Walter Oliver**

In the summer of 1990, 80-year-old Walter Oliver hoped to begin documenting the ship's final voyage and her crew's close escape from Singapore. At least that was the plan. Standing in the way was not writer's block, lost memory, or the fear of revisiting the 1942 inferno, but the weather. It was a glorious day on the West Coast, so he shoved a crumpled pack of Player's into his shirt pocket, grabbed his wartime notes — the ones scribbled onto V-for-Victory letterhead — and set off for the Port of Vancouver.

Fourteen years had passed since he began chronicling his life at sea, and although he had taken breaks, the story never left him, even following him from his apartment to Joyce Station. Settling onto the SkyTrain, Walter pictured in his mind's eye the faces of the Brits, Canadians, Chinese, and Liverpool Irish he had been to sea with, not how they might look today but how they appeared then.

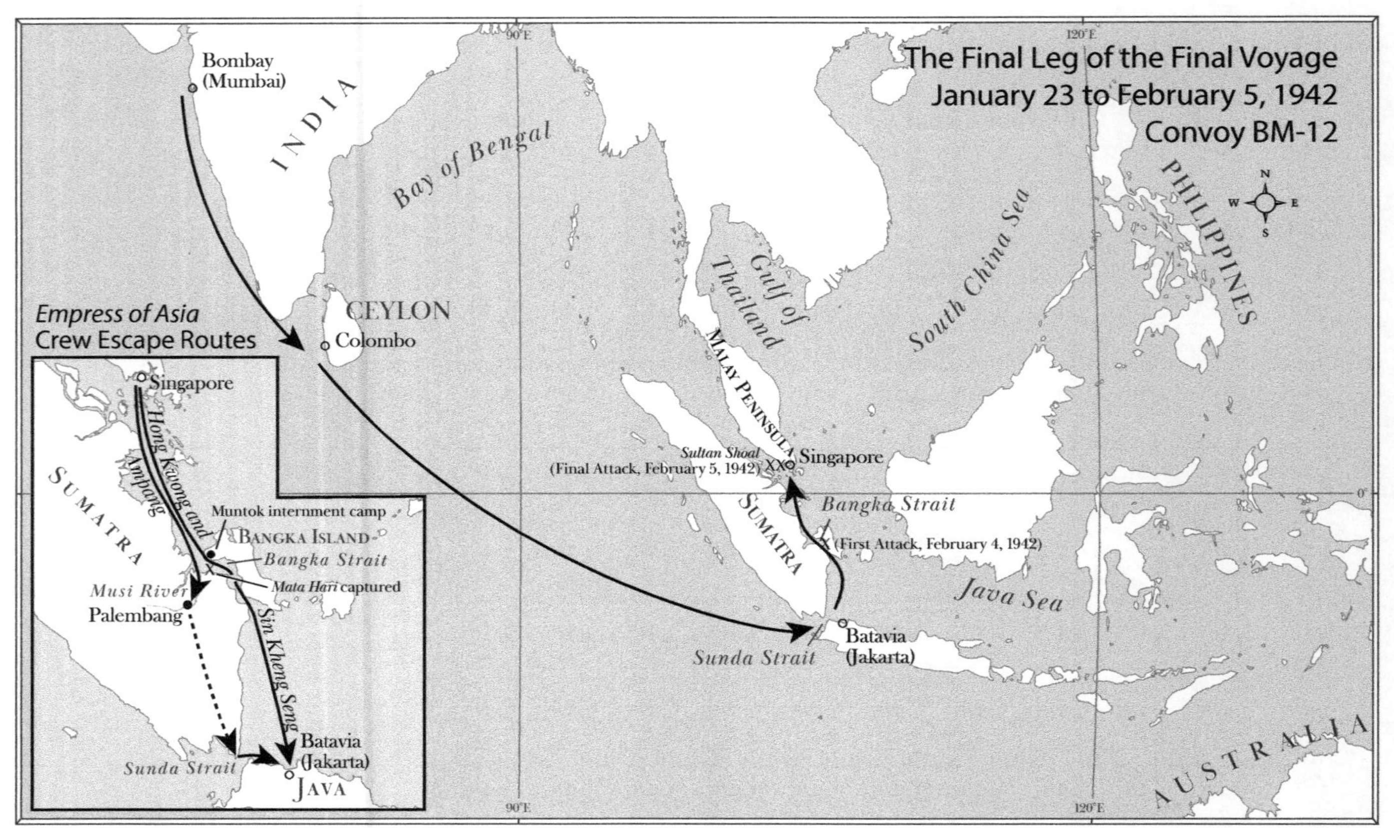

The Final Leg of the Final Voyage
January 23 to February 5, 1942
Convoy BM-12
Bombay
(Mumbai)
INDIA
Bay of Bengal
CEYLON
Colombo
90°E
120°E
0°
N
W
E
S
PHILIPPINES
South China Sea
Gulf of Thailand
MALAY PENINSULA
Sultan Shoal
(Final Attack, February 5, 1942)
Singapore
SUMATRA
Bangka Strait
(First Attack, February 4, 1942)
Java Sea
Sunda Strait
Batavia
(Jakarta)
AUSTRALIA
Empress of Asia
Crew Escape Routes
Singapore
Hong Kwong and
Ampang
SUMATRA
Muntok internment camp
BANGKA ISLAND
Bangka Strait
Mata Hari captured
Musi River
Palembang
Sin Kheng Seng
Sunda Strait
Batavia
(Jakarta)
JAVA

For the most part, the Canadians, especially the West Coasters, remained tight, as did the Chinese from Hong Kong, who worked hard but kept to themselves. In 1941, not one of them was under 30, and they lived and worked with quiet dignity, performing duties without difficulty or resentment. Storekeeper Wu Chiu was the oldest at 48, while the youngest, carpenter/joiner Chan Kam, was 31. In between were Chan's brother, carpenter/joiner Chan Lam, and rustyman Chan Tin Yau in charge of paint stores. Back in Hong Kong, the brothers' wives shared the same small apartment. Wu Chiu, born in the United States, had worked as a police officer in New York City before moving to Hong Kong.[1]

As for the Liverpool Irish, Oliver refrained from denigrating them, being clearly sympathetic to the working class. Perhaps the harsh, stereotypical treatment of the ship's Irish "black gang" triggered memories of the hewers, loaders, and trapper boys who emerged from pitheads at Easington Colliery, with faces so black it was hard to tell them apart. "He warned me as a young person never to cross a picket line regardless of the consequences," recalled his son, Nelson. "On the coastal waterfront at Vancouver, where we lived, there were many families struggling to get by, and I was warned [to] not co-operate or inform on someone who rustled a log or caught a fish illegally. The police, he said, were there only for serious crime."[2]

Scanning the harbour, Oliver lit an unfiltered smoke and considered how lucky he had been to meet so many fine or distinct characters at a time when lives could end so easily. His life had spanned war and peace, through sweeping social and technological change, and he sensed how quickly it was accelerating.

○ ○ ○

In Liverpool, new faces emerged in early November 1941 as the *Empress of Asia* prepared for what was to be her final voyage. Although the galley, bridge, and deck departments still had many old-timers such as trimmer-turned-storekeeper James Jackson and sanitation engineer John Drummond, the young bloods included Maurice Atkins and 18-year-old Ordinary Seaman William "Mac" McKinnon. Youngest of the lot, though, were the two 15-year-olds — laundryman Edward Hughes and Leonard Butler, officers' boy and member of the catering crew. Both were imbued with a sense of adolescent invincibility; however, they would soon realize how dangerous it was in the Merchant Navy.

The engine room and stokehold suffered a mass exodus when three-quarters of the crew left and were replaced with more Liverpool Irish and five husky trimmers described by Geoffrey Tozer as tough "French-Canadians": Roger Denommee, R. Laberge, J. Cyrill Leger, Hector Millette, and Robert Tougas. All under 20, the men arrived November 5 with limited deep-sea time, mostly spent in the stokehold of the former German cargo ship *Chemnitz*, captured in 1939 and renamed the *Saint Bertrand.*

Young and impressionable, Geoffrey Tozer was taken with the brawny lads. "They were a very tough bunch and the Scousers were scared to death of them," he recalled. "The French-Canadians had quite a reputation … and a lot were good at the old Quebec art of kickboxing. There were stories about them being put aboard us to take some of the wind out of the sails of the Liverpool trimmers, but I think we also got them because we were a Canadian ship."[3]

Denommee, the oldest of the lot, had left his father's farm at Newbrook, northeast of Edmonton. The tall, powerfully built lad had recently begun shovelling coal when a stern message arrived from his grandfather advising him to "get the f--k off the sea because it was too dangerous."[4]

In the *Asia*'s galley, Anthanese Lorrain, originally from Oka, Quebec, knew nothing about shovelling coal but shaved 10 years off his age of 60 and signed on as a kitchen porter in which his wage would help support his wife and 16 children in their Alberta homestead. In November 1940 aboard another ship, Lorrain was employed as a galley cook but managed to escape death after the old French freighter *Lisieux* foundered in an icy gale. Deemed as unsuitable for transatlantic service, the ship was 280 miles east of Newfoundland in Convoy SC-13. Men died in the water while others succumbed to hypothermia in storm-battered lifeboats,[5] so he felt much safer being on a larger ship. As a French-speaking porter, Lorrain mostly stuck to the galley and saloon, although his culture, language, and Alberta background was in common with the five trimmers. The influence five men had on a stokehold of 118 firemen and trimmers could not have been huge but given Tozer's impressions it likely disrupted the mainly Liverpool Irish culture.

More notably, given the previous engine room troubles, it appears no administrative or structural change occurred to correct the situation. Still lacking was strong leadership to maintain discipline and resolve important departmental issues that could adversely affect the entire ship. At the same time, it was increasingly difficult to find skilled, dedicated men to work below. "The social structure aboard the *Asia* was [similar] to all British ships from the beginning of the age of steam," and to Tozer there appeared to be "a great deal of class distinction between departments and the various ranks within the departments. There was very little love lost or social interaction between them.... Each group lived and ate in a separate location.... Sounds awful, but that's the way it was and the strict hierarchy was considered essential for the good running of a ship."[6]

He recalled that Captain John Bisset Smith "kept very much to himself," and when it came to socializing, his tight circle was limited to the chief engineer, chief officer, or chief purser. Deck and radio officers and pursers formed their own group, while the bosun, carpenter, and storekeepers constituted another. Then there were groups for the quartermaster, lookouts, boy seamen, and deck crew. This extended to the engine room where officers, non-officer ratings, oilers, stokers, and trimmers formed cliques. "On the lowest rung of the social ladder were peggies [messmen] with senior stewards and cooks sticking together, followed by lower cooks, storekeepers, cabin stewards, and waiters — all pretty much keeping to their own kind."[7] The peggies cleared tables in the firemen's and seamen's messes and sanitized the heads. "They and the Galley Boy formed the lowest strata of society aboard a ship." Among the junior stewards was 17-year-old Bill Taylor, who despite his age had already scrubbed hallways and polished silverware crossing the Atlantic on the *Empress of Britain*, earning £3 19s per month.

Cadet Maurice Atkins was aware of the contrast between cadets who worked for Canadian Pacific and those on British ships. The latter were "cheap labour" and sometimes even had to pay to work the ships. "The CP was very, very good and treated us junior officers [cadets] well — as gentlemen."[8] Among the deckhands there was little interaction between Canadian and British crew, a conclusion that was supported by diaries and postwar interviews. None of these recalled much warmth, even though they worked side by side.

Geoff Hosken, then 25, remained efficient and dependable. Most of his Canadian shipmates knew his father had been a popular, well-connected sea captain, and this likely gave the young able seaman the confidence to challenge authority and push himself to becoming a good merchant sailor despite his lack of sea time — one previous voyage and his 1932 treasure-hunting adventure to Cocos Island.

Like many his age, he displayed occasional bouts of bravado that did not always end well. Furthermore, while working hard at various tasks within the hierarchy, it was clear he preferred taking orders from fellow Canadians.

In the small cabin shared with three mates on the Shelter deck, Hosken relaxed despite the lack of privacy. He enjoyed the camaraderie and poker that could yield extra cash for shore leave. What seemed out of place for his chosen life on the rough seas was the flimsy school exercise book kept as a diary. Stamped in big black letters on the back cover was a list of six danger points for schoolchildren, starting with "Don't run across road without first looking both ways!" and ending with "Don't forget to walk on the foot

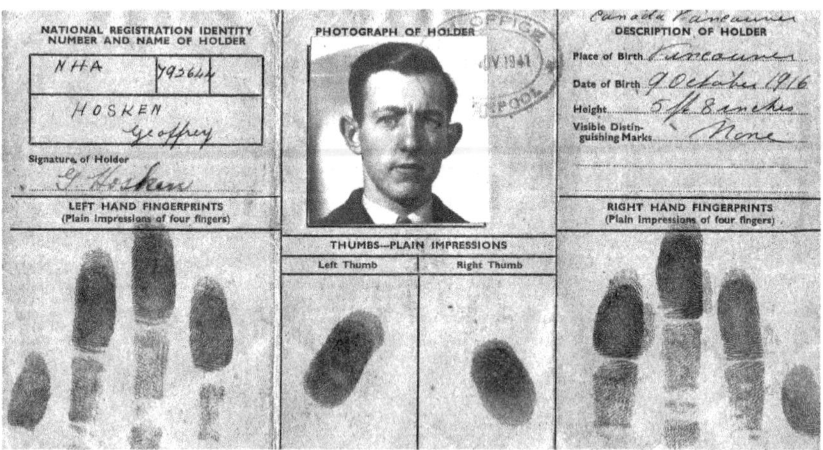

NATIONAL REGISTRATION IDENTITY NUMBER AND NAME OF HOLDER

NHA 793644

HOSKEN Geoffrey

Signature of Holder G. Hosken

LEFT HAND FINGERPRINTS (Plain impressions of four fingers)

PHOTOGRAPH OF HOLDER

THUMBS—PLAIN IMPRESSIONS

Left Thumb | Right Thumb

DESCRIPTION OF HOLDER

Place of Birth Vancouver

Date of Birth 9 October 1916

Height 5 ft 8 inches

Visible Distinguishing Marks None

RIGHT HAND FINGERPRINTS (Plain impressions of four fingers)

Able Seaman Geoffrey Hosken's ID book.

path, if there is one!" Hosken recognized the irony but was not concerned what others thought of his little blue-green scribbler that without fail was up to date.

On November 12, after a chaotic embarking of 2,410 troops, the *Asia* joined Convoy WS.12z bound for Durban. The vessel was overcrowded with soldiers who "had no idea how to sling a hammock" in such tight spaces. "A lot decided they were not happy and decided to leave, so armed guards were put on the gangways, but some even went down the hawser to the quay," recalled Bill Taylor.[9]

The convoy sailed with 16 ships and six escorts. Eleven carried troops while five were laden with cargo, including stores destined for Russia. Freighters occupied the vulnerable outer columns while the troopships held station in between. Only two carried more personnel than the *Asia* — the *Duchess of Bedford* off the *Asia*'s bow had 3,055, while the *Empress of Japan* on her starboard side carried 2,872. Overall, the number of personnel heading to war zones in North Africa, the Mideast, and Far East was over 20,000 with everyone hoping the escorts and ship's crews would keep them safe.

Anticipating rough seas, Hosken spent the first days clearing "scuppers," the holes in the bulwarks that allowed ocean spray and rain to drain. In the afternoon, during his time off, he mended a suitcase for a Royal Air Force officer who paid five bob (shillings), which he added to his poker winnings. On the 18th, while the ship rocked and rolled through stormy weather past the Azores Islands, he and a few mates leaned into wind and spray to corral lifeboats and tighten the gripes holding them against the davits. The importance of securing the boats was not lost on anybody. "One ship … nearly collided with us around 8 o'clock, also another had an explosion on board … but it wasn't enemy action."

While Captain Smith and Chief Engineer Herbert Owen were technically on duty at all times, about 60 percent of the crew maintained four-hour watches, which meant four hours on, four hours off standing lookout or performing deck duties. A sailor did not want to be caught napping, but scanning the grey horizon and keeping an eye on nearby ships was monotonous. In the sweltering heat, flakes of ash drifted down from the funnels, smearing skin and clothes. With water rationing in effect, Hosken sat in bathwater that barely touched the back of his knees and still caught hell for "using as much as a bucket of fresh water."

On the 23rd, the convoy split in two with faster ships taking lead. "Subs in the district all day & there has been quite a few depth charges set off, some being very close to us," wrote Hosken. By 8:00 p.m., the *Asia* was struggling and during the next two hours could not maintain 16 knots, dropping farther astern with one escort. She became a sitting duck, and by morning, the lead ships were out of sight.[10]

Although he may not have said much at the time, it was easy for Taylor to see what led to this. Some "firemen decided to break into the canteen and sample the stock [and] were then unable to work. I did not hear or see any disciplinary action being taken against them, they appeared to be able to do as they wished."[11]

Whatever the case, Hosken was glad to hear the anchor chains roll out for a coaling stop at Freetown — sweeping scuppers topside in the slight breeze beat going below to assist the coal barges, work that left him "wringing wet with sweat."

Also toiling through the oppressive heat were carpenter/joiners Chan Kam and Chan Lam, responsible for opening and closing the two steel exterior doors on the side of the ship leading to coal bunkers. "Their job had nothing to do with carpentry," recalled Atkins. "Once the coal was loaded, they had to make sure the doors, which were a perfect fit, were secured — watertight — with bolts. Both were quiet men, and most of the time did not talk too much to white men. They chatted amongst themselves, and I think they felt happier that way."[12]

Hosken observed more stupidity, nasty behaviour, and arrests marring their arrival. As the pilot boat came alongside to deliver the man responsible for guiding the ship into harbour, "soldiers or firemen threw bottles down at it." Next, a couple of firemen leaped into the sea and were nearly carried away by the tide.[13] Then, to complicate matters, on the night before the *Asia* embarked for Cape Town, a couple of monkeys purchased by firemen at a bazaar were stolen aboard. "Evidently the animals could be re-sold at a handsome profit," Tozer was quick to point out. "But on this particular trip there was a bit of a panic over some monkey-spread virus and the local military police joined in the search and found two animals. The monkeys were put down and two or three crew members logged & fined quite heavily."[14]

The shocking news from Pearl Harbor reached the convoy the same day as the surprise attack — December 7. For many of the long-time crew, it triggered memories of the September 1940 "accidental" bombing off Japan. Surely now the Americans would become fully engaged in the war.

While off the southwest coast of Africa, troops had no way of knowing how ambitious and timed the Japanese plan was to not only hit Hawaii but move with lightning speed on Asian and other Pacific possessions of Britain, Holland, and the United States. The first near-simultaneous attacks included full invasions of the Philippines, Hong Kong, and the Malay Peninsula, the last eventually extending south to the highly vulnerable north side of Singapore Island. But as far as the *Asia*'s crew and her troops were concerned, their destination was the Suez, and their conversations mostly focused on Pearl Harbor.

On the Malay Peninsula, Japanese forces had advanced with bloodcurdling speed from landings in Thailand and northern Malaya. Supported by artillery and tanks and equipped with bicycles, the Imperial Japanese Army rolled south while its superior aircraft bombed and strafed airfields and other ground positions. What was more, the Japanese were utilizing a weapon the Allies assumed would not be useful in Malaya — tanks.

Controlling the peninsula was crucial. However, the major port of Singapore, a vital creation of the British Empire, was top prize, and contrary to assumptions, Singapore was far from impregnable. Conquering it would give Japan strategic advantage with control of vital sea routes and natural resources, including oil, minerals, and rubber. It would also set the stage for further attacks on Sumatra, Timor, Burma, and Java, and not far away, across the Timor Sea, sat Australia and Darwin.

The assumption that Singapore was impregnable came from how the island had been fortified with batteries and gun emplacements during the interwar years, done to ensure it remained Britain's bastion of defence in the Far East. For several years, "Fortress Singapore" had been a deterrent against aggression that could also threaten Australia and New Zealand. However, while serving as chancellor of the exchequer in 1924, Churchill advised Prime Minister Stanley Baldwin that war with Japan was highly unlikely in their lifetimes. He was not the only one, but the assumption led to reduced defence spending on the island.[15]

Measuring approximately 26 miles east to west and 14 miles north to south, the island was connected to the mainland via a causeway over the narrow Johore Strait. Positioned across the island were a handful of Royal Air Force airfields, while the north side was home to the British naval base, centrepiece of His Majesty's Far East defence strategy. The responsibility for overseeing the defence of the peninsula, Singapore, Burma, and Hong Kong belonged to Air Chief Marshal Sir Robert Brooke-Popham. However, on the ground in Singapore, Lieutenant-General Arthur E. Percival commanded the Malayan and Singapore defences.

Overall, those defences aimed to thwart a seaward attack with little attention paid to establishing strong perimeter lines, particularly on the north side along the strait. The feeling shared by many was it would be impossible for an advancing army to resupply and maintain a fighting edge while cutting through dense jungle, only to come up against the strait. There were old racist attitudes that the Japanese were inferior human beings and incapable of jungle warfare, even though they had proven effective in Indochina. The island's main land-based seaward defences, facing south over Singapore Strait, were 15-inch guns, three of which could traverse 360 degrees but were not very effective against a concerted ground attack.

While the Japanese infantry swept south, aircraft delivered devastating blows. On December 10, the Royal Navy lost its two capital ships — HMS *Repulse* and HMS *Prince of Wales* — sunk with tremendous loss of life only two days after sailing from Singapore. At the same time, brave crews of the Royal Air Force operating from airfields on the peninsula and island were fighting aggressively, sacrificing themselves to attack enemy ground positions, airfields, and ships. However, their aircraft proved no match for the faster, more agile, and plentiful Japanese fighters, including the Zeros flown by the Imperial Japanese Army and Imperial Japanese Navy.

That was the situation facing the Allies in Malaya when the convoy rounded the southern tip of Africa, 5,900 miles away as the crow flies. Behind the scenes in the offices and war rooms, the Japanese onslaught triggered emergency responses. For the British Ministry of Shipping, it meant immediately accommodating troop-reinforcement and resupply voyages to Singapore, decisions that determined the fates of thousands of soldiers and ships' crews.

After pounding through the Cape of Good Hope "rollers," the *Asia* and the rest of the convoy struggled into Durban where on the 18th the *Asia* narrowly avoided catastrophe while careening toward the larger *Empress of Japan*. Municipal officials in the teeming port extended greetings along with free transportation, dances, and concerts. Hosken estimated the overall number of military personnel at 50,000, with pubs being the most popular hangout.

Without a standard uniform, merchant sailors were vulnerable in civilian clothes because they looked as if they had not enlisted. Brawls and toe-to-toe encounters with knives and fists led to injury and arrest. "Went pub crawling for the evening & got on board safely," breathed Hosken with a sigh of relief on the 18th. "Got into bit of a brawl." There was more trouble the next day. "My mate got into another fight with soldiers because they were insulting us for being in civvies. And yet they [the soldiers] haven't been anywhere yet," he added.[16]

Besides bruised knuckles, split lips, black eyes, and hangovers, communicable disease spread easily where sex was for sale. Other relationships formed sprung from dances and concerts, but they, too, were fleeting. Christmas Day was a depressing time for fellows, especially family men like Chef Archie Gee. It was also the day, recalled Oliver, of the Allied surrender at Hong Kong.

As part of the rejigging of convoys to reinforce Singapore, Convoy WS.12z reformed with additions and subtractions. It now consisted of three convoys that departed Durban on December 24. The first, WS.12za, sailed for Aden and was expected to arrive on January 4; the second, WS.12zb, was scheduled to arrive in Bombay on January 6; and the third, DM1, was expected in Singapore in mid-January. The *Asia*, meanwhile, stood out as the orphan. After disembarking her troops in Durban, she commenced her 4,380-mile voyage to Bombay on January 1, alone, minus a few firemen who had been arrested, birched, and jailed.[17]

The mood, like the food, deteriorated across the Indian Ocean. Tempers flared in the heavy, wet heat, and on January 10, Hosken remarked how "somebody put poison in the firemen's soup, but only the cook and a couple of others were affected. Nothing fatal that I know of." Then, rather suddenly, Bill Taylor looked out and saw a "gray wall coming straight for us" and "everything went dark." It was a rogue wave, and the "ship hit into it and rose up as if it was climbing then came down with a crash and shudder. I went down to the dining saloon and saw half the dining tables had ripped the bolts from the deck … everything was chaos … pots and pans everywhere.… The forward watertight doors below the bridge were twisted like tin foil and several lifeboats were damaged."[18]

Hosken was chipping paint off the storm rails when land was sighted on the 15th. He had no idea how long the ship would remain at Bombay, but after exchanging poker winnings and pay for rupees, he toddled ashore to purchase a cheap suitcase, better food, and cold brew. Immediately, he noticed how the city's "very modern & well-built" facade slipped into congested streets and alleyways with "a terrific amount of beggars and filth" and "very few white people."[19] Suddenly, he had entered a world opposite to what he knew. The overwhelming majority of citizens were the same type of people often ridiculed in his hometown, and they stared at him.

Chef Gee's mind was on his daughter, Beverley, who would turn nine on the 20th, and so he arranged a celebratory cable, hoping it would reach her in time. How nice it would be to be home again at the kitchen table with his wife and daughters, son, Arthur Jr., and Fritz, his faithful German shepherd at his feet.[20]

On January 22, with a crew of 413, the *Asia* embarked 2,235 troops of the British 18th Division, plus military stores such as tanks, anti-tank guns, small arms, and ammunition meant for Chinese volunteers in Singapore. Commanding the *Asia*'s troop contingent was 51-year-old Lieutenant-Colonel James Dean of the 125th Anti-Tank Regiment. Among senior and non-commissioned officers were seven men who would be dead within two weeks, including 27-year-old Warrant Officer Second Class Ewen McKerchar.

Middle son to John and Christina McKerchar, Ewen, like both brothers, was born into the railway life at one of the stations on the Ballachulish line

Warrant Officer Second Class Ewen McKerchar with his father, John.

in Argyll — not surprising since his father worked for the railway. Life was good, and Ewen had a fiancée and a job as an automobile mechanic in a garage promising him a partnership.

Ten days before the *Asia* embarked her troops, the division's 53rd Infantry Brigade, 135th Field Regiment Royal Artillery, and 287th Field Company Royal Engineers had reached Singapore, but incredibly without artillery or transport. Weeks at sea had left men weary. They had exercised and drilled when and where they could, but many had not seen combat and so were not in top form. Nevertheless, the brigade was committed to battle as part of Malaya Command and were supplied with local weapons. That same month, the Royal Air Force in Singapore received a much-needed shipment of Hawker Hurricanes, which arrived in crates for assembly.[21]

Another convoy transporting the bulk of the 18th Division did not reach Singapore until the end of January. It consisted of American transports *Wakefield* and *West Point* as well as British ships the *Duchess of Bedford*, *Empress of Japan*, and *Empire Star* — vessels that soon saved many lives.

Among the soldiers crowding into *Asia*'s hot and stuffy lower troop quarters were Sergeant Stanley Catherall, Lance Corporal Harold Clare, Sergeant Benjamin Hilton, and 23-year-old Private Leonard Loudwell of the division's 18th Battalion Reconnaissance Corps. Three other twentysomethings were settling in: Lance Corporal Norman Foster of the Royal Signals, Trooper John "Jack" Hodgson, 5th Battalion, Loyal Royal Regiment, and Private Bert Needham of the Light Aid Detachment. Of those men, Hodgson was dead in 13 days while the others became prisoners of war.

The organized chaos at quayside bore no resemblance to the bright, patriotic displays in propaganda films. Among the tired, hurry-up-and-wait expressions were scowls of indignation directed at the rusty old ship some felt was ready for the scrapyard. Other men seemed alarmed or curious. Would she be safe and fast enough? Had she been cleaned and fumigated or was she overrun with rats and cockroaches, and what about the grub?

Looking on, Captain John Watts of the 18th Division shook his head and hoped for the best. Although the destination was unknown to the troops, much speculation followed them up the gangways.

In a report written after he was taken prisoner in Singapore, Watts stated the men began boarding early on the 22nd, although most accounts claim the long process began on the 23rd. It started with Dean's 125th Anti-Tank Regiment filling the forward troop deck. Embarking later, the 251st Field Park Company, 197th Field Ambulance, 18th Division Workshops, and 18th Division Reconnaissance Battalion (5th Battalion, The Loyal Regiment) were allotted space in the after end of the transport. Officers and warrant officers had better quarters, settling into the former liner's first- and second-class cabins.

The troops, observed Watts, "were fully conversant with life on a troopship" and "settled down almost immediately."[22] Dean, however, was seeing red and initially refused to board his troops owing to the ship's filthy condition, but orders were orders.[23]

For many, it was better than the dusty training camp at Ahmednagar, 124 miles east of Bombay, where daytime highs averaged 86 degrees Fahrenheit in January. "For three weeks we trained in the boiling sun, sweating about two stone [28 pounds]," recalled Catherall who reached Bombay on New Year's Eve, 65 days after he boarded a troopship at the port of Avonmouth. At Ahmednagar, "we were housed in ... huts made of mud and straw. What an experience! We slept on Charpoys, a wooden framed bed with a string base and inhabited by millions of bugs which came out at night and made sleep impossible."[24]

Once aboard, it seemed all ranks griped about things they did not like or could not find but wisely steered clear of crew, realizing it was best not to interfere with watches or duties. It had also become obvious how crucial it was for the ship — even though equipped with DEMS — to maintain speed and "station" in convoy. To John Watts, this combination was essential defence "against dangers lurking on, below, and above the oceans."

Along the pier in hot, unrelenting sun, baggage and stores were swung aboard quickly, although Watts was critical of the procedures. "This was due entirely to the standard of the crew who were by no means in the same category of men" who worked other transports. "Furthermore, the stokehold staff were of very poor quality...." While his impressions were accurate, the crisis in Malaya, Singapore, and Sumatra increased pressure on the urgent delivery of troops and supplies. Once again, there was little time to shuffle or remove cargo and uncover rat and cockroach harbourages for fumigation, a process that involved placing iron pots of burning sulphur to produce sulphur dioxide throughout the ship or using hydrocyanic acid. The poison worked, but the breeding rats and roaches knew where to hide. It is likely that Watts did not know the stokehold was undermanned and that the process of loading stores depended on crew and local stevedores.

The *Asia*'s defence capabilities, which no longer included her Bofors gun, was also troubling. "In view of modern air attacks, naturally this is not sufficient protection for a ship this size," Watts wrote, adding that Lieutenant-Colonel Dean appointed a gunnery officer and ordered the "siting of auxiliary weapons throughout the length and breadth of the ship."[25] This extra firepower included 14 Bren guns and crews from the 125th Anti-Tank Regiment and 10 Bren guns with crews from the 18th Division Reconnaissance Battalion. "All these guns were sited to fire in arcs, each gun covering the other with a complete A.A. [anti-aircraft] pattern to cover an attack by aircraft from any angle."

Installing this extra armament was difficult owing to the age and design of the ship, especially on the covered Promenade and open Fiddley. "To make this more inconvenient, a very great amount of space on [the Fiddley] was taken up by three funnels and a conglomeration of ventilators, davits and other in-board accessories which very much reduced the size of the arc of fire of all weapons with the exception of those mounted astern."[26]

To the east, the situation was worsening. Newly gained airfields and the absence of fast, reliable Allied aircraft and guns made it easier for the Japanese to commence low-level bombing of Singapore. On the peninsula, Japanese infantry continued to attack south with a take-no-prisoners fervour, finding ingenious ways to hit and outflank opposition. Within two weeks, four divisions were poised to cross the Johore Strait, setting up what Churchill later described as "the worst disaster and largest capitulation of British history."[27]

Born in Highbury, North London, Norman Foster had no idea where the *Asia* would take him. He had grown up without his mother, Florence, who died when he was three. Child-rearing fell to his father, Frank, who operated a motorcycle shop, and his sister, Dorothy. When Frank remarried, the family moved to Jersey in the English Channel where Norman finished school and landed a job as a bank clerk. At the outbreak of war, he added a year to his age, hoping to join the Royal Air Force. However, without a reply, he altered course and trained as a wireless operator in the Royal Signals.[28]

Jack Hodgson experienced a tougher start in the miners' town of Egremont in northwest England, bordering the Irish Sea. He was the eldest of three boys raised by his parents, John and Hannah. Years of inhaled mining dust prematurely caused John's death, forcing Hannah and her sons to vacate their lodging to make room for another miner and his family. Jack

Norman Foster served as a wireless officer with the Royal Signals.

then worked as a machinist in Lancashire before he and his brothers entered the service, spread among the army, navy, and air force.[29]

Leonard Loudwell was also steeling himself for the voyage. From West Norwood in southeast London, he, too, wondered what lay ahead. The young private had joined the army in December 1939, but his facial scar did not result from training. He had suffered a nasty spill on a bicycle before the war, and to his employer, the injury appeared too unsightly, so they fired him lest store customers be offended.[30]

Bert Needham was another tough boy, and if he had met Foster, the two might have talked motorcycles. A handsome fellow with jet-black hair and even white teeth, Needham may have proudly revealed the nasty scar from his derring-do on the "Wall of Death."

Growing up in a tiny house referred to as the "Iron Yard" at Bridge Gate in Derby, Needham watched helplessly as his mother, Ethel, in the late stages of pregnancy, battled the Spanish flu. A little girl arrived, but the pandemic claimed both her and her mother in December 1918. Bert's father, William, a tea company salesman, was overcome by depression and completely ill-equipped to raise his boy alone, so at age two, Bert ended up being shifted from one relative or household to another. "Great Britain was struggling with the aftermath of war and Bert was just another mouth to feed with very little money coming in," explained his niece, Jennifer Padley.[31]

At age 10, Needham moved in with an aunt and uncle, but when the couple's marriage failed, neither pursued custody. The strangers who purchased the house allowed him to stay in the cellar with a cot and gas ring, an arrangement that lasted a year. Out of school at age 13, he worked for a butcher and then two years later drove a truck carting sand, which allowed him to save for a motorcycle.

Needham's passion for engines and speed made a connection at a local fair where he and his buddies encountered the motorcycle "Wall of Death," a tall cylinder barely 20 feet across but wide enough for tornado-like stunt driving at 30 to 40 miles per hour. While his pals stepped back, Needham stepped up and had a go without a crash helmet, gashing his leg on final descent but landing a job with George "Tornado" Smith.

Then 17 months after receiving his call-up papers, Needham married his cousin, Edna. Before revving up his bike and returning to camp, he told her he would return the next morning, but when he failed to show, she went looking only to learn he had shipped out. Wartime secrecy meant no one could tell her anything; it would be three and a half years before they saw each other again.[32]

From Bombay on January 23, the *Asia*, emitting the usual columns of dark smoke, met a small convoy of troopships: the 5,955-ton Dutch ship *Plancius*, the 17,083-ton *Félix Roussèl* with its oddly square funnels, and the 11,275-ton *Devonshire*. Designated BM.12, the vessels tracked southeast along the western edge of the Indian Peninsula and eastward toward Ceylon. "It was quite evident by the 2nd day we were proceeding to the Far East, and this was confirmed on the fourth day of the voyage to be Singapore," noted Watts, adding "in all probability the ship" would be "attacked from the air."[33]

The shorter route to Singapore would have been due east through the Malacca Strait. However, with the Japanese firmly established on the Malay

Peninsula and in the air, it was far riskier than going the long way around to the Sunda Strait between Sumatra and Java. Regardless of the route, Walter Oliver, who stood near the aft-gun position with binoculars around his neck, accepted what most felt: "This convoy appears to be heading into the most dangerous area in the Far East."[34]

In the lounge beneath the dome, Lieutenant-Colonel Dean, the troop contingent commanders, and Chief Officer Donald Smith met for briefings. Although sticky, the weather was calm, and the spacious room made it easy to imagine what it would have been like during better times. Updates and situational reports continually streamed in, modifying decisions and orders for the troops. It was not a question of whether the ship would come under attack, but when and by how many. Everyone needed to know what their assigned duties were, and while some troops raced to defensive positions, the vast majority proceeded in orderly fashion to designated areas below. Meanwhile, guns remained loaded and manned around the clock, while firehoses were connected and run out from hydrants to vulnerable points along the decks.

Supervising the fire inspection crew fell to John Moloney, the *Asia*'s master-at-arms. His team was responsible for the checking in and timing out of 18 punch-stations at prescribed times. The legwork was usually assigned to boy seamen, including Geoff Tozer, who acknowledged it was not so bad while the ship was at sea, but it was "a very spooky job when she was in port — empty. I remember the last and spookiest spot was the morgue — the final station."[35]

As tension mounted, it was vital for everyone to remain vigilant, especially on watch. Heat, anxiety, and lack of privacy tried men's patience, turning familiarity and situations otherwise ignored into outward contempt. Hosken was feeling this tension between himself and bosun William McKeown. The former felt he was being overworked and underappreciated while the more experienced McKeown, a 64-year-old from Liverpool, had little time for anything above the usual griping.

While the ship was off Sumatra, Hosken asked to be reassigned from general deck duties to four-hour watches as a lookout. "He said OK, but wanted to know why & I wouldn't tell him [why]," explained the able seaman. "The reason is because he's got his Liverpool clique against me. Also, if he won't let me do the jobs that need doing up there & that I wanted to do, I don't want any part of the job. He hates me for being Canadian & so forth. So I am safer in the watch under the circumstances."[36]

This frustrated McKeown who, as the fear set in, was trying to manage several young, inexperienced sailors. From Hosken's point of view, the bosun was assigning him work that appeared less significant given the ship's location. Overall, the episode was indicative of the sour feelings between some Canadian and English crew on the *Asia*.

A seaplane aboard the light cruiser HMS *Emerald* was a welcomed sight on Tuesday, January 27, the day before BM.12 rendezvoused with Convoy DM.2, composed of the *Troilus*, *Malancha*, *Dunera*, *City of Pretoria*, *Warwick Castle*, *City of Canterbury*, and the familiar *Empress of Australia*.

Far to the north, Allied forces on the southern tip of the Malay Peninsula withdrew from Johore to Singapore Island. Then, on January 31, sappers blew up the causeway to slow the enemy that had already identified the island's weakest point of attack.

Remarkably, on that same day, Hosken faced a different threat before running for his life — chased from the mess by an apoplectic peggy wielding a butcher's knife. "He would have definitely run me through, too," complained Hosken, who did not point to any cause. "He's liable to get me in my sleep, but that is his only bet. I'll get him when we get back to England in my own way."[37]

The *Emerald* was relieved by two sloops — HMAS *Yarra* and His Majesty's Indian Ship *Sutlej* — plus the Royal Navy's heavy cruiser HMS *Exeter*. Supplying additional firepower were destroyers HMS *Jupiter*, HMAS *Vampire*, HMS *Danae*, and the Dutch ship HNLMS *Java*, but as minutes ticked by, the fear of attack grew. Below in the oppressive troop quarters, men were nervous and on edge but reasonably confident the escorts and the Royal Air Force would protect them. Farther below in the stokehold, however, many knew there was little chance of surviving a mine, torpedo, or boiler explosion. It helped to be lucky, but survival depended more on how much time there was to escape. Up near the guns, Hosken took the threat seriously. "Haven't seen any enemy yet. In the afternoon we had action stations, that is fire stations, guns, and A.R.P. [air raid precautions]. Myself and three others are below deck in the magazine ammunition supply."[38] Forward on the lower bridge, young Atkins was also on high alert, keeping an eye on the clinometer while remembering the old *Conway* motto "Quit Ye Like Men, Be Strong."

Early next morning as the *Asia* entered Sunda Strait, Ordinary Seaman Jack Ewart was in the crow's nest, just over 100 feet above the water. That was when he heard what he thought were aircraft engines. With his heart skipping a beat, he picked up the phone and reported it to the bridge but was told it was likely the engines of another ship off the starboard beam.

When the ships reached the northern end of the strait, several broke off and headed east to Batavia (today's Jakarta) to support the Dutch colony. The *City of Canterbury*, *Plancius*, *Devonshire*, *Félix Roussèl*, and last in line — the slow-moving *Asia* — steamed northward toward besieged Singapore along with four of the original escorts, *Exeter*, *Danae*, *Yarra*, and *Sutlej*. Of the troopships, the *Asia* carried the most military personnel, followed by the *Devonshire*'s 1,673.

High above, a formation of twin-engine Japanese fighter-bombers scoured the grapefruit-coloured sky and calm sea, paying attention to the islands and straits that forced convoys into single file.

16

VOYAGE OF DOOM

There was this [enemy aircraft] coming down our bow with his machine-gun blazing, and I never felt so helpless in my life.

— **Cadet Maurice Atkins**

Walter Oliver, to a fault, believed in not pointing fingers, privately or publically. He figured it was not his job to cast blame or explain *why* things turned out the way they did. It was better to leave it to historians who would no doubt interview survivors and pore over reports and oral histories. His job, he thought, was to treat the final part of his memoir as a diary, limiting himself to when and where it all unfolded.

○ ○ ○

Far below in the *Asia*'s mess and troop quarters, British troops sat around playing cards and swapping stories. It was early February 4, and while some rested — or tried to — lying in hammocks, others sat bolt upright, either too nervous or complaining about the heat or deplorable conditions. Looking around, Sergeant Robert Wigham of the 125th Anti-Tank Regiment was among those who recognized the filth but found glimpses of a once-glorious ship. "When you saw the saloons and the wooden paneling in there, you knew the *Empress of Asia* had been a lovely ship," he commented years later.[1]

While the ship continued north toward Singapore, Lieutenant-Colonel Dean met with the chief officer and captain on the bridge where quartermaster Ernest Punter had been at the helm. Dean and fellow officers were satisfied plans for disembarkation were shaping up, and the troops understood where they had to be in the event of an attack and were sufficiently familiar with boat and lifebelt drill. A big concern, however, involved the ship's structure: if it were bombed and set afire amidships, communication between the bridge and stern could be lost, making it extremely difficult to move men from one end to another.

As night wheeled toward dawn, the DEMS crew and those on the midnight to 4:00 a.m. watch enjoyed a sprinkling of starlight above the calm, dark sea. Off the *Asia*'s bow, 174 miles to the north, was the entrance to Bangka Strait. Oliver, who had earned the coveted master's certificate, could draw on personal experience while passing through the 134-mile channel. He knew the convoy was vulnerable and that it would be best to pass quickly. Wedged between Bangka Island and the east coast of much larger Sumatra, the strait was angled like a dog's leg, tightening to under nine miles at its narrowest point.

Sanitation engineer John Drummond was expecting the worst, so he packed a few valuables into his patrol jacket, hung within reach on the back of his cabin door. He had keepsakes such as his dog-eared discharge book, identity papers, camera, gold watch, and fountain pen, though if disaster struck, he would have to leave behind his latest paintings and music compositions.

Meanwhile, Hosken added an alarming entry to his little blue-green scribbler: "Wednesday, Feb. 4, 3 A.M.... Lagging behind convoy as usual and cruiser [*Exeter*] keeps signalling to pick up speed for we are endangering the whole convoy."[2]

Hours later, sunlight spilled across the Java Sea, offering a golden start to a morning with excellent visibility and light breeze. This did little to alleviate

Hosken's fears, since by then the *Asia* was well behind the closest troopship off her bow, the ex-French 17,083-ton *Félix Roussèl*. He and others recognized how visible the *Asia* was with her funnels and reflective dome — even with grey wartime conversion the Japanese could easily identify the old liner from her years on the transpacific run.

Moving steadily at 12.5 knots, the *Asia* entered the south end of the strait. Shortly before 11:00 a.m., as Hosken set about preparing mooring lines in anticipation of coming alongside at Singapore's Keppel Harbour, he detected above the ship's throbbing turbines a sound Jack Ewart thought he heard the night before: the drone of aircraft engines.[3]

Boy Seaman Geoff Tozer was also topside, working and enjoying the weather. He noted a few high clouds and all seemed fine — land was visible on both sides of the strait and the convoy was 326 miles south of Singapore. But high above were 27 enemy aircraft sliding in from the north — nine to each V formation, taking stock of the convoy's movements.

Recollections vary on what happened next, but Oliver, positioned near the stern with his binoculars, observed that a single plane from the formation attacked from an "altitude of about 3,000 feet. One stick of three bombs dropped straddling the ship. The marksmanship was excellent. The first fell fifty feet off our starboard side amidships, the second the same distance off the port side, the third being wide."[4] Most recollections state the attack occurred between 11:00 and 11:30 a.m. Tozer and Hosken agreed the alarm for action stations blared closer to 11:30.

Captain Watts of the British 18th Division thought the troops reacted well. Many had been biding their time on the upper decks, and while those assigned to defensive measures or gun crew positions remained topside, the rest "proceeded in an orderly & quick manner to their appointed places [below] & sat down at their mess tables. This movement was carried out in an exemplary manner" considering "the majority … had been since the outbreak of war without a glimpse of the enemy and were without battle experience." This, he added, was "in spite of the natural desire of everyone to gaze into the Air & remain on the decks to have a first-hand glimpse.…"[5]

Len Gibson of the 125th Field Regiment, Royal Artillery, was taking in the sunshine near the bow when he looked up and saw planes. No one

Captain John Watts of the British 18th Division.

panicked, and any remaining troops on deck obeyed orders and headed below. "Our ship being the last one and the poorest one — at the end of the convoy — they (the Japanese bombers) used us for target practice. The *Empress of Asia* bounced up and down and I watched spouts of water on either side."[6]

Cadet Atkins also had a good view from the bridge next to his "dog buddy," Ordinary Seaman William Roderick McKinnon, who at 18 was a year older. Both continued to report directly to much-admired Second Officer Cecil Crofts. "Relatively speaking, McKinnon was an old guy," quipped Atkins years later. "He was my dog buddy because if I wanted something done, he did it, and stayed with me most of the time, ready to help, and we were about to experience some dramatic moments." The upper bridge afforded an unobstructed view in three directions, while its outer wings enabled a look aft. "The attack came out of the blue … and the first thing we knew there were explosions all around us, but they didn't score any hits."[7]

Hosken "saw nine planes coming over at quite a height. Then there were eighteen.… I started forward to get lifebelt and helmet. Was going along the

Ordinary Seaman William McKinnon was part of the ship's requisition voyage and remained on board for the Suez and Singapore voyages. He was only 17 when he signed on in 1941 as a boy seaman.

starboard alleyway on A Deck [Bridge deck], just about the purser's office when bang, I thought the ship was broken in half."[8]

Running aft to assist with the three-inch gun, Tozer peered over his left shoulder and spotted two enemy aircraft descending to roughly 3,000 feet. In his wartime diary, he recorded both released bombs above the *Asia*'s starboard side with five hitting the water, three falling aft, and two landing off port side. The anti-aircraft gun responded, "but the range was set too low" and the aircraft veered off within seconds. He speculated the planes were not low enough for the six large-calibre Oerlikon guns and 10 Hotchkiss light machine guns — six of which were within machine-gun nests on top of the ship behind the bridge.[9]

Oliver witnessed red tracer bullets, which had obviously lost velocity, bounce off the aircraft while Watts observed several shots from the three-inch gun exploding "not far short" of the planes. It was clear the aircraft were out of range, but to Watts the greatest challenge was to suppress the impulse of Lewis machine gunners to open fire indiscriminately. "They had been told for months before hand, both when they were manning guns at home and aboard previous troopships that it was out of the question to effectively engage aircraft at more than 1,000 feet. This being their first experience of bombing, it would be superfluous to emphasize the impatience shown by the crews to fire at the enemy and the disappointment of all when a suitable target did not present itself...."[10]

Hosken made it to the three-inch gun but was sent below to retrieve shells from the ammunition bunker: "Two of us went down — it was very scary. Had some boxes of shells hoisted up and we were told to come out ... and keep a lookout by the gun." He remembered five enemy bombs striking the water. "Two were near hits. No. 2 and 2A lifeboats splintered by shrapnel. One piece landed beside [Ordinary Seaman] Mike Costello and one near the gun...."[11]

Sidestepping empty firehoses connected to the hydrants on the Bridge deck, Oliver felt the ship lurch and shake with concussion, but his initial assessment was that aside from split doors and cracked port windows, the damage was limited.[12]

Sanitation engineer John Drummond grabbed his patrol jacket and bolted from his cabin. Immediately, he went looking for damaged water pipes, noting that one bomb had dropped "within twenty feet of the ship, forward on the Port Side" with the concussion "smashing some of the large, square port glasses [windows] in the saloon dining room" where it also split iron tables in two. He recalled seeing nine bombers followed by another six driven off by "heavy A.A [anti-aircraft] fire and two of our own fighters [Royal Air Force Hurricanes], but this gave away our position as we were only 24 Hours sail from our destination."[13]

In the meantime, Gibson and Watts had noticed a slight buckling of steel plates above the waterline on the port and starboard sides.[14]

Donning helmets and observing from the bridge, Captain Smith, Chief Officer Smith, and Lieutenant-Colonel Dean watched as the *Asia*'s DEMS crew returned fire, bolstered by some of her assigned gun crew and gunners from the 18th Division. The captain stated in his official Commander's Report that the attack occurred at about 11:00 a.m. In addition to listing the *Asia* as the last ship in the convoy, he confirmed she "had been allotted this position on account of our steaming difficulties," adding that "the ship almost invariably dropping astern of station when fires were being cleaned. The speed of the convoy was 12½ knots."

The captain's report does not record the number or type of aircraft, just that the *Asia* was the only ship attacked with five near-misses. "The ship was badly shaken but received no material damage except that two … lifeboats were pierced by shell splinters. Shell splinters were also found about the decks."[15]

A separate account from the captain recorded 18 bombers at a height of 5,000 feet, and one dropped "a large number of bombs which all fell together" missing the ship. He judged the nearest to be 10 feet away. "The bombs exploded striking the water, sending up columns of water … which descended on the deck. We received the brunt of this attack, no other bombs falling near the other ships."[16]

Watts recalled a "great deal of good was gained … because after months of boredom … they [the troops] had now seen the enemy." He commented if the Japanese bombers struck again at a lower height, they "would experience the stored up energy of the 'man at the gun' and destructive spray of the weapons.…"

The *Exeter* and other escorts drove off the enemy with anti-aircraft fire, and within seconds, the last of the twin-engine bombers were heading south with Royal Air Force Hurricanes "on their tails," added Watts. With the all-clear, the *Exeter* pulled out of her lead position to check, like a mother hen, on her "chickens."[17]

As the dust settled, the two Smiths and Dean ventured onto the Bridge deck to assess the damage. Dean surmised the bombs were an incendiary type, a theory he did not have to prove to the chief officer, whose "white shoes were covered with a black substance thrown up by the explosion."[18]

Interviewed years later as one of the last surviving crew members, Atkins was unequivocal when asked about Hosken's diary entry regarding the ship losing speed in the Bangka Strait — the *Asia* maintained speed and position as last in line. "We were used to being tail-end Charlie and so it got so we didn't pay much attention to it, but by that time the old man [captain] persuaded the chief [engineer] that somehow or another you better keep steam up."[19]

Hosken, Watts, and many others saw it differently, pointing blame for loss of speed on the stokehold. "The bloody firemen came up when the first bomb was dropped, but the priest told them to go back down and [they] did," Hosken wrote that day. "They are nearly all Roman Catholics. Twenty-five soldiers are [now] down in the stokehold to each watch also. All the port [windows] have been ordered closed & it makes it very stuffy. It was a very scary episode & hope never to experience it again. Four of our fighters [Royal Air Force aircraft] patrolled the area but never attacked the Japs. I believe they won't sacrifice the planes unless the Japs actually start diving. Shows you the scarcity of British planes." Hosken did not identify the priest, but the man was likely a chaplain attached to the 18th Division. He also did not specify how many firemen left.[20]

But for Watts the exploding bombs in the water on either side of the ship "brought these people up through the iron gratings on to the Fiddley Deck from the depths of the Stokehold. This worried us tremendously" because the absence of men from the stokehold for a "half hour" … reduced steam pressure, and the ship "dropped considerably behind the main convoy."[21]

British captain William Harris, who was topside with a 25-man emergency fire squad, was also highly critical. "We were lagging behind the rest of the convoy because we had had some trouble with the [engine room] crew on board, and yes I like to say this because they were a lot of bastards," he noted in a postwar interview. "I expect you couldn't understand they weren't very willing to make any sacrifices and do their jobs because they were really the sweepings of the Liverpool streets, almost impressed men, I guess, because the real stokers of this ship had always been Chinese. And there were not very good conditions in the stokehold, but these fellows refused to work, and they came up … and we stoked the ship into Singapore."[22]

Besides remaining adamant about the ship's speed through the Bangka Strait, Atkins had absolutely no recollection of firemen abandoning the stokehold on February 4. "I'm sure if we had slowed down … Donald Smith would have mentioned it in his official report to the company. Everything seemed perfectly normal." It is "unnatural — in my opinion — for a deck hand — an able seaman — to know anything about what is going on in the engine room."[23]

Captain Smith did not mention the incident in his report, and Oliver, for reasons stated, steered clear of the controversy while chronicling his time on board, although his son, Nelson Oliver, said his father was aware of the problem.[24]

It is important to consider that in filing official company reports following a disaster, ships' masters often remained circumspect when it came to details that could jeopardize an insurance claim. As far as the behaviour of the firemen is concerned, it is worth remembering how terrifying that moment would have been — far below the waterline in an undermanned stokehold with a younger, much less experienced crew that seemed to lack sufficient discipline

British ambulance driver Jack Greenberg was topside when bits of shrapnel sprayed the ship. As instructed, he "ran down into the middle of the transport, but the Liverpool Irish crew ... they ran up — they were frightened — they ran up, and we had to go down and stoke the ship." Greenberg did not become an emergency volunteer stoker, but he recalled several troops did. He also remembered the *Exeter*'s captain or one of her officers using a loud hailer, imploring the *Asia* to increase speed. Gibson of the 125th Field Regiment, who preferred sleeping on deck rather than below, was startled awake by the same message, which in essence stated: "*Empress of Asia*, get a move on or we are leaving you."[25]

During the forenoon, the convoy split into two sections. Two of the faster ships, the *Devonshire* and *Plancius*, escorted by the *Exeter* and *Sutlej*, pulled away from the other three escorted by the *Danae* and *Yarra*. "Twilight until daylight ... everything remained quiet, with the convoy proceeding towards Singapore at twelve knots in clear tropical weather," noted Oliver.[26]

With no sign of the enemy on the evening of the 4th, Watts commented as "darkness set in we celebrated our good fortune at not being hit and [so] retired to our beds with the comforting thought the morrow would bring shore-based fighter escort and final protection as we docked at Singapore." At sunrise, the weather remained clear with placid seas. Unfortunately, the morning brought "many problems and catastrophe … the major one of which was the conduct of the stokehold crew."[27]

Overnight, a contingency plan emerged. In co-operation with the ship's captain, a crew of volunteers from the various units would report to the stokehold at first sign of trouble. "This [idea] was immediately put to the men of all units aboard; they knowing full well the situation in the engine room [and] volunteered in such numbers that we were able to choose the men who were accustomed to working at high pressure with a shovel." Watts witnessed a few officers volunteering to command and assist should the occasion arise, including a reconnaissance corps officer with marine engine experience who immediately went below to become acquainted with the machinery. "Such a ready response from the men for a task so strenuous, let alone dangerous in the event of a hit, was extremely gratifying."[28]

Meanwhile, the ship's catering department under Francis Wright remained steadfast in the galley. Chef Gee and the cooks had little time to think about home while the peggies cleared and washed dishes and mopped up spills. However, anyone hoping to use the heads would be disappointed because, explained young Bill Taylor, the lack of steam pressure made it impossible to flush.[29]

By morning, Watts was acutely aware that the *Asia* was "still lagging behind … and the gap growing. On the horizon, two black columns of smoke … gave us our first indication of the direction in which Singapore lay, but there were no sounds of battle, although we were now within 20 miles of our destination."[30] At 6:30 a.m., the troops left their messes to collect kit, stores, and rifles from the armoury in preparation of going ashore within three hours. "The men knew the danger, were expecting it, & it seemed from Officers on watch on the Bridge they were expecting it with enthusiasm." When the much-anticipated fighter escort failed to appear at 9:45, officers and men figured that "from the air, the situation must be quiet."[31]

By 10:00 a.m., the convoy's trailing section composed of the *City of Canterbury*, *Félix Roussèl*, and *Asia* approached the lighthouse on Sultan Shoal, 13 miles southwest of Keppel Harbour. The ships reduced speed to embark harbour pilots, but the one assigned to the *Asia* never made it aboard.[32]

"The ship is moving at eleven knots and is stern ship in the convoy," lamented Oliver, who frequently checked his watch. "At 10:05 our position is five miles from Sultan Shoals [*sic*]. At 10:28, nine enemy planes sighted flying at high altitude. Action stations immediately sounded, but the planes passed overhead without incident and the all clear given. At 10:37, action stations called as twenty-seven enemy planes are flying low and directly towards us. We are now in for a dive bombing attack; it is the worst and most accurate of all bombing."[33]

For the second time in 24 hours, the troops scrambled below. Settling into his unit's designated place in the belly of the ship, Sergeant Wigham sat with his mates and braced for a frightening experience. The men were in the dark with no idea how the attack was unfolding.

Watts, on the other hand, had a clear view: "They were flying in the same formation as the day before — our fighters were still absent. The formation crossed the convoy apparently without taking any notice … and it was thought … they must be British machines.… This false sense of security lasted one or two minutes … as suddenly our escort vessels opened rapid fire at five enemy bombers, the first of which already screaming down in his dive."[34]

First hit but not lost, the *Roussèl* managed to get a harbour pilot aboard. The bomb struck her port side, forward of the bridge, recalled Oliver, although Tozer thought the blast occurred about amidships, while some of Watts's men witnessed "great pillars of smoke from her stern." The *City of Canterbury* was also targeted, suffering damage to her steering gear from a near-miss.[35]

From the *Roussèl*, the enemy turned on the *Asia*, which according to Oliver, "at least gave the other two [troop]ships the chance to make port. The attack was fast and furious, one plane followed immediately by another. As each plane attacked … the bomb could be seen leaving the rack next the explosion followed by the concussion. All these aircraft attacked from forward in line with the bow at incredible speed."[36]

On the bridge, Captain Smith watched as a swarm of aircraft, flying at high and low altitudes, attacked from all directions. "All ships of the convoy and the escorting light cruiser [*Danae*] and a sloop [*Yarra*] opened fire at once on the invaders. Bombs started falling all around the *Empress of Asia* and it was evident the ship had been singled out.... Terrific concussions shook the ship and standing on the bridge it was difficult to know what were near misses and what were direct hits."[37]

Close by, Atkins and McKinnon figured they were about to die. "There was this [enemy aircraft] coming down our bow with his machine-gun blazing, and I never felt so helpless in my life," recalled Atkins. "He was heading straight for the ship — coming over the bow. All I could see was the machine-gun shells hitting the deck and bridge, and felt for sure one had my name on it."[38]

In the struggle, it is understandable why recollections vary on the timing of the attack and the number of bombs striking the *Asia*. Tozer was certain the first bomb slammed through Chief Steward Francis Wright's cabin and then exploded in the pantry, and that bombs two and three crashed through the glass dome above the officers' lounge before the next one hit aft of the

A basic depiction of the *Empress of Asia* illustrating, approximately, where three of the bombs struck the ship.

bridge. His next entries state: “Firemen left stokehold. No [water] pressure fire getting bad. Smoke streams along starboard side hid planes. Gunners fire at noise.” Hosken, on the other hand, records that “one steward got a belly full of shrapnel” while “one fellow got 7 machine-gun bullets.” The former was 23-year-old pantryman Douglas Elworthy, who had been with the ship since her requisition voyage. Also wounded was Chief Purser Brian Moran, who suffered severe burns.[39]

In Oliver’s notes, the enemy scored its first hit at 10:48 a.m. — seven minutes after the attack began — when an incendiary bomb crashed through the Fiddley deck between No. 1 and No. 2 funnels, exploding in the lounge, spreading flames, and inflicting casualties, mostly among the troops sheltering below decks. “The fire starting in the lounge quickly spread … and although fire parties sprang into action, there was no water available apparently due to damaged water mains.”[40] Among those killed were Captains Richard Dixon and Arthur Lawson, both of the 5th Battalion, The Loyal Regiment, and British Warrant Officer Ewen McKerchar.

Watts witnessed a bomb pass over the bridge before smashing through the Fiddley deck and exploding below: “To those Officers on the Bridge the missile was plainly seen, black & growing larger & it appeared as if it would strike the Forward Well Deck [Bridge deck on the bow]. The Ship’s Gunnery Officer & all light automatic weapons were still engaging the plane as it pulled out of its dive, the former firing a light automatic weapon out of the windows of the Lower Bridge.” The observant Watts also described an oil bomb weighing about 100 pounds falling with a “swishing sound.” Within seconds, two more struck, and in “a very short time the ship was belching smoke from a short distance behind the bridge to the third funnel.”[41]

Atkins, ordered by Chief Officer Smith to take a crew and activate the hoses at the hydrants amidships, discovered, to his horror, zero water pressure.[42] “Something had happened — in all probability the main pump or the fact that the hoses had been severed at a main artery by the first bomb,” documented Watts. “This was really a catastrophe & fire, unopposed, spread relentlessly … fanned by a following wind which enabled the flames to travel at a remarkable pace [and so] windows, bulkhead doors & ports were immediately closed [as this] was the only method of checking the fire left open to us. Shortly before this was done, we could see Officers & men, blackened & burned, being passed & assisted through the open windows of Cabins & Recreation Rooms on the Promenade Deck."[43]

Despite this, the *Asia* fought back. "From the start … our guns put up an extremely good and steady barrage," wrote Oliver. "The barrels of the Bren guns became so hot they had to be discarded and replaced…."[44] Hosken and Costello dashed to the ammunition bunker for the three-inch gun and nearly fell over each other passing shells up to the gunners. "I believe our 3″ gun shot about forty rounds. The ship felt like it was turned upside down with guns going off and machine guns. [When] we came up … we looked amidships and she was ablaze…. Every time one [a bomb] dropped we would crouch down. It was very terrifying." Ordered to report to the starboard lifeboat stations, Hosken could only reach the motorboat, since "all the rest were ablaze."[45]

"The dive-bombers swooped and leveled at about 400 feet to drop their bombs," recalled 21-year-old Costello, who had joined the Merchant Navy in Vancouver as a teenager. Between plumes of smoke, "we could see the pilot clearly in his cockpit and the red ball [circle] on the wing."[46]

Soon one of Lieutenant-Colonel Dean's worst fears materialized when explosions and fire knocked out communication between the bridge and stern, denying safe access and making it difficult to assess events as they unfolded. Below decks, Drummond joined a crew that had managed to aim a near-empty hose at a coal bunker fire. But it was a lost cause, and the sanitation engineer headed topside to assist with lifeboats.[47]

The smoke was so thick on the Fiddley deck that the Bren gunners had to relocate to the upper bridge railing facing the bow. "The concussion of all this firing set off many of our own rockets," observed Oliver. "Some … flew horizontal around the bridge deck, adding … to the confusion. The rockets with the parachutes attached also took off on their own, at the wrong time, and landed with all the wiring back on deck."[48]

By now, *Asia*'s Oerlikon guns were running out of ammunition and two Hotchkiss guns jammed, leaving the ship woefully lacking in firepower with just her three-inch gun and small arms remaining. The third bomb that crashed into the ship's officers' quarters behind the bridge spawned a fire that now raced up the walls into the ceiling. On the bridge, smoke filled the wheelhouse and the floor buckled from heat. Toxic levels of gas formed in the engine room, and pieces of burning ceiling dropped onto the men, prompting the captain to issue an evacuation order. But even with that, some members of the "black gang" remained at their posts.

By 11:25 a.m., the captain knew the fires were out of control. "The saloon was burning fiercely, also the pantry, and cabins on A Deck [Bridge deck]." He also understood that the "fire pumps were working at full pressure," confirming the problem lay with the water mains. Every use was "made of fire extinguishers, fire buckets, and fire axes, but the flames spread with extreme rapidity...."[49]

Incredibly, troops remained in their messes as instructed. Many did not know fire was ripping through the upper decks. "Imagine for yourself," suggested Watts, "men sitting on their tables in the depths of the ship, some below the waterline, helpless should anything hit their particular section, listening to the reports of guns & the thunder & action of exploding bombs on and around the vessel." The men only became aware of the conflagration when plumes reached them through the ventilators, and "the heat of the approaching fire could be felt on the steel bulkheads."[50]

Sergeant Wigham was among them. No one, he recalled, sent word explaining that while the ship remained watertight, she was a floating inferno. More alarming, loud banging from the bow followed by the ear-splitting sound of steel on steel caused men to think the *Asia* was breaking up. The ship was by then without power, and they had no idea she was drifting toward a minefield. What they heard was the heavy clattering of corroded anchor chains as the portside anchor was deployed at 12:15 p.m., followed by the starboard one 15 minutes later. Captain Smith recalled that the ship gradually swung round to a position three-quarters of a degree east of Sultan Shoal Lighthouse and was anchored. All in all, "it was a pretty horrible experience," noted Wigham.[51]

But the men did not panic, although curiosity caused some to climb the stairs to the watertight door to find out what was happening. As Wigham reached the top, the door creaked open, and when the men emerged topside, they immediately realized why they were below. The fire ignited ammunition stored around the various gun positions. "There was ammunition exploding and buzzing around people's ears all over the place, but nobody came and told us that. It was our baptism of fire."[52]

Watts was quick to praise the actions of Chief Engineer Herbert Owen, who entered the noxious engine room to reduce steam pressure in the boilers — his experience, coupled with time as chief engineer on the *Russia*, served him well. "The Chief Engineer accomplished practically the whole of this dangerous & all most [*sic*] impossible task & furthermore reached his fellow Officers on the Bridge, in spite of the heat & falling embers of the now blazing ship."[53]

As the smoke and fiery debris swirled above the ship, the wounded, walking wounded, and unscathed mustered on the exposed fore- and afterdecks, hoping the relentless dive-bombers would leave them be. Lying among the injured on the foredeck were two army officers who received "the full blast from one of the incendiaries, one being burnt from head to waist plus open wounds," recalled Oliver.[54]

By noon, the lower bridge and chart room were ablaze. While the damaged stairs and ladders to the upper bridge and top of bridge were inaccessible, ropes tied to the non-burning port side allowed 36 men to escape. Those on the starboard wing could not see through the smoke to the ropes, so they faced a difficult choice: Jump 29 feet to the foredeck or 52 feet into the sea.[55] Cecil Crofts, among the last to escape, turned to Bill Taylor and told him he decided on the former. The young waiter warned him he would break his legs on impact, but Crofts went for it. "I watched him fall and hit the deck. He did not get up again."[56]

With the *Asia* on her chains facing northwest, the troops emerged from below to abandon ship, warned to stay clear of the minefield. On the bridge, Captain Smith did as expected and tossed the steel box containing classified documents over the side.

Exposed on the forecastle, men waited to go over the port side where Atkins and McKinnon worked furiously to lower long loops of thick mooring line to the water. Their action saved lives by allowing men from the forecastle to slide some 40 to 50 feet into the sea, although some suffered nasty rope burns to hands and legs. When it was their turn to go, Atkins and "Mac" slid together on opposite strands of a loop. As they did so, the crackling sound of burning wood, popping glass, and exploding ammunition rose above the roaring flames, devouring the bridge, Promenade deck, and deckhouses. "We swam around a bit — had our lifejackets on — and a boat came by and picked us up," noted Atkins. "I have no recollection from then until we reached shore."[57]

Jack Ewart and seaman Tim Cameron came to the aid of Cecil Crofts. With a puttee leg binding and makeshift leg splint, *Asia*'s second officer was lowered over the side. "I must have fainted because I hit the water then and the next thing I remember was two soldiers trying to hand me onto a raft," recalled Crofts. While the raft rescued other survivors, a Malayan torpedo boat came alongside and pulled off the wounded, including Crofts — all were taken to Sultan Shoal to await further rescue.[58] By then, Cameron and Ewart had boarded the *Danae*, which Captain Smith had joined by motor launch. In the meantime, Boy Seaman Tozer was portside, attempting to launch a lifeboat, when he glanced down and noticed badly wounded pantryman Elworthy clinging to a swamped boat.[59]

Exacting First Officer Leonard Johnston, who had admonished Atkins and Art Le Patourel for their loud music, also entered the water and made for shore. First Radio Officer Richard Farrell did the same but had a tougher go. The radio room had been badly damaged — his buddy, Second Radio Officer Patrick Harkins, died in the attack, although a third, William Lindsay, survived. Badly burned, bleeding, and in shock, Farrell destroyed his code books before going over the side — one of the last to leave. He jumped into the water, and when he surfaced, spotted a strange object heading his way. His first thought was a shark, but it was another man needing help with his life jacket. "Dad helped him to get it on correctly and they went their own ways," recalled his daughter, Deirdre. "They never met again."[60]

Scottish-born Fifth Officer Hugh McLean and Chief Barkeeper Bernard Bree were also injured. While helping to lower a packed lifeboat, Bree mangled a hand in a rope block (pulley). Costello, who was assisting him, reached for his knife, but not before another member of the crew, wielding a butcher's knife, severed one side of the lifeboat falls. The boat tipped, dumping its passengers into the water several feet below. Costello recalled: "It was just a sea of heads all around the ship. It was a bloody mess...."[61] Somehow, Bree and McLean reached shore, although with his injury, Bree's fate was sealed.

For many, the choice was to jump or climb down a rope ladder. Taylor, a non-swimmer, figured it was better to jump from the bridge than be consumed by flames, although he was haunted by visions of sharks attacking spoiled freezer meat discarded two days earlier in the Sunda Strait. As far as he could tell, the bombs had frightened them away. Wearing only his shorts and life jacket, Taylor jumped, feet first, plunging so deep he thought he would hit bottom until he suddenly bounced up like a cork.[62]

Wigham recalled how "the boats were no damn good, only one kept afloat while some burned in the [lifeboat] falls. We were told 'anybody who could swim — you've got your lifejackets make for the [Sultan Shoal] and leave the rafts and things for the lads who can't swim." The soldier followed his training. He removed his heavy boots and stockings, rolled up his gaiters, and placed everything into his boots that were left in a neat row next to other boots. His one mistake was not removing his steel helmet. The only item of value, aside from his life, was his pipe, tucked into a pocket beneath his life jacket. Wigham started down a rope ladder that looked like the newest bit of equipment on the ship. Halfway down, the weight of the descending troops caused it to stretch, and those who could not swim feared the worst. Before jumping, Wigham took a deep breath and pulled down on the collar of his life jacket to prevent it from slamming into his neck on contact with the water. He was a good swimmer, but as he plunged into the sea, the chinstrap on his helmet shot up around his eyes and ears.

Discarding the troublesome helmet, Wigham swam toward the shoal, rescuing a tall bombardier who could not swim. "He hung on to me like a six-foot streamer and I swam and swam. It was for over an hour, heading for the lighthouse." The crew of a naval cutter plucked both from the water, along with another fellow drowning in his life jacket, and after delivering them to the crowded shoal, returned to rescue more, along with several private craft. After landing in Singapore, Wigham felt around for his pipe; a puff or two would have taken the edge off, but it had jumped from his pocket when he hit the water.[63]

Non-swimmer Len Gibson also reached the city, wearing only a shirt. "I had never been able to stand cold water to be able to learn to swim." However, he discovered the water was warm, and to make himself lighter, removed trousers and socks. Swimming for his life, he wondered how long it would be before his cork life jacket became waterlogged. Advised to swim toward the pall of smoke above Singapore, some 10 miles away, Gibson was among hundreds in the water. He witnessed men showing others how to swim but often found himself alone. "All kind of thoughts came into my mind, wondering about how my parents would feel if they could see me in this situation … so I had to reach that smoke if I hoped to get back to them."

More and more injured and bedraggled men came ashore at Singapore, checking to see who else had made it. For the British troops, whether reasonably healthy or in agony, it meant waiting until they were dispatched to an army camp or hospital. Dripping wet, Gibson stood with only the shirt on his back and his saturated pay book in his pocket. When several well-dressed women arrived to offer assistance, the modest Gibson had to think fast. "The wind was rather playful and so I had to sit on my shirt [tails until trucks] came and took us away."[64]

Previously, around 12:45 p.m., according to Tozer's diary, Captain Wilfred Harrington of the Australian sloop *Yarra* manoeuvred his warship next to the *Asia*'s port quarter. It was a highly risky manoeuvre due to the minefield and unexploded ordnance on the troopship. Tozer, Drummond, Higgs, Hosken, Third Officer James Donnelly, Chef Gee, and Boy Seaman Owen Gillett, along with Wu Chiu, Chan Kam, Chan Lam, and Chan Tin Yau, were just some of the 1,800 crew and soldiers rescued by the courageous sloop, although estimates vary.[65]

With hundreds of survivors squeezed onto and below the *Yarra*'s narrow decks, the men had to sit still to help trim the ship. "We were being taken care of on the decks while she was still contacting the enemy with ack-ack fire," recalled Senior Second Engineer Herbert Stainton, a long-time CPSS employee.[66] Without the *Yarra*, many could have perished. Higgs recalled seeing "hundreds of spent ack-ack brass cartridges stacked like cordwood," and the determined *Yarra* was still pounding away when Hosken stepped

His Majesty's Australian Ship *Yarra*, the brave little sloop that rescued many.

aboard, clutching a canvas briefcase containing his identity papers and blue-green scribbler.[67]

"Before abandoning the [*Asia*'s] afterdeck, the ammunition for the six-inch gun and the remainder from the three-pounder had to be heaved overboard away from the mines. Only two lifeboats were intact," recounted Oliver. "Both ... were launched; one made it, one capsized."[68] With both legs badly scorched, he boarded the *Yarra*, but not before checking the *Asia*'s lower decks where a dedicated member — one of the Liverpool Irish — was sweating profusely from the heat and smoke. When asked why he was still there, the answer was he had not received the order to evacuate. As Oliver descended a ladder onto the *Yarra*, his binoculars wedged between the ladder and the *Asia*'s freeboard. The strap was ripped from his neck, but there was no time to retrieve it.

Crofts eventually reached Singapore on an Australian warship. "That was when I was lucky," he recalled, noting that an army doctor came aboard and "was putting all the wounded out with shots of morphine, but I wouldn't let him put me out.... I was the only one in the ambulance who was awake and believe it or not that saved my foot because I was taken from the ambulance into the operating room ... and believe me at that time they were just cutting off legs and arms [because] they didn't have

A rare photo showing the *Empress of Asia* burning off Singapore, February 5, 1942.

time to do anything else ... afterwards there were stacks of arms and legs in the hallway."[69]

By 1:00 p.m., the *Empress of Asia* remained a burning hulk that could be seen for miles. From the *Danae*, Captain Smith transferred to HMIS *Sutlej* and made a sweep around his ship "to make sure no one had ... been overlooked."[70] At 2:30 p.m., the captain and his chief officer again circled the ship before returning to Singapore to organize firefighting crews with floats and pumps. The plan was to remove some of the valuable military stores, including machine guns and ammunition, but when the crews approached the ship, the wind picked up and it became too dangerous.[71]

Remarkably, of the 2,648 men on the Singapore voyage, only 16 soldiers and four crew died during or soon after the attack (see Appendix 1: *Empress of Asia* Casualty Lists, 1914–42). At least 238 were injured, many severely. The enemy lost two, possibly three aircraft.[72] "Packed like sardines [on the *Yarra*], we crawled into Singapore while our ship was blazing in the distance," recalled *Asia* engineer William McArthur.[73]

Over the next several months, the War Office's Casualty Branch interviewed survivors in an attempt to locate missing men or update casualty details. Investigators in November 1942 spoke to Trooper G. Eaves of the

18th Reconnaissance Corp, who mentioned the "disciplinary trouble in the boiler room." He also helped confirm that Captains Dixon and Lawson were in the officers' lounge when the bombs hit. "If unhurt they should have been able to escape. But if stunned or otherwise injured by the bombs, they were unlikely to get out and would certainly be burnt to death." Eaves also stated he watched 20-year-old Trooper Edward McGrath disappear during a shark attack.[74]

The *Empress of Asia* was finished, slowly sinking, consigned to a shallow watery grave in one of the world's busiest ports. However, the suffering was not over.

17

ESCAPE FROM SINGAPORE

It was evident we had passed from a bombed and burning ship to a bombed, burning, and besieged island.

— **Captain John Bisset Smith**

Even if it were possible, 80-year-old Walter Oliver would never go back to reassure his 31-year-old self that everything was going to be okay. He would no more do that than expect the ghost of his future self to reassure him now in the final years of documenting his life at sea. As a practical sailor, he understood the importance of charting course, but what made it worthwhile was the lack of a guarantee, whether in the Strait of Magellan or off Singapore.

For the moment, a steaming cup of tea next to his typewriter — and his smokes — pulled him back to Singapore when he stepped off His Majesty's Australian Ship *Yarra* into a horrific situation, minus his binoculars.

○ ○ ○

Bolstered by tea, biscuits, and cigarettes, the *Asia*'s crew huddled at Keppel Harbour beneath the acrid smoke and thunder of guns. Four hundred and thirteen crew arrived with the ship, but some were missing, including First Radio Officer Richard Farrell, who unbeknownst to any of the crew, had found shelter in a Singapore bank, and deck hands Irvine Hampshire and Harry Worrall of the DEMS crew.

Among those accounted for: laundryman Edward Hughes, who was the youngest at 15 years, four months, and the oldest, bosun's mate William McKeown, pushing 65. Twenty-three were 17 or younger when they signed on, and at least 24 were in their fifties or sixties. Most sat tight-lipped, working on settling their nerves while others helped the wounded or escorted them to Singapore General. There was also no word on the status of barkeeper Bernard Bree, who had likely lost fingers lowering one of the lifeboats; injured Second Officer Cecil Crofts; and the more severely wounded pantryman Douglas Elworthy. Chief Purser Brian Moran's oil burns looked terrible, far worse than Walter Oliver's scorched shins, and so he was led to hospital by Oliver and Second Purser Harold Angus.

Encouraged by nursing staff to have his legs tended to, Oliver did not stay long, electing to rejoin his dishevelled shipmates, many of whom appeared accepting of the situation. Largely unscathed were First Officer Leonard Johnston, nicknamed "Bertie," James Donnelly, Drummond, Ewart, Gee, Higgs, Hosken, and James Jackson. Quartermaster Ernest Punter, who had been at the helm on the way in, was also there, as was Tozer. Close by, an island unto themselves, were the four quiet and dependable Chinese: Chan Tin Yau, Wu Chiu, and brothers Chan Kam and Chan Lam. "For tonight only," wrote Oliver, again opting for the present tense, "the deck and engine-room officers are staying at the Tanglin Club."[1]

The Tanglin, Singapore's exclusive private members club, was not their first stop. From the waterfront, "we were taken in trucks to a compound two miles north of the docks,"[2] recalled John Drummond, but two hours later, most of the men had entered the Seaman's Mission where they rinsed the oily grime off their bodies. By dusk, the hungry group entered the Tanglin, although others headed off to the YMCA or St. Andrew's Mission for Women and Children.

Maurice Atkins did not remember the waterfront but recalled being shoeless and drifting off to sleep on a Ping-Pong table with a life jacket as mattress before entering the Tanglin. "It [the Tanglin] was the be-all to end-all. When we went in most of us were in damp oily clothes — old clothes — and the high-class of Singapore were sitting there. They took one look at us and I could tell they were disgusted, wondering how we could be allowed in. They were having a dance or a party — all sitting there with their black jackets and long dresses."[3]

If Drummond was offended, he did not say. His opinion was more favourable, noting the men received a good meal and a bottle of beer each before settling onto cots for the night. Oliver just appreciated leaning against the bar "swapping tales of the day" with shipmates, overhearing someone say that during *Asia*'s brief deployment as a Second World War troopship she had logged 47,224 miles, although the distance was closer to 57,787.[4]

Stepping off *Sutlej* with burned and skinned hands, Captain John Bisset Smith accepted an invitation to spend the night in army barracks, although to "get to this camp we had to drive through an area of thick smoke and burning godowns [warehouses]. The whole island lay under a pall of smoke that rose from innumerable fires. The noise of gunfire from both sides of the Strait of Johore was deafening, and the viaduct [causeway] ... already blown up. It was evident we had passed from a bombed and burning ship to a bombed, burning and besieged island."[5]

Hosken and Tozer spent the night on the concrete floor of St. Andrew's Mission Hospital, hoping to sleep through an air raid. In the morning, during another aerial attack, Hosken and a few mates lucked out when they sought shelter in a customs house that offered food. Not far away, hunkered down in the bank, First Radio Officer Farrell had little or nothing to eat while staring blankly at useless piles of cash.

On the morning of February 6, Oliver learned from the hospital that one of the British officers severely wounded on the ship had died overnight and that Cecil Crofts would need to be fitted with a plaster cast. The compound fracture to his lower left leg required insertion of a wire through the bones of his heel.[6] In a letter written to Jack Ewart after the war, Crofts confided "they operated on me [but] they didn't make too good a job as now my left leg is three-quarters of an inch shorter than my right. I had a plaster cast on it for eleven months and I lost the use of my leg for a while but now it is pretty good and I get a forty per cent disability pension."[7]

Assisting with wounded at the understaffed and overrun hospital was *Asia*'s own surgeon, Dr. William McGinley, who had signed onto the ship in November 1941. A graduate of Queen's University, Belfast, he had been a pathologist in Sheffield before replacing the ship's previous surgeon.

Fed and washed, Oliver and others bedded down on cots in the musty cellar of the Seaman's Mission. "The air raids are becoming more frequent as the Japanese … come closer to the island. Not a pleasant prospect," he thought, shutting his eyes until something dashed across his woollen blanket. Whatever it was, he tried blocking it from his mind, but by morning, the men were staring at "two of the largest rats any of us had ever seen. These monsters had been our nightly roommates.…"[8]

By the next day, most of the crew, save for the captain, chief officer, and chief engineer, had been moved through heavy rain from one mud-filled army rest camp to another. No. 7 Reinforcement Camp consisted of huts and slit trenches. "It is a little too close to the Main Defence Line for comfort … rumour has it the Japanese are enlarging their strength across the strait facing the northwest side of the island. This makes an all-out attack on the island imminent."[9]

To Oliver, it appeared Japanese aircraft had unfettered access — flying over at will to bomb the island's aerodromes, something that had been developing since December and was now forcing the Royal Air Force to operate from Sumatra. "The harbour is another favourite target … and it is the reason most of the work gangs have discontinued unloading ships and leaving freight around the docks."[10]

By evening, word reached camp that 110 of the *Asia*'s firemen, greasers, and trimmers, in addition to trimmer-turned-storekeeper James Jackson and two others, were being repatriated. The men had been rounded up on February 6 and taken aboard the damaged *Félix Roussèl* and *City of Canterbury* bound for Britain. Among the evacuated engine-room crew were hard-working Liverpool Irish firemen Patrick Brown and James Cranny, plus greaser Wesley Glover from the Great Lakes crew. Held back were a further 13 who would prove helpful in the days ahead.[11]

Those remaining were shocked and more than a little outraged that certain stokers were among the first to evacuate. However, there was little time to dwell on the decision, since No. 7 was in urgent need of more slit trenches.[12] Beginning around 10:30 a.m. and continuing throughout the day, some 15 artillery shells passed overhead, followed by an all-night barrage.[13] "Nothing was dropped around us, but the ack-ack was going pretty steady," noted Hosken. "Everybody is very nervous and shaky."

While some clung to hope, it was not looking good. The enemy, with a lot fewer men than Singapore's defenders, seemed unstoppable, and that day, Japanese forces had invaded Burma. "This all adds up to Japan's conquest continuing at an alarming rate," wrote Oliver.

With overnight artillery support on February 8 and 9, Imperial Japan's 5th, 18th, and Imperial Guards Divisions began storming across the Johore Strait. "We are being shelled and bombing remains steady during daylight hours," wrote Oliver. "An ammunition dump was hit ... causing further explosions ... fires ... plainly visible on each side of the line." It was not long before the village of Ama Keng fell and the Tengah Aerodrome — west of the camp — was under attack. The enemy, which had also suffered numerous casualties, had gained a foothold on the north side of the island along the Kranji River, west of the destroyed causeway.[14] From Hosken's diary: "They [the Japanese] blasted our troops with trench mortars & came across the water, which is about 2,000 yards, under the arc of the shells.... Most of the soldiers I've talked to are very discouraged and they've really been through hell."[15]

Visiting the camp for the first time was the *Asia*'s captain with a request from Singapore's director of civil medical services. By then Smith, along with his chief officer and chief engineer, had moved from the army barracks to the famous Raffles Hotel. The request was for volunteers from the *Asia*'s catering crew to serve as cooks and orderlies at Singapore's beleaguered hospitals — an idea put forward by Dr. McGinley.[16] According to Captain Smith, the proposed agreement stipulated that remuneration would be through private arrangement and anything offered would be supplementary to seamen's wages. Furthermore, in the event of injury or death, compensation would match the scales established by the Ministry of Shipping.[17]

Amid the sound of guns and flying projectiles, 133 quietly accepted the deal. The group included 15-year-old Len Butler, assistant butcher Harry Hesp, Chef Archie Gee, laundryman Eddie Hughes, waiter Bill Taylor, and Chief Steward Francis Wright, as well as the old man of the group — Quebec-born Athanase Lorrain, now a survivor of two sinkings. All went into it hoping they would be doing something useful.

Life at the camp remained dangerous, though. Oliver guessed that by late on February 9, the enemy was just 10 miles from Keppel Harbour. "Spent a lot of time in our slit trenches again," scribbled Hosken that night. "Very heavy gun fire this evening. I must say we're not getting much sleep."[18]

In the morning, Oliver watched as the first Japanese shell struck the camp, spraying shrapnel into huts and over the men in the V-shaped trenches. The "situation is grim and deteriorating hourly, and only time will tell."[19] Most of the heavy shelling and aerial attacks targeted the waterfront as tens of thousands of panic-stricken civilians attempted to flee burning homes and businesses to reach evacuation vessels.

Singapore was falling, and at 4:00 p.m., the remaining ex-*Asia* deck and engine-room crew received instructions from British naval authorities to board trucks for the harbour where they were to crew three shallow-draft coastal freighters.[20] This marked the start of an escape that stretched around the world, but as the trucks pulled away from the camp, they left behind the 133 volunteers who had not yet left for work in the hospitals. In less than a week, those men, along with 10 other ex-*Asia* members, including Dr. McGinley and the hospitalized Cecil Crofts and Chief Purser Brian Moran, would become prisoners of war.

For years, that moment stayed with Tozer, who felt that with "a little thought and planning on the part of the Master and Chief Officer of the *Asia* ... the stewards [hospital volunteers] could have either been sent out on the troopship [as the firemen] or taken with us on the coastal freighters. I can still remember the look of bewilderment on the faces of a group of young stewards, sitting on sandbags around a gun emplacement, at the rest camp as they watched us climb aboard the army lorries on our way to crew the ships we had been assigned to. I also remember feeling that the Canadian senior officers, and we as Canadians, had let them down."[21]

Given the speed of the Japanese advance, anyone left at the camp was extremely vulnerable. "While we were waiting ... hospital trucks, motor-cycles, & equipment lorries [loaded with barbwire] came whizzing past both ways," noted Hosken. "I got talking to a soldier posted on the road ... and he said the Japs were only three miles down the road and would be here by 4 in the morning."[22]

The day also brought news that pantryman Douglas Elworthy had died from the shrapnel wounds suffered on the ship. "He was a Canadian steward & a fine one too," lamented Hosken before boarding a truck that evening. Amid the rush to escape, First Officer Donald Smith may have been the only crew to attend Elworthy's funeral at Bidadari Cemetery on February 9.[23]

After reaching a hotel used as a clearing station, the men awaited transportation to the docks where they acquainted themselves with the coastal vessels. The boats were nothing special, but solid enough for the *Asia*'s experienced crew to operate and repair if necessary. By name, they were the oil-fired 213-ton *Ampang* and the coal-burning *Hong Kwong* and *Sin Kheng Seng*, 207 and 200 tons, respectively. Longest of the three was the 118-foot *Ampang*.

The following morning, each crew began taking inventory, although there was not much to list. For the *Hong Kwong*, it was a matter of acquiring, by hook or by crook, charts, an Aldis lamp, and enough fuel and supplies to reach Batavia, 544 miles to the south.

Boarding the *Ampang* were 33 ex-*Asia* crew, among them Donald Smith, Leonard Johnston, Herbert Stainton, Arthur Le Patourel, William McKeown, William McKinnon, and Tozer, plus a handful of evacuees and stragglers from the Royal Navy and British Army. Among the non-*Asia* crew were: British lieutenant Patrick Spicer, his wife, Elizabeth, and a fellow listed only as Mr. Pratt. While Smith was the ranking officer, Spicer was technically in command, although Johnston became de facto leader. Stores were sparse, and Tozer was aware of just two bags of rice abandoned by the former crew. "There was not much water & just a few provisions the Spicers had brought so we had no food that night."[24]

The *Hong Kwong* would depart with 36 ex-*Asia* crew members — apart from Walter Oliver and John Drummond, the list included trimmer Roger Denommee, Chief Engineer Owen, quartermaster Ernest Punter, Chan Tin Yau, Wu Chiu, and deckhands/DEMS gunners Alfred Thurtel and Spencer Towler. Lieutenant S. Lockwood, Royal Navy Volunteer Reserve, would be in charge of the boat, although Oliver would assist with decision-making. Discovered later would be eight British military stowaways.

The *Sin Kheng Seng* listed 40, including five British naval stokers. Among her ex-*Asia* crew were Captain Smith, James Donnelly, Tim Cameron, Ernie Higgs, Geoff Hosken, Michael Costello, Owen Gillett, Maurice Atkins, and Jack Ewart, plus carpenter Chan Kam. Smith, as ranking officer, was in charge, but it fell mostly to Donnelly to take the coaster to Batavia.

Enemy aerial attacks continued to pound the waterfront, getting worse by the minute, and the prospect of navigating through heavily mined waters congested with dozens of vessels intent on escaping with civilians was viewed with alarm. While loading coal, Oliver's crew jumped for cover as machine-gun bullets ripped into the wharf and warehouses. There were, however, a few smiles when the men broke into a padlocked shed stocked with canned meat, fish, vegetables, cases of beer, and even a little stout.

While the *Sin Kheng Seng* loaded coal, 27 bombers released their payloads. "Every bomb was a direct hit except for a streak of them that cut across the water," observed Hosken. "Sheds and everything along the docks went up in smoke. It was absolutely horrifying."[25] Fearing for their lives, he and other crew members raced into high gear, loading 15 tons of coal in seven and a half hours, while along the wharf military personnel pushed new, recently delivered vehicles into the harbour to prevent them from falling into enemy hands.

Japanese infantry had advanced from the northwest, capturing the high ground at the town of Bukit Timah, some five miles from the harbour, intent on also taking the vital water reservoirs supplying the city. Frightened, shocked, and injured, civilians straggled toward the docks where Oliver noticed soldiers attempting to do the same, appearing to have lost contact with their unit: "Our instructions from the Naval Authorities were we are not, under any circumstances, to let any of them aboard."[26]

Once bunkered, the coasters were directed to remain at anchor inside the harbour's breakwater, although the delay was excruciating. "Waves and waves of planes have been bombing the city, docks, and anchorages since daylight," wrote Oliver. "It is low- and medium-level bombing, easier to take than dive bombing. Many small craft tied up three abreast alongside and inside the breakwater close to where we are were bombed and set afire." Fortunately, they were not holding passengers. In the midst of the violence, the crew of the *Hong Kwong* witnessed a mother and two daughters, minus luggage, in an old sampan desperate to board a larger vessel.

Awaiting departure on February 11 was the 12,656-ton, 524-foot-long cargo liner *Empire Star*, her holds jammed with approximately 2,400 passengers and crew. The evacuees included nearly 1,600 Royal Air Force, British Army, and Royal Navy personnel; Australian Imperial Force personnel; and dozens of Australian army nurses. The diesel-powered ship also carried 10 members of the *Asia*'s crew, including two stewards and four engineers — immediately assigned to the engine room. One of the engineers, William McArthur, a dapper man from the Scottish town of Dunblane, came with diesel certification. Already a survivor from two previous ships lost to enemy action, McArthur was a no-nonsense sort of fellow, who before boarding the *Empire Star* helped render useless vehicles and machinery being left behind; while in the *Asia*'s engine room he always kept a large wrench handy in the event of trouble, and indeed, had swung it more than once.[27]

Shortly before 3:00 p.m., the *Ampang*, *Hong Kwong*, and *Sin Kheng Seng* weighed anchor and fled in single file past the breakwater and into the harbour at a top speed of six knots. Passing the *Empire Star*, Hosken stared at her, knowing she had taken on former *Asia* shipmates. "They gave us their good luck and we kept going straight out to sea ... clearing the harbour one stick [of bombs] came 30 or 40 feet from us."[28]

At one point during the frantic exodus, a massive explosion sent shock waves across the water followed by towers of black, fiery smoke caused by the destruction of the large fuel depot on Bukom Island, deliberately blown up to prevent enemy use. To the west sat the hulk of the *Empress of Asia*. "Once First World War auxiliary cruiser, then a Queen of the Pacific, latterly a troopship, and now a burnt out hull," observed Oliver. Still, it was a relief to be sailing again, even under risk of death.[29]

Tozer and others were too busy or frightened to spend much time looking at her as bombs exploded in the water around them. The terrific heat generated by burning coal had likely buckled her hull plates, and so rather than sinking right away, the *Asia* settled with some dignity, her funnels, bridge, and upper decks now much closer to the waterline. It was hard to say if she was floating or grounded, but the old Greyhound of the Pacific would never sail again.

° ° °

Back in Singapore, the catering crew staffing the hospitals faced the sickening realization of a losing battle. Casualty lists grew by the second, and across the island, reports similar to those heard earlier from the peninsula told of soldiers, Chinese volunteers, and civilians brutally executed after surrendering.

At Tan Tock Seng Hospital, the *Asia*'s assistant butcher Harry Hesp, working as an orderly, felt lucky to be alive after a bomb exploded in his ward the first day he started work. He helped carry twice-wounded survivors to Singapore General where more gruesome scenes unfolded. "The dead and dying were everywhere.... They were being brought in on rickshaws, motorbikes, and anything else. The fighting was only a quarter of a mile away." While carrying the dead into hospital mortuaries where the remains "were three and four deep," he entered stairwells streaming with blood and passed a room crowded with young, shell-shocked soldiers in straitjackets. With sniper activity increasing, it was too dangerous for outside burials.[30]

Bedridden in the same hospital, with his left leg elevated and immobile, Crofts was still without his plaster cast. He could hear screams and bombs exploding outside and wondered when the enemy might show up.

Young Len Butler smelled the carnage before he even reached the hospital. The staffing was not good. Many local workers had left to rejoin families, while Allied military nurses, who had worked tirelessly, were under orders to evacuate, although many did not want to leave. Entering a large ward, he noticed beds full of the critically wounded, and beneath each, a somewhat less serious case. It was nauseating to sop up blood and help pick maggots from festering shrapnel and bullet wounds. Beyond the wards, the burned and bleeding cried from hallways, laundry rooms, kitchens, and balconies, and like Hesp, Butler would never be able to shake from memory images of the dead in the non-refrigerated morgue.

Both were at work on Valentine's Day when Japanese soldiers entered Alexandra Military Hospital. With fixed bayonets, soldiers went room to room killing unarmed medical staff and patients, including one man under anaesthetic on an operating table. This was followed by the cold-blooded murder of dozens of people after they were marched out of the hospital. While helping a wounded soldier remove his clothes at Singapore General, Hesp could not get the patient to hand over two hand grenades. "He wouldn't give them to me.... He said one of them is for them when they come and the other one is for me."[31]

○ ○ ○

Holding their breath as they passed single file through the minefields to distance themselves from concentrated aerial attacks, the crews of the coasters made reasonable progress. On the bridge, Oliver set his watch knowing the sun would set before 7:30. For navigation, the crew of the *Hong Kwong* would have to rely on old methods because the pivot on the boat's magnetic compass had failed, likely through concussion. Steering by the stars and gauging shoreline positions by "rule of thumb" would have to be the norm. At least the weather was clear.

Sunrise February 12 exposed more than the Durian Strait south of Singapore with the islands, and low silty coast of Sumatra, to the west. The overcrowded *Empire Star* had joined the exodus overnight, and while clearing the strait came under attack by dive-bombers. Direct hits killed 13 and severely wounded 37. As nurses tended to the wounded, others raced amid falling bombs to extinguish fires and get the ship, which had lost power, underway again. "The second officer in charge of the larger anti-aircraft gun lost an arm," recalled Oliver.[32]

Not surprising then that during the chaos, the eight stowaways who had managed to sneak aboard abruptly emerged from the *Hong Kwong*'s cargo hold. There was little to do except carry on, although the vessel suddenly had 45 mouths to feed.

As they settled in for their second night, the crew's attention quickly focused on the *Ampang* converging on her starboard side. The semaphored message was that she had no charts and would have to follow the *Hong Kwong* to the Musi River. She was running low on fuel, so her plan was to leave the other boats and head upriver to the oil refineries at Palembang, Sumatra. In darkness, however, the *Hong Kwong* lost all contact with the *Ampang*. Attempts at locating her by switching on stern lights and blowing a whistle were to no avail.[33] On the afternoon of February 13, the *Ampang*, shortly before reaching the Musi, hailed the *Sin Kheng Seng*, which dispatched a small boat to deliver charts. "The old man [Captain Bisset Smith in the *Sin Kheng Seng*] told them [the *Ampang*] to bugger off because the planes would be back," remembered Hosken.[34]

Now on their own, the *Ampang*, with Tozer as quartermaster, devised an emergency plan that likely saved lives. The vessel's previous native crew had left behind clothing, including "coolie" hats, sarong-type skirts, and loose white tops. "At any sign of a Japanese approach two of the crew were to don these clothes and take over the bridge while the rest hid [under a tarpaulin] on the afterdeck," recalled Tozer. At about noon, a two-seater Japanese floatplane appeared, and the plan was set in motion. Wearing the native garb, Chan Lam was particularly brave, putting himself at great risk while remaining visible to the aircraft.

However, while the men huddled beneath the tarpaulin, one of them — a military man nicknamed "Mad Hatter" — attempted to get out on deck with a rifle and shoot the plane. The fellow was so empowered by shell shock that it took three men to hold him down and pull him back under the tarp. At one point, the aircraft was so close the men could see the goggles on the faces of the pilot and observer, who had his machine gun trained on the vessel. Fortunately, Chan Lam remained cool, and the Japanese fell for it. "The whole thing only took about five minutes, but it was the longest five minutes of my life," admitted Tozer.[35]

The two other coasters, meanwhile, continued south and were making progress until 10:00 a.m. when a lot more enemy planes arrived, flying in the direction of Palembang. Oliver witnessed one attack "two oil tankers anchored at the mouth of the river. We are unable to see if any of the bombs hit. Then, a solitary [plane] flew low right over us. We had everyone under cover … and the plane left us, probably deciding we were a native crew." Drummond wrote that everyone except Lockwood, who was steering the boat, took cover, and when the danger passed, the lieutenant took a long swig of whisky and then handed the bottle down to the grateful crew.[36]

During this time, dozens of vessels had been passing the coasters while others moved more slowly. On the *Sin Kheng Seng*, the crew watched a bomb explode on the afterdeck of a ship that had overtaken them. "We passed her and could see they were beaching her," wrote Hosken. But surviving the gauntlet and limping into Batavia with her wounded was the severely damaged *Empire Star*.[37]

On the night of February 13, Oliver's boat passed a hospital ship "lit up like a Christmas Tree" appearing to be at anchor. A fleet of fast-moving ships — too hard to identify in the dark — easily overtook them.[38] By midnight, under harsher weather, the *Hong Kwong* reached the southern end of Bangka Strait and barely survived a gale that forced all hands "to go to port or starboard to keep from turning over."[39] Hours behind on the *Sin Kheng Seng*, Hosken noted it "rained very hard during the night. Ship slowed down to about 3 knots. The coal seems to be all slack & no life in it."[40]

Having finally found the silty mouth of the Musi around 6:00 p.m., the *Ampang* picked up a pilot and began heading upriver to the refineries. She had just started in when the Musi closed to traffic. Unknown to her crew, enemy aircraft were about to seed the sky with paratroopers intending on capturing Palembang and the refineries. As Tozer described, "It was a very dark night, raining and quite cold. During the trip we anchored several times owing to breakdowns and poor visibility. There were no lights for the binnacle so we steered by flashlight which was very dim."[41]

Senior Engineer Herbert Stainton remembered the vessel had no cargo or ballast and "we had to trim the ship to counteract bends in the river by moving passengers from one side to the other."[42] When the *Ampang* dropped anchor in the middle of the river opposite the refinery at 8:35 a.m. on February 14, she was down to 12 gallons of fuel. The crew was looking forward to a little ration break, but as they discussed the situation, their words were lost beneath air raid sirens, the scream of Japanese fighter aircraft, and the drone of heavy air transports.[43] Looking up, Stainton saw at least 200 paratroopers with chutes supporting two or three men.[44] Tozer figured there were at least four times as many.

With no choice but to abandon ship, the party reached shore by lifeboat and then fled on foot through the outskirts of Palembang. Wondering whether they would survive the next few minutes let alone the day, they commandeered an old bus and sped off with engine-room storekeeper

Senior Second Engineer Herbert Stainton.

George Fraser amped up behind the wheel. They were heading south, aiming to catch the ferry at Sunda Strait to Java. However, within an hour, the bus rolled to a stop, halted by British soldiers who urged the party to get out and walk to a railway station, likely because the vehicle might attract enemy aircraft fire. The "dedication and guts" of the soldiers calmly remaining at their posts while the island was under attack impressed Tozer. "They were sure that [General Sir Archibald Wavell, commander-in-chief of American-British-Dutch-Australian Command] was about to arrive and lead them to victory," he recalled.

The group remained alert while trekking through the jungle known for large carnivorous wildlife and poisonous snakes, all the while listening intently for enemy aircraft or other activity. Twice they had to scatter and dive for cover as machine-gun bullets ripped through the canopy. Reaching the station, they ducked into a warehouse and discovered, to their delight, cigarettes and brandy. It was still February 14 when they settled into a railway baggage car, hoping to arrive at a village for the night.[45]

Tozer woke up the next morning in a thatched hut. Staring outside, he rubbed his eyes, convinced he was dreaming. A few yards below him was "a whole bevy of naked teenage girls bathing in the stream — a sight that would have gladdened the heart of any adolescent boy at any time in history."[46]

Elizabeth Spicer, the only woman among the evacuees, also left an indelible impression, especially on Tozer. "Mrs. Spicer was a very good looking blonde in her mid-twenties. She was a Vancouver girl who had married Patrick Spicer a few years before." In civilian life, Patrick, who was not to survive the war, had been a rubber planter. "She had learned to speak Malay and her ability to talk to the local population in their own language [between Palembang and Batavia] was a big contributing factor to our successful trip overland after we left the *Ampang*."[47] Stainton was also very appreciative, noting she had the presence of mind to pack food before leaving Singapore: "Our hunger was appeased by Mrs. Spicer who produced some tinned goods which helped to make meals for the people on our boat."[48]

○ ○ ○

Lieutenant Patrick Spicer and his wife, Elizabeth.

To the east, the *Hong Kwong* and *Sin Kheng Seng* straggled into Batavia the following day, while to the north the British surrendered Singapore. "All of a sudden one night the shelling stops — all is peaceful and quiet," remembered Len Butler, who along with other hospital staff was herded into a lecture hall. "Later on at night … there was a thumping on the door … and somebody went to open it. In marched a Japanese officer with an escort of four soldiers [with rifles and] fixed bayonets.… The officer was very smart — never said a word — just marched to the centre of the room … had a quick look … then marched out. Placed a guard on the door and that was it. Next morning they herded us out … into a big field [the Singapore Pedang] near the harbour.… Soon they got us into columns and we started marching to a place called Changi."[49]

○ ○ ○

Unexpectedly, Chief Officer Donald Smith's fluency in Scottish Gaelic proved useful — even on the southeast coast of Sumatra — when he convinced a somewhat reluctant Scottish ferry captain to transport the party across the Sunda Strait to Java. From there, the *Ampang* party boarded a train and reached Batavia around 2:00 p.m. on February 16; by then Palembang had fallen.

So close, but they had been lucky. Hundreds of evacuees never made it to Batavia. Bombs and bullets killed and wounded many, destroying or rendering useless many ships, including the small freighter *Vyner Brooke*, knocked out near Bangka Island. Among the captured, more than 20 Australian nursing sisters were marched into the water that day and machine-gunned, in what has become known as the Bangka Island Massacre. One survived.

The date of the massacre coincided with the surrender and capture of the 1,020-ton HMS *Mata Hari*. A former cargo ship requisitioned as an armed patrol vessel, she had left Singapore with 483 on board, including crew, and many women and children hoping to escape. Lit by searchlights from Japanese warships, the *Mata Hari*'s captain decided he had too many civilians to fight back. He ordered the destruction of all documents and important equipment, since the vessel had no choice but to steer under Japanese escort to Muntok on the northwestern tip of Bangka Island. Chief Barkeeper Bernard Bree, still nursing his mangled hand, as well as William Marlow, Thomas Roberts, Herbert Smallwood, and Lawrence Thompson, were on board and among hundreds who became internees. Many met their deaths in Sumatra while thousands more captured in Malaya and Singapore suffered similar fates.

Over and above the tens of thousands of Allied military personnel taken prisoner during the fall of Singapore along with scores of civilians, the *Asia*'s prisoner-of-war list totalled 154 and included the injured Crofts and Moran, Chef Gee, kitchen porter Lorrain, Dr. McGinley, Butler, Hesp, and Hughes. Private Bert Needham, the "Wall of Death" rider, had been fighting through the Malayan jungle when shrapnel struck him in the chest.

Luckily, the impact was light, and after extracting the smouldering metal, he dressed the wound himself and carried on but was later captured. Lance Corporal Norman Foster, Private Leonard Loudwell, and Sergeant Robert Wigham, minus his pipe, were marched off to prisoner-of-war camps, along with Captain John Watts, who wrote that his time on the *Asia* was "a trial which has no parallel."[50]

For those who escaped on the *Ampang*, *Hong Kwong*, *Sin Kheng Seng*, and *Empire Star*, perilous ocean routes lay ahead, fanning out in opposite directions from the Dutch East Indies.

18

HOMEWARD

He was a fine seaman in his own quiet way, perhaps the bravest man I have ever known.

— **Boy Seaman Geoff Tozer**

In his 85th year, the last of his life, Walter Oliver did not see himself as especially wise. He felt worn out and confused by the pace of progress, although he could easily feel the restorative powers of a decent handshake or warm smile. It is also true he accepted life as a fight, beginning at birth and carrying on with every breath until the last. The secret came in knowing when it was good to leave a bad situation, and more importantly, to believe it was acceptable to do so. "He who fights and runs away will live to fight another day" was advice he followed from the ancient Greek statesman Demosthenes.

As he sat down to describe his homeward journey from the Far East, Oliver remembered the challenge of arranging passage from Batavia on rusty flat-bottomed coastal boats not designed for deep-sea travel. Coupled with those frustrations had been the fear of more Japanese attacks, as well as depressing thoughts on the darkening fates of crew left in Singapore, and the missing *Ampang*.

o o o

Rain slickened and with naval escort, the *Hong Kwong* steered through the minefield and into the crowded harbour in Batavia on the morning of February 15. Surprisingly, a "Dutch agent boarded with eight cases of beer," recalled Oliver. "He had not much to say except he was of the opinion that unless the American Pacific Fleet came to the rescue Java is doomed. This was wishful thinking when so many capital ships had been sunk or damaged at Pearl Harbor."[1]

Later that evening in continuing rain and lightning flashes, Captain John Bisset Smith, Jim Donnelly, and crew anchored the *Sin Kheng Seng* just off Batavia. By morning, she was out of coal and needed a tow into harbour. En route she "dipped" her ensign to a few navy ships and sloops, including the brave little *Yarra*, then minutes later those on board were relieved to spot the *Hong Kwong* loading 45 tons of coal — her crew bent on carrying on to Darwin, Australia. At least that was the plan before word came that the northern Australian port was no longer open to shipping. Regrettably, this was after they had already loaded the coal. The *Hong Kwong* was not going anywhere.[2]

Being stranded was made far worse with Dutch restrictions on outdoor movement — enforced with a 6:30 p.m. curfew. Hosken described their temporary home as a bank crowded with grizzled and distressed seamen from around the world: "The Dutch officer in charge of the building gave us quite a mean talk … also said his guards would shoot if we tried to leave."[3]

Maurice Atkins believed the strict measures were in response to some bad behaviour: "Some of our guys couldn't resist the women and they went [out] and the local people were not happy and so they locked us up. It was very comfortable there, very nice, but we were placed behind [steel] bars" until moved the next day to a monastery or Catholic hall.[4] In the midst of this, on February 16, the crew from the *Ampang* arrived — minus the

Ampang — but with an adventure to share. Facing them all were more days of uncertainty, but it seemed a miracle that the crews from all three coastal vessels had reached Batavia.

The question of what to do next was answered after two crew members paid a courtesy call to the British consul. "Get out of Batavia" was the advice. "If worse came to worse," noted Oliver, "we could try sneaking the ship [*Hong Kwong*] out of the harbour in the middle of the night, go through the Sunda Strait, then head for Ceylon. This, however, would be a last [resort]."[5] But there would be no chance of that; on February 18, the vessel was seized and her crew removed and taken to a holding camp by Dutch authorities. Oliver discovered this when he returned to the dock after meeting with naval authorities. The *Hong Kwong*'s seizure impacted the fates of many — none more so than two ex-*Asia* crew who had returned to the harbour only to find the vessel gone. Left behind and on their own, Able Seaman Herbert Kennedy and deck boy James Towers eventually became prisoners of war. Kennedy survived, but Towers became a casualty of forced labour on the notorious Burma Railway in 1943.[6]

As the ex-*Asia* crew worked to secure passage to Australia, including offering themselves for crew responsibilities, the strange but true news came that Captain Smith and Chief Engineer Herbert Owen and two other engineers had left on the Dutch steamer *Plancius*, bound for Colombo. Some had also heard that Chief Officer Donald Smith had done likewise, but that rumour was false. He remained in Batavia, but three other crew members who had reached Batavia on the Straits Steamship Company's *Ipoh* also joined the *Plancius* when she sailed on February 16.

○ ○ ○

Back in Singapore, *Asia* hospital volunteers Len Butler, Archie Gee, Harry Hesp, Bill Taylor, Francis Wright, and others had joined the throng of civilians marching out of the Padang, the sprawling outdoor complex home to cricket, tennis, and lawn bowling venues. The captured soldiers, meanwhile, were headed to other prisoner-of-war camps, and many ended up on the Burma Railway.

The procession of guarded civilian prisoners walked under the blazing sun along the East Coast Road toward Changi Jail near the present-day airport on the east side of the island. The march took several days, with stops at Katong and at least one other place, and Hesp was among many who fell victim to severe prickly heat rash. Along the way, he witnessed the rounding up of Chinese singled out for execution, ditches strewn with bloated bodies swarming with flies, and severed heads on posts serving as a warning to looters. "It was all death," he recalled, believing that relatives or friends of the dead had dared not go outside to claim and bury the remains.[7]

Bill Taylor managed to scrounge three tins of beetroot that helped to sustain him. "Quite often we heard rifle shots at the rear of the column. I, in my wisdom, kept to the front. The day got hotter … and being fair skinned my face started to blister." He estimated it was at least 20 miles to Changi. "[The jail] was most imposing; high exterior walls and as you entered by the main gate, you went through a narrow area and through another set of gates on the inner wall. There were guard turrets on the walls, not a very inviting place."[8]

Metal steps led Hesp to the third floor where a nine-by-twelve-foot cell in B Block was shared with the manager of a rubber plantation and a rather "cultured gentleman" who owned a palm-oil plantation. Female prisoners occupied a separate block. To the best of Hesp's memory, there were four blocks or sections containing dozens of cells each, and in the centre of each cell was a stone block with a stone pillow. In one corner was a primitive toilet minus a seat. There was no heating, and cellmates competed for the right to sleep on the raised block, which meant others slept on the floor. Bedding came in the form of rice sacks, which aggravated Hesp's heat rash and contributed to infestations of bedbugs that crawled from the walls and floor, especially at night.[9]

Work involved the planting of sweet potatoes or tapioca in arid soil that could barely support growth. Parties of men uprooted trees for firewood, since the jail had stopped using oil to fire the stoves. Food in the form of overcooked rice or weak soup was just edible if one did not stare at it too long. Overall, Changi, where the ex-*Asia* crew had earned the nickname "Asia Boys," was better than the camps overcrowded with soldiers. Lance

Corporal Norman Foster, who had escaped the *Asia* only to be captured, worked the docks at Keppel Harbour until moved to Ban Pong, Thailand.[10] Bert Needham also ended up there, joining other Allied prisoners of war and up to 250,000 Southeast Asian civilians in forced labour. "Every twenty miles or so the camp had to be moved to keep up with the [Burma] Railway's progress. The camps weren't fenced [because] the surrounding jungle was the fence and escaping prisoners never got away. Each prisoner was given a pint of rice a day...."[11]

In his early days at Changi Jail, Len Butler hauled sacks of rice and chopped wood, likely because he was young and relatively healthy. To cool down and wash the vermin and grime off their bodies, the prisoners were permitted to bathe in the ocean near the camp, but one day while staring at the sandy beach, Butler noticed human remains, realizing only later he had been washing in a place where many Chinese prisoners had been executed.[12]

As *Asia*'s senior officer at Changi, Cecil Crofts willed himself to get better and hobbled around with his leg finally in a cast, occasionally cursing the inconvenience. Lack of vitamins caused his eyesight to deteriorate, but he could stand or lean against something while working in one of the kitchens. For Crofts, his men came first. There was no way to know how long he and other lads would remain at Changi, so it was important to secretly compile a list of the 143 men, complete with dates of birth and addresses of next of kin — information shared with a Royal Navy representative. There was little outside news, although some likely trickled in on smuggled or makeshift radios concealed within the camp.

Chef Archie Gee, who was pushing 50 when he entered Changi, did whatever he could to stay alive for his family back in Vancouver, since there was no way for him to explain what had happened. He could not imagine what anguish and worry Flora and his children had been going through, but from Cell 2-2-14, Gee kept faith, sharing the small cold room with two British-born civilians — Murray Jack, who up until capitulation had been employed by the Registrar Supreme Court in Singapore, and Arthur Allen, traffic manager at the local Ford Motor Company plant. The chef treasured his pocket King James Bible, which he read to himself at night thanks to a battery, a piece of wire, and a tiny light bulb he had squirrelled away. With its dim light, he found comfort in chapter and verse, interrupted only by snoring and the footsteps of guards. It was risky because if there was anything that could get a prisoner executed, it was possession of a radio receiver or anything resembling one. The Japanese listened carefully for transmissions, and cells were "tossed" often. During one raid, guards stormed into Gee's cell and turned the place upside down, but when they flipped his sparse bedding, the battery, light bulb, and wire somehow remained hidden.

○ ○ ○

At dusk on February 19, sanitation engineer John Drummond and electrician George Meilke found themselves aboard SS *Jalaratna*. She was the slowest in a convoy that departed Batavia, and by the 20th, after Japanese forces landed 620 miles to the east at Bali, the 3,942-ton vessel was passing through the Sunda Strait for Colombo. Two other cargo vessels with shipmates aboard were heading in the same direction. The name of the first was the *Hai Lee*, and on board was Maurice Atkins and several other ex-*Asia* crew members. The second, the 3,204-ton *Wuchang*, carried at least three from the *Asia*, plus tons of munitions.

However, most out of Batavia were ocean bound for Fremantle in Western Australia, crammed aboard the banged-up *Empire Star*, the 7,475-ton *Marella*, and the flat-bottomed coastal vessel *Whangpu*, which was overloaded with refugees. On the *Empire Star* were five from the *Asia*, including engineers who assisted during the hellish voyage out of Singapore. Ernie Higgs, Geoff Hosken, Chan Kam, Chan Lam, Donald Smith, and Herbert Stainton were among the 41 ex-crew members who boarded the *Marella*, while several offered their assistance on the *Whangpu*, including Walter Oliver, Leonard Johnston, and Geoff Tozer, who in addition to maintaining a diary had taken to writing poetry, such as this one entitled "Lookout at Sunset":

> Into the mist horizon the sun dissolves
> Turning the western clouds dull red, then pink
> Till grey descends, then black and

one by one the stars come out to blink — a night warning.
Except on the horizon. Far — Where the sun has set
The cold grey blue of evening lingers, Yet
On the fo'c'sle head the lookout then begins to think —
That the days bring nothing but the sunset
But a light leads, tho the sun may Sink.
Half an hour later it is dark
He calls himself a fool and starts to think
Of some girl in Liverpool.

To the northwest, it took the *Jalaratna*, the *Wuchang*, and the *Hai Lee* 11 days to reach Colombo. It was mostly good weather, but all three were under constant threat of attack. Those aboard the *Wuchang* had a dramatic story to tell after they reached Colombo on March 3, one that made them thank their lucky stars they had crossed on a flat-bottomed boat. "The first intimation we had of the attack was two torpedoes speeding under our shallow craft," stated Able Seaman Alexander Murdoch in a letter home, shared with the *Dumfries and Galloway Standard* beneath the headline "U-Boat Scared by Seamen." A submarine, he explained, surfaced near the vessel to "survey" her kill, but her commander must have been surprised when the *Wuchang* became the aggressor and steamed straight for his sub. "I was at the wheel and I was determined that the submarine should go up with us if the bombs which we were carrying were exploded by torpedoes or shellfire," he added. "The submarine submerged as quickly as she could and we made off at full speed."

Expecting the sub to resurface, Murdoch and crew covered the *Wuchang*'s half-dozen life rafts with a tarpaulin, through the centre of which they positioned a flagstaff at an appropriate angle. "When the periscope poked up astern of us there was an efficient-looking four-inch 'gun' leaning over our stern," continued Murdoch. "The attacker took note of this and vanished."[13] While the crew swore it happened, some, including Atkins, wondered why a submarine would waste torpedoes on such a small flat-bottomed boat. One explanation was perhaps the sub knew about the munitions. Regardless, Tozer remembered that for years the tale made its rounds in a Vancouver seamen's club where the yarner typically claimed it as "the gospel truth."[14]

It was over two weeks before Atkins and others embarked on the *Nieuw Amsterdam* for Durban, arriving on March 28, and a further 12 days later before the party could secure passage on yet another ship, the *Narkunda*, for the four-day voyage to Cape Town. Meanwhile, Captain Smith, Drummond, and seven other ex-*Asia* members departed Bombay on March 14 aboard the *Strathnaver* and alighted in Cape Town 12 days later. It had been a long and arduous crossing, but at least the men now resembled regular passengers, having acquired new luggage and clothes, including tailor-made suits and dress shoes purchased in Bombay before they left.

All was not 100 percent agreeable for Drummond, however. An entry in his journal implies relations between the ex-*Asia*'s captain and himself had soured considerably. "We occasionally met J.B. Smith and Owen," he wrote. "Captain only spoke to me once, asking for information." Drummond also noted how some of the *Asia*'s firemen, who had joined the *Strathnaver* in Bombay, were quickly ridiculed by British and Greek sailors and merchant sailors after the firemen lost a series of sporting competitions. More concerning was the health of a female passenger diagnosed with smallpox on March 25, which forced the vaccination of everyone on board a day before the ship reached Cape Town. This meant that those wishing to go ashore faced a mandatory 14 days at a quarantine camp, although somehow Drummond managed to be released after only five days on the ward.[15]

○ ○ ○

Already in Fremantle, the *Empire Star* was joined by the *Marella* in late February. For the *Empire Star*, her arrival was particularly bittersweet, since two weeks earlier, shortly before reaching Batavia for repairs, she had solemnly committed her dead to the sea.

During the *Marella*'s crossing to Australia, the bad blood between Hosken and McKeown continued. According to Hosken, as he rested beneath a saloon table, McKeown urged Chief Officer Smith into suggesting work for the able seaman, even though officers had no legal authority over crew since the loss of the *Asia*. "He [McKeown] certainly doesn't like me," wrote Hosken. "If mate [Smith] ever wants to hear me, I'll give him an earful...."[16]

Far more serious concerns were troubling the *Whangpu* as she neared the end of her 12-day crossing. Oliver had noticed rations getting low, and as bad weather approached, it became necessary to break into the lifeboat supplies. "At the latter part of the trip ran into a westerly gale. These Chinese riverboats are flat bottomed and although quite seaworthy, they can certainly roll. The passengers, however, are a hardy lot."[17]

Ravaged, with hungry passengers and several mysteriously missing boat cats, the *Whangpu* struggled into Fremantle on March 4. "As is the custom in all Australian ports the cats have to be rounded up and locked away [on board]," recalled Oliver. "We left Batavia with five. Now no cats [after searching the ship]. We found out they had all gone into the stew pot."[18]

○ ○ ○

Oceans away in Cape Town, 13 ex-crew sat tight until April 21 to board the Cunard White Star Line's 81,235-ton *Queen Mary* bound for Rio de Janeiro and New York City. Atkins was in awe. The *Queen* was massive — an oil-burner that made the *Asia* seem like a "piece of the past."[19] From one of the cabins, Drummond good-naturedly updated his journal, sketched, and committed scenes to memory, perhaps to paint later. And reminiscent of better times on the *Asia*, there was a grand piano — a Steinway — in the spacious lounge, which Drummond "had a shot at ... when nobody was around."[20]

Entertainment included afternoon picture shows, beginning with Joan Crawford and Margaret Sullavan in *The Shining Hour*, and two days later, the aviation drama *Night Flight* starring John and Lionel Barrymore, Clark Gable, and Helen Hayes. Such distractions likely helped passengers cope with the omnipresent threat of U-boats — at the time, the Kriegsmarine was winning the Battle of the Atlantic, with more than 600 Allied ships sunk between January and August. Sending such a massive troopship to the bottom would be a priority.[21]

The *Queen Mary* spent only a day in Rio. It was too risky to linger, and it was her speed, after all, that helped guard her against U-boats. So, fuel and water were briskly brought aboard and then on April 28 she steamed at 26.5 knots straight for New York City. Hot and clear weather turned cooler before a U-boat scare on May 6. "Some excitement this morning when we passed close to six lifeboats tied together one upside down," noted Drummond. "A submarine must be in the vicinity, passengers have likely been taken off. We are going full speed now; shortly after a submarine periscope is sighted. We kept going full speed, and at 7 PM another periscope was sighted and all the crew ordered to Action Stations. We zigzagged a lot, but nothing happened."[22]

Early next morning, more than three months after the destruction of the *Asia*, the *Queen Mary* majestically sailed into New York City. Train travel to Montreal with connections to Vancouver followed for Drummond, Atkins, and several others from the West Coast.

○ ○ ○

From Fremantle the *Asia*'s crew split yet again. Only one man, engineer David Somerville, elected to stay with the *Whangpu*, which became a submarine depot ship. Six, including Johnston and Tozer, filled vacancies on the *Mangola* bound for Sydney, while the majority remained with the *Marella*, reaching Melbourne on March 9. Oliver and three other mates spent a couple of extra days at Fremantle before shipping out to Sydney. When the *Mangola* party reached Sydney, the men happily embarked on the *Queen Elizabeth*, the slightly larger sister of the *Queen Mary*, steaming for New York City. Those arriving in Melbourne exchanged the *Marella* for a flophouse then a hotel before boarding the Royal Mail Line's 25,689-ton *Andes* bound for Halifax. "There have been a few games of deck hockey but that was stopped because it is a little dangerous," wrote Hosken, who confessed to having "quite a few beers and a bottle of wine" the night before, which made him "pretty high." He also noted his knee was bad, "caused from the wine, I suppose. Two of our boys are isolated in hospital with V.D."[23]

Off Halifax on April 17, the men on the *Andes* lined up and dropped their trousers for the usual "short-arm inspection" conducted by a Royal Australian Air Force doctor, then at 3:00 p.m. the next day, Hosken and the other Vancouver lads boarded a westbound train. "Had quite a good sleep in upper berth, & meals are very good compared to our regular diet before."[24]

o o o

Walter Oliver opted for a different route. Without any other ex-*Asia* crew, he boarded the 6,676-ton *Salamaua*, a twin-screw diesel capable of only eight knots, and crossed into the South Pacific, past volcanic islands into a realm that made him think of HMS *Bounty* mutineers. The cargo ship charted an easterly course to Fiji, then north across rising seas past Hawaii, and eventually through the Strait of Juan de Fuca, arriving in Vancouver by late June. After 10 days' leave, he hopped aboard a train to Quebec City and resumed the dangerous life of a wartime merchant sailor on the North Atlantic.

Geoff Tozer's time with Oliver on the *Asia*, *Whangpu*, and much later while crossing the Atlantic on the *Gatineau Park* left the young seaman from the Okanagan in awe of the fourth officer. "He was a fine seaman in his own quiet way, perhaps the bravest man I have ever known. He always seemed to know exactly what to do and how to do it."[25]

With wartime censorship in effect, details on the *Asia*'s destruction and fates of crew did not appear in the press until April when the men began arriving home. However, even then the news was unofficial because the Admiralty waited until May. One of the earliest reports, published on April 18 by the *Vancouver Daily Province*, appeared after the ubiquitous wartime dateline "An East Coast Canadian Port." It stated that survivors from the *Asia* were safe in Canada, heading home, while many others remained in Singapore as prisoners of war. Boy Seaman Owen Gillett described where and how the ship met its end and named the *Yarra* "as one brave boat."[26] A follow-up piece on April 28 named several survivors, covering their warm homecomings at the Vancouver train station. Among the men named publicly were Chan Kam, Chan Lam, Mike Costello, Jack Ewart, George

Fraser, Owen Gillett, Ernie Higgs, Geoff Hosken, Chief Officer Smith, and Herbert Stainton.

Ernest Punter and James Jackson had also made it home. Punter stepped off the train in a second-hand suit purchased for him in Halifax by the Red Cross. He had worn it to impress his parents waiting at the station. However, as Norm Hutton recalled, when the sailor reached home, there was a bill from the Red Cross for $13.50.[27]

○ ○ ○

As some families embraced returning sons and fathers, many other households endured long, lonely days praying for loved ones left in prison camps.

Quartermaster Ernest Punter arrives home in Vancouver wearing his second-hand suit.

For the prisoners themselves, the separation was excruciating. Held captive at Changi, Richard Farrell expressed this while writing home in November 1942: "Dear Wife & Bairns. So far I have received no letters from you and I am feeling very anxious. If possible write via International Red Cross.... I hope you received my previous card ... I am ok and in good health.... Looking forward longingly to the day I return home."[28]

In Vancouver, Chef Archie Gee's family had a longer wait. "It was over a year until my mother [Flora] got any word on whether Dad was dead or alive," remembered Gee's daughter, Beverley, who had turned nine just two weeks before her father entered Changi Jail. "I think it was a phone call, who it was from I don't know. Mom used to send five dollars every month to the Red Cross to send a parcel to him, and I know he got one in 1945.

First Radio Officer Richard Farrell as a prisoner of the Japanese.

She also sent postcards, something like twenty-five words on each … not very much."

Deliberately not included in postcards from home was the horrific death of Archie's 19-year-old son, Archie Jr., on January 16, 1943, although one mailing from Flora included the cryptic words: "Archie Sea." Like his father, he was a merchant seaman, working the freighter *Northolm*, running supplies along the coast from Alaska to California. The vessel sank in treacherous waters off the northern end of Vancouver Island and young Archie was among the missing. "At … that point we didn't know whether Dad was dead or alive, and even if he was in a prisoner of war camp it wasn't the greatest thing to think about. It would have been nice if we knew he had escaped Singapore, but that didn't happen. And then there was the death of my brother. For about two weeks Mom couldn't eat at all. I was sent to the store to buy a can of consommé. That was the only thing she could keep down. With Dad missing she had relied heavily on my brother. When he died it just about took her, too."[29]

Without a body, authorities listed Archie Jr. as missing and likely presumed drowned, so Flora and her two girls clung to the hope he could have reached shore and been suffering from amnesia. "It was the kind of hopeful dream families hang onto after such tragedies," offered his grandson, Dave Thackray. "But it was pretty clear he had perished."[30]

Meanwhile, the "Asia Boys" at Changi thought it best not to think of home. The food had worsened with smaller portions, and then the men were transferred to nearby Sime Road Camp where the food was not much better or worse and where disease claimed more lives. Butler contracted malaria, but at least he dodged typhus. As the men's health deteriorated, the work, including the digging of tunnels and draining of swamps for an airport, pushed them to the breaking point. Cutting down coconut and rubber trees and lifting logs onto trailers made from stripped-down, engineless trucks led to injury and exhaustion, and with diminishing rations, it was hard to find the last sparks of energy. Beriberi, malaria, and pneumonia were common, but so, too, were jungle ulcers. Meanwhile, possession of an illegal radio receiver still carried a death sentence. Butler remembered a group of men marched away for interrogation, some not seen again. With wood in short supply, a reusable coffin, built with a false bottom, discreetly released remains into the ground.

○ ○ ○

While interned in Sumatra, barkeeper Bernard Bree found it hard to write home to Nell. His injured hand and failing eyesight meant that fellow prisoner Lawrence Thompson wrote the letters for him. "Just a line to let you know I am alive and well," noted one dated March 15, 1942. "Hoping this finds you & the family in the pink. Remember me to everybody. Sorry I have not been able to write this myself owing to my eyes going back on me … and not at all able to get suitable glasses, but there is nothing to worry about. Hoping to see you all in the near future. I remain your affectionate hubby, Bert."[31]

On November 14, 1944, Bree died of malaria in Sumatra. Three other ex-crew he had travelled with also died there: waiters William Marlow and Thomas Roberts and kitchen porter Herbert Smallwood. Buried in Sumatra, their remains were later interred in Batavia.

Civilian Internment Camp
Palembang Sumatra 15/3/42

Darling Wife,

Just a line to let you know I am alive & well. Hoping this finds you & the family in the pink. remember me to everybody. Sorry I have not been able to write this myself owing to my eyes going back on me & not at present being able to get suitable glasses but there is nothing to worry about. Hoping to see you all in the near future

I Remain your affectionate Hubby

B. M. BREE

Bert

The back of a March 1942 postcard sent by prisoner-of-war Bernard Bree to his family in England.

In a postwar interview, Butler did not provide a date but spoke of seeing "white tracers in the sky." They were, he recalled, American reconnaissance planes. "One day in the morning we were out working and the air raid sirens went. We had heard them before and so we didn't take notice and about five or six miles away was the naval base at Singapore and we looked up and saw these big planes, flying perfect formation. Big, four-engine planes. They went across and all got away, none shot down."[32]

Life got worse until finally all work stopped. "We thought something was strange. Then all of a sudden into our camp walked six European soldiers with reddish berets. They walked through … with Japanese escorts with fixed bayonets and went to the hospital and looked around. We thought they were Germans. It was only when they were gone that a newsletter went around saying they were British paratroopers, dropped at Changi airport. The Japanese had surrendered. [Later] another plane came over tossing leaflets. We realized then the war was over. That was it."[33]

Butler, Crofts, Gee, and Hesp were among many from the *Asia* who had endured a total of three years and six months as prisoners of war until liberated in late August 1945. When it came to considering claims, the Canadian government determined the incarceration lasted 1,288 days, from February 15, 1942, to August 25, 1945. Interestingly, it was a year and 10 months into that captivity that Crofts and other Canadian internees received the standard Christmas greeting and good wishes from Prime Minister Mackenzie King.[34]

At Keppel Harbour, Butler boarded the Dutch ship *Nieuw Holland* and reached Liverpool on October 15, 1945, via the Suez Canal. By then he was coughing blood and weighed 116 pounds. Crofts, Hesp, Gee, and other emancipated crew also endured thousands of miles of ocean travel to reach home. Unknown to Crofts, his mother had died the day the very first contingent of *Asia* survivors reached Canada in the spring of 1942.

Gee's long journey began September 14, 1945, when he and several shipmates boarded the *Tegelberg* destined for Liverpool. The ship arrived on October 11, and Gee's sister, Rosa, who lived in the United Kingdom, met him in Liverpool and later carefully broke the news that his only son had died in 1943. In addition to losing Arthur Jr., Gee had missed his mother

Annie's funeral in May 1945, so steps to ensure the health of the chef's beloved German shepherd, Fritz, had grown more important. When Gee finally arrived home by train from Montreal on November 19, the old dog was in his usual place under the kitchen table. Sensing his master's presence, Fritz did not find the energy to stand, but Gee's daughter distinctly remembered the wag of his tail.

o o o

Returning home to Brentwood Bay on Vancouver Island in 1942, Maurice Atkins had trouble fitting in around his parents, finding he had changed from the youngster who had left home. He soon headed to San Francisco and went to sea again, eventually winding up in Liverpool where he signed onto the *Empress of Canada* skippered by former *Asia* captain George Goold.

Shortly before midnight on March 13, 1943, with a small moon breaking through a heavy ceiling off the west coast of Africa, the first of two torpedoes fired by the Italian U-boat *Leonardo da Vinci* struck the British-registered *Canada*. On board were 1,530 passengers, including 500 Italian prisoners of war, Polish military personnel, British government officials, and a crew of 362.[35] Although asleep in his cabin when disaster struck, Atkins moved quickly when Goold gave the "abandon ship" order. As *Canada*'s third officer, he helped launch the only operable motorboat before the next torpedo struck at 12:50 a.m.

During the next several hours, as all moonlight disappeared, Atkins and a couple of mates pulled survivors, including a pregnant woman, out of the sea. With the ship's starboard-side boat deck almost level with the water, Goold stepped off, and after locating a swamped lifeboat, joined the effort to pick up survivors. Nearly 43 hours later, after a Catalina flying boat circled the scene, a destroyer and two corvettes arrived. Three hundred and ninety-two perished.[36] This occurred just 13 months after the *Asia*'s destruction, and so Atkins was twice lucky.

o o o

Like Atkins and other ex-*Asia* merchant seamen, Walter Oliver continued to help maintain oceanic supply lines to embattled England. He had married the year the war ended — to a woman he had met in New Westminster, British Columbia — and soon entered fatherhood. As the years rolled on, he continued to go to sea before settling into work closer to home, ending with his stroke on the *Pitt Lake Express.* In the fall of 1995, the years caught up to him. He was not well — down with a cold that turned to pneumonia. Before entering hospital and leaving his one-room apartment in the Columbus Tower for the last time, Fourth Officer Oliver could point to his memoir, typed with two fingers over 19 years. It was not about the writing; it was the history, fulfilling a promise to himself, the *Asia*, her crew, and his family. On November 3, Walter Oliver took his last breath and returned to his first love.[37]

EPILOGUE: SIGNING OFF

Things are better at sea these days. I wouldn't want anybody to go through the slugging I did.

— **Geoffrey Hosken**

Following the destruction of the ship and narrow escape of her crew on coastal vessels, OBEs (Officer of the British Empire) were awarded to Captain J. B. Smith and Chief Officer Donald Smith. MBEs (Member of the British Empire) were bestowed upon Chief Engineer Herbert Owen, First Officer Leonard Johnston, and Fourth Officer Walter Oliver.

Well after the *Empress of Asia* slowly disappeared beneath the waves, she became the focus of salvage operations. One of her massive anchors was retrieved in 1998, and since 2015, it has been on permanent display at the Singapore History Gallery of the National Museum of Singapore. In 2020, the Maritime and Port Authority of Singapore issued a notice of work to remove the wreck, and local officials report it has since been cleared as the densely populated city pursues land reclamation and waterfront redevelopment.

○ ○ ○

Maurice Atkins made the most of his postwar life, travelling across Canada with the Department of Transport until he retired and established his own company, Toronto-based M.D. Atkins Marine Services Limited. Before he died on June 23, 2022, the old *Conway* graduate remained gracious and generous with his time and insisted on presenting this author with the life jackets that saved his life while escaping the *Asia* and *Empress of Canada*. "I've carried them all my life," he said. "I'm 96. It's time to get rid of them."[1]

Cecil Crofts also returned to sea. In an April 1966 letter to fellow crew member Jack Ewart, he summarized his work after the war. "I rejoined the C.P.R. I went skipper on one of the Park ships they were operating and after one year she was sold to Greek interests and I remained there as skipper on a year's leave of absence from C.P.R. I then returned to my company and was appointed first officer on *Empress of Scotland*. We were still trooping in the Mediterranean — evacuating Palestine [in 1948]. Our run was Liverpool-Gibraltar-Malta-Port Said-Haifa [and return]. I came back to Montreal and joined the *Beaverburn* as chief officer and we were running [between] Montreal and Liverpool. I quit C.P.R. and went skipper in a Greek ship operating around the world — a seven-month voyage." Crofts later joined the British Columbia ferry service, skippering the ferry from Texada Island to Powell River. He died on October 29, 1967. In his letter to Ewart, Crofts thanked his former mate for clarifying how he suffered severe rope burns under his armpits while being evacuated from the burning ship — with his broken leg. "I must have fallen out of the bowline you tied around me, but at the time I was probably passed out."[2]

In Walter Oliver's opinion, Crofts was an example of an incredibly brave and experienced seaman who understood — especially after injuring himself during the ship's final moments — that rash decisions propelled by courage sometimes impaired a person from considering all possibilities.

Lionel Dale Douglas, the Empress captain who reached Vancouver on the four-masted *Silberhorn* in 1897, announced his retirement in 1940. Since joining the Empress fleet in 1905, he had amassed 43 years of ocean travel.

He was 84 when he died in hospital in Vancouver on September 14, 1962, from a ruptured abdominal aortic aneurysm.[3]

John Drummond, who loved to paint nautical scenes, compose music, and "tickle the ivories," died of coronary thrombosis on August 25, 1950, at War Memorial Hospital, Williams Lake, British Columbia.[4] Visitors to his son Bill's home in Sechelt, British Columbia, were impressed by the *Empress of Asia* decor — keepsakes inherited from his father. In addition to Chinese furniture and several paintings by John, there were elephant carvings on the mantel, purchased in Colombo during his long journey home in 1942. Best of all, the sanitation engineer left his remarkable wartime journal.

Jack Ewart joined the Royal Canadian Navy and served on merchant ships as a DEMS. With important contributions from mates Ernie Higgs and Geoff Hosken, he pulled together various notes and wrote a detailed history on the loss of the *Asia*, mostly focusing on the escape. The account was published in the *Red Duster* in the summer of 2001. Ewart died on April 22, 2002.[5]

Archie Gee died of a heart attack at Vancouver General on July 1, 1964, and was survived by his wife, Flora, and two daughters. He was 70, and his occupation was still listed as chef. His daughter Beverley Thackray, interviewed for this book, died in April 2023, but his grandson, Dave, hangs on to the memories.[6]

A.J. Hailey, along with Captain Samuel Robinson, was — for many years — a senior skipper of the Empress ships. Master of the *Asia* from 1919 to 1921, and again from 1927 to 1930, "Bricks" had seen it all. He died at age 83 in Vancouver on July 15, 1958. The purposeful captain, who broke transpacific speed records, was a known animal lover. His bulldog, Bully, joined him on the *Monteagle*, and he kept a cat on the *Empress of Canada*. On the *Asia*, he was accompanied by his canary, Sun Yat-sen.

Ernie Higgs was in a care home when he died on May 24, 2011, following 16 years of progressive dementia. He was 90. His daughter, Joan, said even up to the end of his life he remained "dapper, well-dressed, and fit-looking" and enjoyed all genres of music, including classical, but was a huge fan of the Beatles. "We were lucky his genial nature remained throughout his mental demise," she noted.

Following the loss of the *Asia*, Higgs remained in the Merchant Navy, helping to transport war supplies to Britain, Australia, New Zealand, and India. Postwar, he achieved his master's foreign-going certificate and became chief officer on the weather ship *Stonetown*, and on Canadian Hydrographic Survey vessels, and later on shore as a marine insurance adjuster. "We were of the middle class, but on the humbler, lower end," noted Joan. Higgs saved enough and made time to build a beautiful sailboat, christening her the *Pahquee* (a plant used to make medicine to promote healthy blood). A high point was sailing her with his family to the 1962 Seattle World's Fair.

Geoffrey Hosken was 82 when he died in Nanaimo, British Columbia, on October 14, 1998. He had outlived his brother, Flight Officer (Navigator) Charles Douglas Hosken, killed when his Halifax bomber went down over

Ernie Higgs's sailboat *Pahquee* was a dream come true.

Holland in 1943. The loss of his brother and the bomber's five other aircrew did not stop him from joining the Royal Canadian Air Force, and in January 1946, he earned the British Empire Medal for meritorious service. At the time of Geoff's death, his occupation was master mariner with the BC Ferry Services. It is also fortunate that his little blue-green scribbler survived and is one of the best accounts of wartime life on the *Asia* during the war.[7] In an interview in May 1979, Hosken declared: "Things are better at sea these days. I wouldn't want anybody to go through the slugging I did."[8]

James Jackson was 89 when he was murdered on Monday, April 8, 1974. The former trimmer was in the beer parlour of the Vanport Hotel when he got up from his table and shuffled through heavy cigarette smoke to the washroom where someone stabbed him to death. Charges were laid, but the case ended in acquittal.

Geoffrey Hosken and his wife, Marjorie, strolling in downtown Vancouver.

James Jackson with his long-time friend and fellow *Asia* crewmate, George Fraser (left), and his grandson, Kenneth Gaskell, in Vancouver.

Herbert James died at age 67 on March 19, 1943, in Vancouver. He remained a formal, immaculately dressed fellow, and when he retired in 1935, he was captain of the *Empress of Russia*. During a sea time spanning more than four decades, he — as his grandson, Robert Gibson, noted — "witnessed a quantum change in shipping and everything else." A man of many talents, James collected stamps and coins, loved gardening, and designed and fitted out a beautiful large-scale model, the composite of a passenger ship named RMS *Africa*. Best of all, perhaps, James left a detailed diary chronicling the *Asia*'s First World War operations, details that would otherwise be lost.[9]

Ernest Punter, after the loss of the *Asia*, served as bosun and then third officer on several other foreign-going ships during and immediately after the war. These included the *Earlescourt Park*, *Queens Park*, *Mission Park*, and *Lake Cowichan*. Before his retirement in the 1950s, he worked as a longshoreman. His nephew, Norm Hutton, remembers the story of him being in convoy bound for Liverpool when the ships were surrounded by fog that hovered 20 feet above the water, making it difficult to spot U-boats but easy for the enemy to locate ships with their periscopes. "He said that throughout the morning, all you could hear were explosions, and when the fog lifted, there were no other ships in sight. He never found out how many escaped or were sunk."

Punter collected coins from every port — had a bag full of them. He also kept a canvas bag with a sailmaker's sewing palm, a ball or two of seaming twine, and a piece of beeswax. "He was adept at stitching up torn sails, pillows, curtains, or whatever, as well as rope splicing and rigging." The former *Asia* quartermaster died in hospital on April 2, 1975. His ashes were scattered at sea.

Samuel Robinson was 87 when his heart gave out on September 5, 1958. Despite the frustrations he endured during the First World War Red Sea operations, Robinson remained a highly successful and prominent sea captain, especially following the Great Kantō earthquake. As a senior skipper of the Empress fleet, his diaries, written in pencil, along with his large collection of photographs, are a tremendous but mostly untapped resource for maritime scholars. Thankfully, they both survive at the City of Vancouver Archives and Vancouver Maritime Museum.

Donald Smith died at age 90 on April 24, 1976, at Vancouver General. After the war, he continued to stay in touch with Prince Philip. During the Braemar Gathering in 1963, the Duke of Edinburgh stepped out of the royal box to greet the Scottish-born ex-naval captain. By then, the silver-haired Smith was 78.[10]

John Bisset Smith died on June 26, 1971, at the same hospital as Donald Smith. He had been ill for a long time, but the cause of death was listed as "senility," a consequence of cerebral arteriosclerosis, a disease that occurs when arteries in the brain become hard, thick, or narrow due to plaque.[11]

Hilda Mary Smith never married. The assistant matron of nursing died of breast cancer at King's Daughters' Hospital, Duncan, British Columbia, on October 8, 1965. The former Almonte, Ontario, native was 60. She had worked as a nurse for more than half her life — 37 years.[12]

Geoffrey Tozer had many more years on the high seas. Throughout the war and beyond, he served on several Park ships, including the *Gatineau*, *Temagami*, *Cornish*, *Aspen*, and *Earlescourt*. He was a cadet when he joined the *Gatineau* but rose to third officer and became second mate on other ships. The last merchant vessel listed in his Continuous Service of Discharge book is the *Lake Manitou*, which he signed off on in January 1951. From there, Tozer went to the Mackenzie River/Great Slave Lake waterways where he worked for the Yellowknife Transportation Company and became a captain. In addition to being at home on the sea, he continued to write and read poetry and jot down philosophical insights, perhaps indicative of how the sea can make poets of us all. He was 82 when he died of heart failure in Kelowna, British Columbia, on March 12, 2005.

Phylis Williams, who died in her 80th year in 1990, continued to benefit from her "astuteness and captivating social charm." The former hairdresser eventually moved to Ontario working in provincially run homes for the elderly. Initially contributing in a supervisory capacity, assisting the facilities to organize handicraft activities for residents, she became an inspector and helped assess the quality of care. After she retired from government work, Williams relied on her knowledge to establish a group of unpaid senior volunteers to support other seniors navigating through government services. This was followed by more volunteer work, but her son says "she never forgot her sailing experience and time on the *Asia*." While watching the 1987 Steven Spielberg/Tom Stoppard film *Empire of the Sun*, the frail, little lady rose from her seat and said: "I was there. I was there."[13]

ACKNOWLEDGEMENTS

This undertaking began in 2019 on the eve of the coronavirus pandemic, and while it became exceedingly difficult to travel and meet key contacts in person, the research benefited from online connections. Without this, the layering of the *Empress of Asia*'s human history would have fallen well short. Likewise, the book would have been incomplete without the valuable assistance of archivists, librarians, museum curators, and historians.

I want to begin by singling out researcher Nelson Oliver of Port Moody, British Columbia, who hosts empressofasia.com. It was Nelson, son of Walter Oliver, who urged me to write this book, and when I agreed, he remained at my side through a renewed research effort that went beyond the scope of his vast research. Together, we worked like private investigators, doggedly tracking down and establishing contact with people to interview.

Furthermore — in addition to sharing his father's unpublished memoir — Nelson answered my sometimes awkward questions regarding his father and spent hours fact-checking early drafts, assisted by his wife, Sharon. He also introduced me to Jenny Smith, a skilled editor who spent days poring over the drafts. Her early copy-editing and more substantive recommendations were spot on. Both volunteered their time, and so the book is for them as much as anybody else. Behind the scenes, offering additional support on the technical side, were their kin, Peter Muntener and Kevin Smith.

I am also very grateful for the interviews with the last remaining individuals who served or travelled on the ship. Maurice Atkins, Frank Davis, and Arthur Le Patourel patiently answered questions, shared photos, and provided valuable first-hand information. Sadly, all three gentlemen have passed, but it has remained an honour to have the assistance of Maurice's wife, Micheline; Frank's daughter, Kathleen; and Arthur's son, Doug.

The research also produced a variety of diaries, journals, and unpublished memoirs filed away in homes or public archives. Those written by crew include the Commander Samuel Robinson series (album and diaries) at the City of Vancouver Archives, gained with the assistance of reference archivist Kira Baker; the John Drummond Wartime Journal shared by his son, the late Bill Drummond, and material supplied by his grandson, Jim Drummond; the Geoff Hosken Diary, furnished by his daughters, Margot Wallace and Brenda Hosken; the Herbert James Journal and supporting perspectives from grandson, Robert Gibson, assisted by his daughter, Steph; and the Geoff Tozer Diary, letters, and artifacts shared by son, Allen Tozer, with assistance from the Kelowna Museums and archivist Tara Hurley. The Tozer story also benefited from dozens of emails written by Geoff Tozer and shared with Nelson years ago — all shedding light on life at sea for merchant sailors during the war.

Then there are contributions from Aimee Herndon and the Mary Frances McNeal diary; Ester Benjamin Shifren for sharing research — and her book *Hiding in a Cave of Trunks* — detailing family time in Shanghai and prisoner-of-war internment; Michal Singer of the University of Cape Town Libraries for the George Wille Papers; Sarah Russell Spray for sharing photos from the William F. Russell Family Photo Archives, and the diary of William F. Russell; Martin "Bud" Walsh for articles and emails with Nelson shedding light on engine-room operations; Helen Watts for the Captain John Watts Report; and University of Virginia reference librarian Penny White for assistance in providing the American Red Cross Mission to Russia 1917–18 papers and diary of Dr. William Thayer.

Meanwhile, scores of individuals produced excellent material and made themselves available for interviews. Not everything shared could be included, but from the smallest recollection to the longest interview or document, each contribution helped, and any error arising from the research is mine.

Among the many contributors are Robert G. Allan for John Masson; Fiona Allen and her aunt Deirdre for information on Richard Farrell; Tony Allwright for Gerry O'Neill material; Evelyn Atkinson for Warrant Officer Ewen McKerchar; Bill Attwell for interviews and material on Phylis Williams; Cathie Barrett for Trimmer James Jackson; Glen Bowe for steering me to the Falun Historical Society's *Freeway West* digital collection at the University of Calgary; Polly Elder, author of *All This Shall Pass*; Rosalind Brice for Colin Albert Alt; the late Ron Bridge MBE, AFC, chairman of the Association of British Civilian Internees Far East Region, for files on civilian Far East prisoners of war and assisting with research at the National Archives, Kew, London; Al Brown for Patrick James Brown; Gwen Butler for Edward Evans; Jan Care for files on Jack Hodgson; Roy Carver for Clarence Carver; Raymond Chong and Michael Chong for Gim Suey Chong; Frances Clemmow for Dr. Keith Gillison and family; Robert Collyer for Don Collyer; Terry Costello for Michael Costello; Kathleen Davis for Frank Davis; Jill Ewart and Joan Noel for Jack Ewart; Peter Foster for Lance Corporal Norman Foster; Iain Gaskell and Denys Courtnay Gaskell for Gunner Edgar Mole's diary; Darren Gillett, Rhonda Gillett, and Debbie Eaton for Owen Gillett; Al Hailey and Nelle Picard for Captain Alfred James Hailey; Kay Hasler for Bernard Bree; Kevin Hazard for William Patrick Hazard; Joan Higgs and Doris Clemens for Ernie Higgs; Dennis, Lorne, Mildred, and Richard Hugh, and granddaughter Shana Hugh, for Albert Hugh; Norm Hutton for Ernest Raymond Punter; Peter Hvidsten for his book and information on Per Hvidsten; Kathleen Kershaw for Henrietta Kershaw; Ronald Konzen for historic album photos; Paula Linn for Len Johnston; Terry Loudwell for Private Leonard Loudwell; Norma Lumley for Barrett Lumley; Norman A. Mackenzie and Sheena and Alan Beedham for William McArthur; and Michael May for Lieutenant Douglas May and Rosalie Pryce.

Research assistance also came from Chris Morris; Doug McElroy for Herbert Stainton; Steve McKinnon for William Roderick McKinnon; Dale

Mitchell for Clarence Mitchell; Gerry O'Neill for his biography *Favourable Winds*; Jennifer Padley and Rob Padley for Private Bert Needham's story; Barb Paquette for information on Staff Sergeant Willard "Dusty" Perrin; John Pitt-Brooke and Lynda Pitt-Brooke for Reginald Pitt-Brooke; Anne-Marie Reading for the Pietro Colombo photographs; Kenneth Roistacher for Second Lieutenant Herbert Roistacher; Sherri Rush for Leona Kearns; Patricia Slade for Hilda Mary Smith; Kathryn Smith for her article on Frank Lloyd Wright; Chris Stringer and Geoffrey E. Whitfield for information on the *Asia* maiden voyage members returning home on the *Empress of Ireland*; Beverley Thackray and son, Dave, for Archie Gee; Dr. K.G. Tregonning for help with the Straits Steamship Company; Felicity Yost for Charles and Gertrude Yost; and Patricia Wong for Low Shau Wah.

After submitting access requests to Veterans Affairs Canada's Merchant Navy Registry, staff provided dozens of personnel files. At Library and Archives Canada, I located postwar files on Cecil Crofts and was assisted by researcher Glenn Wright with the 1940 "accidental" bombing files. Excellent assistance also came from the Vancouver Maritime Museum Archives, gained with the help of librarian and archivist Ashlynn Prasad; the Wallace B. Chung and Madeline H. Chung Collection at Rare Books and Special Collections, University of British Columbia Library, with assistance from Chelsea Shriver; Royal BC Museum for recorded interviews with Captain L.D. Douglas and John Bird, part of the CBC Collection; Library and Archives staff at Ingenium — Canada's Museums of Science and Innovation — for making available the Commander's General Voyage Reports requested as RG11-043, RG11-082, and RG11-083; and Drew Waveryn for sharing his extensive photo collection. I am also indebted to the *Vancouver Sun*, *Vancouver Province*, *Victoria Daily Times*, and *Victoria Colonist* for being the source of articles written about the ship, passengers, and crew, and it was good to reach out to the Friends of Keewatin while researching the Great Lakes crew.

Overseas, research was supported by archivist Gary Haines at the Royal Air Force Museum, London, who shared an article and answered questions on the Royal Air Force's role during the Battle of Singapore. Another excellent source was the online oral histories kept by the Imperial War Museum, and photos from the Australian War Memorial, Canberra. My gratitude extends to Jimmy Yapp and Veronica Chee at the Singapore National Library Board's magazine *BiblioAsia* for publishing an article I wrote on the ship's final hours, and to Fairfield coordinator Abigail Morris, Fairfield Heritage Project, Govan Workspace Limited, Glasgow.

Closer to home, I am deeply grateful to James P. Delgado for writing the Foreword, and to Robert D. Turner, author of *The Pacific Empresses*, for his assistance.

On the publishing side, it was a pleasure to work with the highly professional staff at Dundurn Press. Kwame Scott Fraser welcomed the project, and the Dundurn team meticulously edited and applied layout and production skills to each page. That team also included chief publishing officer Meghan Macdonald, editorial director Kathryn Lane, managing editor Elena Radic, editor Michael Carroll, art director Laura Boyle, and production manager Rudi Garcia. It is much better because of them. I am also grateful for the excellent maps produced by Julie Witmer of Julie Witmer Custom Map Design.

Finally, on the personal side, I want to thank my wife, Alice, children, and grandchildren for love and support.

APPENDIX 1: *EMPRESS OF ASIA* CASUALTY LISTS, 1914-42

RESEARCHED AND COMPILED BY NELSON OLIVER

Armed Merchant Cruiser Deployment, 1914–16

Wong Wa, trimmer, October 16, 1914, beriberi, Hong Kong Memorial, Hong Kong.*

Lung Sack, fireman, October 22, 1914, beriberi, Colombo (Kanatte) General Cemetery.

Ip Yee, trimmer, October 25, 1914, beriberi, Colombo (Kanatte) General Cemetery.

Wong Quong, trimmer, November 21, 1914, beriberi, Hong Kong Memorial, buried in Singapore.

Lee Tuk, fireman, December 12, 1914, drowning, Hong Kong Memorial, body not recovered.

Yee Yeong aka Hong Bin, able seaman, December 29, 1914, beriberi, Hong Kong Memorial, buried in Aden.

Ahmed, fireman, November 6, 1915, beriberi.

* Ship's log states two firemen died on this date, one of heart failure, the other of beriberi.

Amiruddin Fazal, fireman, November 6, 1915, beriberi, Bombay Memorial, Mumbai, India.

Saiyid Ahmad Raqi, trimmer, November 28, 1915, beriberi, Bombay Memorial, Mumbai.

Geva Rosa, fireman, November 29, 1915, beriberi.

First World War Transpacific Service, 1916–18

Cheong Kong, fireman, September 6, 1916, heart failure, Vancouver.

Ah Fan, saloon boy, May 5, 1917, St. Paul's Hospital, Vancouver.

Tam Cheong, saloon boy, July 7, 1917, suicide by drowning.

Troop Carrier Deployment, North Atlantic, 1918–19

Tsang Fong, cook, June 18, 1918, suicide by drowning, New York.

Edward Fisher, steerage steward, October 15, 1918, pneumonia, Walton Hospital, Liverpool.

Wong Yun, smoke room boy, November 2, 1918, Bellevue Hospital, New York City.

Ho Poa, cook, November 7, 1918, pneumonia, Walton Hospital, Liverpool.

Wong Wong, greaser, November 7, 1918, opium, on board ship, New York City.

Second World War Pacific Service, 1939–40

Harry Morris Mann, boatswain, September 26, 1939, cardiac failure, buried at sea.

Yau Tse, night watchman, June 7, 1940, pneumonia, Vancouver.

Crew Deaths Shortly After the Attack or While Prisoners of War

Douglas Richard Elworthy, pantryman, February 10, 1942, Bidedare Cemetery (Kranji War Cemetery), Singapore.

Patrick Leo Harkins, second radio officer, February 10, 1942, Tower Hill Memorial, London, United Kingdom.

Irvine Hampshire, DEMS deckhand, February 15, 1942, Portsmouth Naval Memorial, Hampshire, United Kingdom.

Harry Thomas Worrall, deckhand, DEMS able seaman, Royal Australian

Naval Reserve, February 15, 1942, Plymouth Naval Memorial, Devon, United Kingdom.

James Joseph Towers, deck boy, June 13, 1943, Kanchanaburi War Cemetery, Thailand.

Herbert Smallwood, kitchen porter, March 29, 1944, Jakarta War Cemetery.

Thomas Henry Roberts, waiter, June 25, 1944, Jakarta War Cemetery.

William James Marlow, waiter, August 11, 1944, Jakarta War Cemetery.

Bernard Medlicott Bree, barkeeper, November 14, 1944, Jakarta War Cemetery, Indonesia.

British Troop Deaths on the Day of the Attack, February 5, 1942

Harold Armitage, private, Royal Army Ordnance Corps, 18 Divisional Workshop Company.

Gordon Anthony Cluley, trooper, Reconnaissance Corps, 18th (5th Battalion, The Loyal Regiment) Regiment.

Richard Wickham Dixon, captain, Reconnaissance Corps, 18th (5th Battalion, The Loyal Regiment) Regiment.

John Hodgson, trooper, Reconnaissance Corps, 18th (5th Battalion, The Loyal Regiment) Regiment.

Mark Edward Kirby, gunner, Royal Artillery, 125th Anti-Tank Regiment.

Howard Reginald Lambert, warrant officer, Royal Army Ordnance Corps.

Arthur Hamilton Lawson, captain, Reconnaissance Corps, 18th (5th Battalion, The Loyal Regiment) Regiment.

Robert Charles Lloyd, private, Royal Army Ordnance Corps, 18th Divisional Workshop Company.

William Longden, trooper, Reconnaissance Corps, 18th (5th Battalion, The Loyal Regiment) Regiment.

Bert Mayor, trooper, Reconnaissance Corps, 18th (5th Battalion, The Loyal Regiment) Regiment.

Patrick Edward McGrath, trooper, Reconnaissance Corps, 18th (5th Battalion, The Loyal Regiment) Regiment.

Ewen Kennedy McKerchar, warrant officer, Royal Army Ordnance Corps, 18 Divisional Workshop.

Clifford John Partridge, warrant officer, Royal Army Ordnance Corps, 18 Divisional Workshop.

Frederick James Randall, major, Reconnaissance Corps, 18th (5th Battalion, The Loyal Regiment) Regiment.

Patrick Regan, private, Royal Army Ordnance Corps, 18 Divisional Workshop.

Robert Barker Wilson, lieutenant, Royal Artillery, 125th Anti-Tank Regiment.

○ ○ ○

Escaping death on February 5, 1942, was British captain John Watts of the 125th Anti-Tank Regiment, Royal Artillery. He died on February 26, 1945, while a prisoner of war and is buried at Kanchanaburi War Cemetery, Thailand.

APPENDIX 2: TIMELINE

Summer 1911: Contracts signed for two new passenger liners to be named *Empress of Asia* and *Empress of Russia* for Pacific Ocean service.
August 28, 1912: *Empress of Russia* launched at Fairfield Shipbuilding and Engineering Company, Govan, Scotland.
November 23, 1912: *Empress of Asia* launched at the same shipyard.
May 31, 1913: *Asia* completed and delivered to Canadian Pacific.
June 14 to August 31, 1913: Maiden voyage of *Empress of Asia* — Liverpool to Vancouver — via the Cape of Good Hope.
June 28, 1914: Archduke Franz Ferdinand assassinated.
July 28, 1914: Austro-Hungarian Empire declares war on Serbia; First World War begins.
August 3, 1914: *Empress of Asia* commissioned for war service as armed merchant cruiser with deployments in the Yellow Sea, the Philippines, the Indian Ocean, and the Red Sea.
August 4, 1914: Great Britain declares war on Germany.
October 22, 1915: *Empress of Asia* released from war duties and resumes Canadian Pacific Ocean Services.
May 22, 1918: *Asia* departs Vancouver to embark Chinese Labour Corps members at William Head Quarantine Station, Vancouver Island. Sails to Liverpool via Panama Canal and New York City.

July 31, 1918: *Asia* commences transporting American troops across the Atlantic Ocean to Britain and France.
November 11, 1918: First World War armistice signed.
January 2, 1919: *Asia* departs Liverpool with repatriated Canadian troops bound for Victoria, British Columbia. On board, the Asia Land Settlement Committee is born.
February 27, 1919: First postwar voyage of *Asia* — Vancouver to Hong Kong.
February 19, 1920: First motion pictures screened aboard *Asia*.
April 17 and June 12, 1921, July 20, 1924: Record crossings of the Pacific Ocean for *Asia*.
September 1, 1923: Great Kantō earthquake.
January 11, 1926: Fatal collision of *Asia* with the *Tung Shing*.
January 21, 1926: Baseball player Leona Kearns washed overboard on *Asia*.
January 24, 1927: *Asia* arrives in Vancouver with Shanghai refugees.
August 1, 1927: Record silk shipment for *Asia*.
February 13, 1928: *Asia* arrives in Vancouver after terrific storm that nearly swept the helmsman and officer of the watch away.
October 1929: Stock market crashes and the Great Depression ensues.
July 5, 1931: Expanded service of *Asia* to Hawaii announced.
August 18, 1937: *Asia* sails for Hong Kong from north of Shanghai with approximately 1,300 evacuees fleeing the war-torn city.
March 18, 1938: Crew member of *Asia* diagnosed with typhoid fever.
April 22, 1938: American pilot Elwyn Gibbon removed from *Asia* for interrogation in Yokohama, Japan.
September 1, 1939: Second World War begins.
September 14, 1940: *Asia* bombed north of Oshima Island, Japan.
January 11, 1941: *Asia* completes her 307th and final crossing of the Pacific Ocean in Vancouver.
January 13, 1941: *Asia* formally requisitioned as a troopship.
February 13, 1941: *Asia* departs Vancouver for the Firth of Clyde, Scotland, then Liverpool.
April 24, 1941: *Asia* embarks British troops and sails from Liverpool bound for Suez.

June 28, 1941: *Asia* departs Suez with European evacuees and Italian prisoners of war. Prince Philip of Greece serving as midshipman.
September 4, 1941: *Asia* reaches New York City with evacuees.
September 16, 1941: *Asia* departs Halifax, Nova Scotia, for Liverpool in HX.150, the first Atlantic Ocean convoy escorted by American warships.
November 12, 1941: Final voyage of *Asia* commences from Liverpool, England.
December 7, 1941: Japanese attack Pearl Harbor.
December 8, 1941: Japanese invade Malaya.
December 25, 1941: British colony of Hong Kong surrenders to Japanese Empire.
January 15, 1942: *Asia* arrives in Bombay and later embarks troops bound for Singapore.
January 31, 1942: The last British Commonwealth troops withdraw across the Strait of Johore to Singapore.
February 4, 1942: *Asia* attacked by Japanese bombers in the Bangka Strait off Sumatra.
February 5, 1942: Destruction of *Asia* by Japanese dive-bombers on final approach to Singapore.
February 11, 1942: Dozens of *Asia* crew members begin escape on coastal vessels from Singapore.
February 15, 1942: Singapore surrenders to Japanese. *Asia*'s catering crew, who had volunteered at local hospitals, among prisoners of war.
April 18, 1942: First *Asia* crew members who escaped Singapore arrive in Halifax, Nova Scotia, reaching Vancouver six days later.
April 30, 1945: Adolf Hitler commits suicide.
May 8, 1945: Unconditional surrender of Germany; Second World War in Europe officially ends, though pockets of German military continue to fight or resist in Europe weeks after VE day.
August 6 and 9, 1945: Atomic bombs dropped on Hiroshima and Nagasaki, Japan.
September 2, 1945: Complete capitulation of Japanese Empire set out in the Instrument of Surrender signed aboard the USS *Missouri*. Prisoners of

war, including those who crewed or travelled on the *Asia*, are liberated, but many face a lifetime of physical pain and trauma.

1998: One of *Asia*'s anchors salvaged from the wreck site.

2015: Anchor of the *Empress of Asia* placed on permanent display at the National Museum of Singapore.

NOTES

1: Heater Boys and Holder-Ons

1 Nelson Oliver Interview, November 1, 2020. Hereafter Nelson Oliver Interview.

2 As the 20th century unfolded, the number of separate yards along the Firth of Clyde grew to more than 200. See landmarktrust.org.uk/searchresults/?q=clyde+shipbuilding.

3 W. Kaye Lamb, "Empress Odyssey: A History of the Canadian Pacific Service to the Orient, 1913–1945," *British Columbia Historical Quarterly* 12 (January 1948): 6. Hereafter Lamb, *BCHQ*.

4 See *Steel Chest, Nail in the Boot and the Barking Dog*, a 1986 documentary film by David Hammond, digitalfilmarchive.net/project/steel-chest-nail-in-the-boot-193.

5 W. Kaye Lamb, *Empress to the Orient* (Vancouver: Vancouver Maritime Museum, 1991), 76.

6 Anne Massey, *Designing Liners: A History of Interior Design Afloat*, 2nd ed. (New York: Routledge, Taylor & Francis Group, 2021). See also oldwestburygardens.org.

7 Martin Walsh Emails, March 5, August 6, 2003. Hereafter Walsh Email or Emails.

8 Lloyd's 1912–13 Registry.

9 University of British Columbia Library, Rare Books and Special Collections, Chung Collection, CC-TX-246-4-5.

10 *Vancouver Sun*, February 21, 1912, 12.
11 Henry Beauclerk/Alice Shaughnessy marriage certificate, June 3, 1911, Charlotte, New Brunswick, Certificate No. 1142.
12 *Vancouver Sun*, February 22, 1912, 12.
13 Robinson Diary, March 15–May 7, 1913, City of Vancouver Archives, Box 539-B-05 fld. 22. Hereafter Robinson Diary.
14 See en.wikipedia.org/wiki/Kingston-upon-Hull.
15 *Vancouver Sun*, March 12, 1913, 10.
16 Robinson Diary, May 4, 1913.
17 *Victoria Daily Times*, August 27, 1913, 16.
18 *Los Angeles Herald*, February 2, 1898; *Victoria Daily Times*, February 2, 1898, 6.
19 *London Gazette*, October 31, 1902.
20 Lamb, *BCHQ*, 5.
21 Massey, *Designing Liners*, 63.
22 University of British Columbia Library, Rare Books and Special Collections, Chung Collection, CC-TX-225-51-9.
23 *Vancouver Sun*, May 22, 1913, 13.
24 Robinson Diary, May 29, 1913.
25 Robinson Diary, May 30, 1913.

2: Dancing Like a Hurricane

1 Walsh Email, March 5, 2003.
2 Walsh Email, June 9, 2003.
3 Walsh Email, June 22, 2003.
4 Robinson Diary, June 25, 1913.
5 Robinson Diary, June 27, 1913.
6 Robinson Diary, June 28, 1913.
7 Jean Hétu, "Pérodeau, Narcisse," *Dictionary of Canadian Biography*, vol. 16 (Toronto and Quebec City: University of Toronto/Université Laval, 2003), biographi.ca/en/bio/perodeau_narcisse_16E.html.
8 Robinson Diary, June 29, 1913.
9 University of British Columbia Library, Rare Books and Special Collections, Chung Collection, CC-TX-225-51-9.
10 George Wille Papers, University of Cape Town Libraries, Cape Town University, J.D. Noonan Diary, Empress of Asia Trip Round the World, BC 646, 1. Hereafter Noonan Diary of Wille Papers.
11 Wille Papers, 1.
12 Noonan Diary of Wille Papers, 2.

13 Noonan Diary of Wille Papers, 3.
14 Noonan Diary of Wille Papers, 4.
15 Wille Papers, 2.
16 Noonan Diary of Wille Papers, 5.
17 Wille Papers, 2. The quote is verbatim for historical accuracy.
18 Robinson Diary, July 17, 1913.
19 Wille Papers, 4.
20 John Scott Cowan, Principal Emeritus, Royal Military College of Canada, "From Flanders Fields to Tokyo Bay," 2.
21 Cowan, "From Flanders Fields to Tokyo Bay," 2.
22 Library and Archives Canada, RG 150, Accession 1992-93/166, Box 2022-3, Lawrence Vincent Moore Cosgrave, en.wikipedia.org/wiki/Lawrence_Moore_Cosgrave.
23 Noonan Diary of Wille Papers, 13.
24 University of British Columbia Library, Rare Books and Special Collections, Chung Collection, CC-TX-246-4-5.
25 Noonan Diary of Wille Papers, 14.
26 Wille Papers, 8.
27 Robinson Diary, July 29, 1913.
28 The tonnage of coal loaded at Nagasaki was typical of amounts loaded onto the *Asia* during her homeward-bound voyage.
29 *Vancouver Sun*, September 1, 1913, 3.
30 *Vancouver Sun*, September 1, 1913, 3.
31 Robinson Diary, September 10, 1913.
32 *Vancouver Sun*, September 12, 1913.
33 Robinson Diary, August 3, 1914.

3: Stripped for War

1 Robinson Diary, July 30, 1914.
2 Herbert James Journal, Nelson Oliver Collection. Hereafter James Journal.
3 National Schools and Admissions Records, Queen Elizabeth's Grammar School, 1887, Barnet, England.
4 Robert Gibson Interview, March 20, 2023.
5 *Vancouver Daily Province*, February 8, 1935, 8.
6 James Journal, August 2–3, 1914.
7 TNA ADM 137/11/4 1914 Folio 154.
8 James Journal, August 8, 1914.
9 James Journal, August 9, 1914.

10 See rtpovey.homeip.net/familytree/fam606.html;http://www.dreadnoughtproject.org/tfs/index.php/Category:Royal Navy Officers Educated at Eastmas%27s_Royal_Naval_Academy; see discovery.nationalarchives.gov.uk/details/r/D8119414; Carl Whiteley, "Captain Walcott," www.royhodges.co.uk/Captain%20Walcott.pdf; TNA-ADM 196/125/207.
11 James Journal, August 11, 1914.
12 Robinson Diary, August 11, 1914.
13 James Journal, August 13–14.
14 James Journal, August 15.
15 James Journal, August 15.
16 Robinson Diary, August 15, 1914.
17 James Journal, August 16, 1914.
18 James Journal, August 20, 1914.
19 James Journal, August 23, 1914; P&O Heritage, Ship Fact Sheet.
20 Robinson Diary, August 29, 1914.
21 Robinson Diary, August 29, 1914.
22 James Journal, September 3, 1914.
23 Robinson Diary, September 4, 1914.
24 James Journal, September 7, 1914.
25 James Journal, September 9, 1914.
26 James Journal, September 12, 1914.
27 James Journal, September 21, 1914.
28 Robinson Diary, September 21, 1914.
29 James Journal, September 29, 1914.
30 James Journal, October 3, 1914.
31 In the words of *Emden* torpedo officer Prince Franz Joseph von Hohenzollern, "We were after her like a fox."
32 James Journal, October 13, 1914.
33 See okthepk.ca/dataCprSiding/news/2011/2012010204.htm.
34 James Journal, October 15, 1914.
35 Percy B. Egan, "An Outbreak of Beri-Beri in the 'Empress of Asia,'" *Journal of the Royal Naval Medical Service* 3, no. 2 (1917): 195–201.
36 Lamb, *BCHQ*, 12.
37 Gunner Edgar Horace Mole's Diary, courtesy Iain Gaskell and Denys Courtnay Gaskell. Hereafter Mole Diary.
38 James Journal, October 21, 1914.
39 James Journal, October 24, 1914.
40 James Journal, November 4, 1914.

41 James Journal, November 7, 1914.
42 James Journal, November 14, 1914.
43 James Journal, November 16, 1914.
44 Mole Diary, November 16, 1914.
45 Mole Diary, November 16, 1914.

4: Guns and Shuttlecocks

1 James Journal, November 27, 1914.
2 James Journal, November 25, 1914.
3 John Keegan, *The First World War* (Toronto: Key Porter, 1998), 218.
4 John Baldry, "British Naval Operations Against Turkish Yaman, 1914–1919," *Arabica* (June 1978).
5 Robinson Diary, December 12, 1914.
6 James Journal, December 13, 1914.
7 James Journal, addendum containing January 30, 1915, letter from Colpoys Walcott to naval authorities.
8 James Journal, February 2, 1915.
9 James Journal, February 3, 1915.
10 James Journal, March 8, 1915.
11 James Journal, March 14, 1915.
12 James Journal, April 10, 1915.
13 TNA ADM-196-42-455 TNA ADM-196-88; see also dreadnoughtproject.org /tfs/index.php/Philip_Howard_Colomb.
14 James Journal, April 14–June 29, 1915.
15 Mark L. Connelly, "The British Campaign in Aden, 1914–1918," *Journal of First World War Studies* 1, no. 3 (2005): 73.
16 Connelly, "The British Campaign in Aden, 1914–1918," 73.
17 F.A. McKenzie, "Shadows in the East: The Defence of India," in *The Great War: The Standard History of the All-Europe Conflict*, vol. 7, chapter 128, eds. Herbert Wrigley Wilson and Sir John Alexander Hammerton, 159–61.
18 Connelly, "The British Campaign in Aden, 1914–1918," 75–76.
19 James Journal, July 7, 1915.
20 James Journal, July 7, 1915.
21 James Journal, July 18, 1915.
22 James Journal, July 21, 1915.
23 James Journal, July 21–22, 1915; the Honourable Artillery Company is the oldest regiment in the British Army and the second-most senior unit of the Territorial Army.

24 The British casualty total is difficult to determine precisely. A July 26, 1915, *Times* of London article quoting the Press Bureau sets the number at "about twenty-five all ranks."
25 James Journal, July 21–22, 1915.
26 Robinson Diary, July 24–25, 1915.
27 Robinson Diary, July 27, 1915.
28 Robinson Diary, July 31, 1915.
29 James Journal, August 1, 1915.
30 Robinson Diary, August 2, 1915.
31 James Journal, August 23, 1915.
32 James Journal, August 22, 1915.
33 James Journal, September 18, 1915.
34 See naval-history.net/OWShips-WW1-09-HMS_Lama.htm.
35 Gibson, March 20, 2023.

5: Atlantic and Pacific

1 Ilya Tolstoy, *Reminiscences of Tolstoy* (New York: Century Company, 1914).
2 Charles W. Yost, unpublished memoirs, 1980, Seeley G. Mudd Manuscript Library, Princeton University, New Jersey. Hereafter Yost, Unpublished Memoirs.
3 Felicity Yost, correspondence with author, April–May 2021.
4 Yost, Unpublished Memoirs.
5 Dr. William S. Thayer Diary, American Red Cross Mission to Russia, 1917–1918, Joseph M. Bruccoli Great War Collection, Accession 10875-c, Special Collections Department, University of Virginia Library, Charlottesville, Virginia. Hereafter Thayer Diary.
6 Thayer Diary.
7 Frank Billings, M.D., "The Work of the American Red Cross Mission to Russia," *Military Medicine and Surgery* 69, no. 20 (November 17, 1917), jamanetwork .com/journals.
8 Thayer Diary, July 6, 1917, 8.
9 Thayer Diary, July 7, 1917, 9.
10 Thayer Diary, July 11, 1917, 11.
11 Thayer Diary, July 13, 1917, 14.
12 Thayer Diary, July 14, 1917, 15.
13 Thayer Diary, July 14, 1917, 15–16.
14 Thayer Diary, July 15, 1917, 16–17.
15 Thayer Diary, July 16, 1917, 17.

16 NA Registers and Indexes of Births, Marriages and Deaths of Passengers and Seamen at Sea, 1891–1922.
17 Dan Black, *Harry Livingstone's Forgotten Men: Canadians and the Chinese Labour Corps in the First World War* (Toronto: James Lorimer, 2019), 36. Original source: *Asia* ship manifests April 12, 1917, June 25, 1917, August 20, 1917, October 15, 1917, December 10, 1917, and March 4, 1918.
18 W.W. Peter, "Yellow Spectacles," Kautz YMCA Family Archives, University of Minnesota, 7–8.
19 TNA, FO 228/2894.
20 Black, *Harry Livingstone's Forgotten Men*, 276.
21 Black, *Harry Livingstone's Forgotten Men*, 276.
22 TNA, FO 228/2894.
23 TNA, FO 228/2894.
24 TNA, FO 228/2894.
25 *Empress of Asia* March–June 1918 logbook.
26 New York, New York, Extracted Death Index, 1862–1948.
27 Progress Charts, Transports, *Empress of Asia*, War Diary/Intelligence Summary, Major William Henry Macdonald, former archival reference no. LAC RG9-III-D-3.
28 LAC, RG 150, Accession 1992-93/166, Box 5090-6.
29 Derek Lukin Johnston, "Memories of the Empress," *Vancouver Historical Society Newsletter* 43, no. 9 (June 2004).
30 Royal BC Museum and Archives, CBC Collection, T0915:0001, John Bird recording. Hereafter Bird, CBC.
31 LAC RG 150, Accession 1992-93/166, Box 746-15.
32 *Vancouver Daily Sun*, January 25, 1919, 1.
33 LAC, Complaints Re. SS *Empress of Asia*, File Part of 1903 Army Headquarters Central Registry. Former archived ref. No. RG24-C-1-a, vol. 898.
34 War Diary/Intelligence Summary, Major William Henry Macdonald, LAC RG9-III-D-3.
35 LAC RG24-C-1-a, vol. 898.
36 Canadian Pacific Facts and Figures, general publicity department, Canadian Pacific Foundation Library, 1937, 194. Hereafter CPFF.
37 Bird, CBC.
38 Bird, CBC.
39 *Victoria Daily Colonist*, September 27, 1919, 13.

6: Boys to Sea

1 *Vancouver Sun*, May 15, 1971, 1; Nelson Oliver Interview, April 5, 2024.
2 Nelson Oliver Interview, October 20, 2022.
3 Oliver Memoir, 3.
4 See durhamrecordoffice.org.uk/our-records/coal-mining-and-durham-collieries.
5 Nelson Oliver Interview, January 11, 2022.
6 Oliver Memoir, 1
7 Oliver Memoir, 10.
8 Walter Oliver Discharge Certificate.
9 Maurice Atkins Interview, January 29, 2021. Hereafter Atkins Interview(s).
10 Atkins Interviews, January 29 and February 2, 2021.
11 See sunnyokanagan.com/ogopogo.
12 Allen Tozer Interview, October 25, 2022. Hereafter Tozer Interview.
13 Canadian Pacific Steamship Letter, November 29, 1940.
14 W. Kaye Lamb, *History of the Canadian Pacific Railway* (New York: Macmillan, 1977), 346.
15 Canadian Pacific Steamship Letter, November 29, 1940.
16 Jill Ewart Interviews, 2021–22.
17 Joan Higgs Interview, December 2021.
18 A messenger rating was used for a while instead of boy seaman.
19 Higgs Interview, December 2021.
20 Nelson Oliver Interview, October 11, 2022.
21 Terence became a master mariner and Second World War Canadian Merchant Navy veteran as well as director of the Vancouver Maritime Museum. He died in 2014 at age 94.
22 Twenty-four men composed the party.
23 *Vancouver Sun*, February 20, 1932, 9.
24 *Vancouver Sun*, December 9, 1931, 28.
25 *The Sun*, Sydney, New South Wales, Australia, August 1, 1927, 1.
26 See geni.com/people/Lt-Col-John-Edwards-Leckie-Strathcona-Horse-Rgmt.
27 David Mattison, "Gosnell, R. Edward," *Dictionary of Canadian Biography*, vol. 16 (Toronto and Quebec City: University of Toronto/Université Laval, 2003).
28 Margot Wallace Files on Geoff Hosken.

7: Salt of the Sea

1 LAC RG 150, accession 1992-93/170, vol. 16-1525.
2 LAC RG 117, Vol. 472 ABZ1968, July 13, 1953, War Claims Commission Correspondence.

3 LAC RG 117, Vol. 472 ABZ1968.
4 *Canadian Skirmisher* Manifest, New York, August 6, 1925.
5 Dave Thackray Interview, March 16, 2021.
6 Beverley Thackray Interview, March 19, 2021.
7 Beverley Thackray Interview, March 29, 2021.
8 Beverley Thackray Interview, March 29, 2021.
9 Beverley Thackray Interview, March 19, 2021.
10 Norm Hutton Interview, March 2, 2021.
11 Peter Kemp, ed., *Oxford Companion to Ships and the Sea* (New York: Oxford University Press, 1976), 679.
12 *Russia* Logbook, March 21, 1934, 20–21.
13 *Vancouver Sun*, April 10, 1934, 7.
14 LAC RG 150, Accession 1992-93/166, Box 2595-2.
15 Oliver Memoir, 64.
16 The *Admiral Graf Spee* was scuttled on December 17, 1939, following the Battle of the River Plate, east coast of South America.
17 Oliver Memoir, 77.
18 Oliver Memoir, 80.

8: Rising Twenties

1 *Victoria Daily Times*, June 13, 1921, 14.
2 *Victoria Daily Times*, June 13, 1921, 14.
3 The *Empress of Canada*'s time was eight days, 10 hours, 53 minutes, *Victoria Daily Times*, June 18, 1923.
4 Robert Turner, *The Pacific Empresses: An Illustrated History of Canadian Pacific Railway's Empress Liners on the Pacific Ocean* (Victoria, BC: Sono Nis Press, 1981), 111.
5 *Canadian Railway and Marine World*, June 1920, 321, canadiana.ca/view/oocihm.8_06968_95/41.
6 Turner, *The Pacific Empresses*, 112.
7 *Vancouver Daily Province*, February 18, 1920, 20.
8 Grandmother Mary Frances McNeal's 1921 Journal, courtesy Aimee Herndon and Sandra Howell. Hereafter McNeal Journal.
9 Sixty percent of the world's raw silk shipped from Yokohama, as noted in CPFF, 239.
10 *Vancouver Daily Province*, April 19, 1921, 7.
11 *Victoria Daily Times*, April 17, 1922, 8.

12 CPFF, 236.
13 *Vancouver Sun*, May 18, 1926, 15.
14 Joel Thomas Livingston, *A History of Jasper County, Missouri, and Its People*, vol. 2 (Chicago: Lewis Publishing, 1912).
15 McNeal Journal.
16 Black, *Harry Livingstone's Forgotten Men*, 443.
17 *Vancouver Daily Province*, June 20, 1921, 22.
18 *Vancouver Daily Province*, June 20, 1921, 17.
19 *Vancouver Sun*, June 24, 1921, 2.
20 McNeal Journal.
21 Nelle Picard Interview, August 8, 2022. Hereafter Picard Interview.
22 *Empress of Ireland* first-class passenger list compiled by Geoffrey E. Whitfield and Craig Stringer.
23 Craig Stringer Email to Nelson Oliver, December 31, 2007.
24 *Vancouver Daily Province*, August 7, 1910, 5.
25 Picard Interview.
26 McNeal Journal.
27 The Grand Hotel, Yokohama, Guidebook, circa 1920.
28 McNeal Journal.
29 *Victoria Daily Times*, August 18, 1921, 1.
30 *Victoria Daily Times*, August 18, 1921, 1.
31 *Victoria Daily Times*, August 18, 1921, 1.
32 *Victoria Daily Times*, August 18, 1921, 3; Marian A. Packham, "MacCallum, Archibald Byron," *Dictionary of Canadian Biography*, vol. 16 (Toronto and Quebec City: University of Toronto/Université Laval, 2003), biographi.ca/en/bio/macallum_archibald_byron_16E.html.
33 Sarah Russell Spray Interview, January 4, 2022. Hereafter Spray Interview.
34 *Victoria Daily Times*, August 19, 1921, 8.
35 Spray Interview.
36 William F. Russell Family Photo Archives, courtesy Sarah Russell Spray.
37 Kathryn Smith, "Frank Lloyd Wright and the Imperial Hotel: A Postscript," *The Art Bulletin* 68, no. 2 (June 1985), 296–310, updated 2014.
38 Smith, "Frank Lloyd Wright and the Imperial Hotel, 296–310.
39 Smith, "Frank Lloyd Wright and the Imperial Hotel, 296–310.
40 *New York Times*, August 16, 1914, 12.
41 Smith, "Frank Lloyd Wright and the Imperial Hotel, 296–310.
42 Richard Gimbel Photo Album, Nelson Oliver Collection.

43 Marla Hardee Milling, "Cornelia Vanderbilt's Brushes with Death," *Blue Ridge County Magazine* (July 1, 2017), blueridgecountry.com/newsstand/magazine/cornelia-vanderbilts-many-brushes-with-death.
44 *Victoria Daily Times*, March 23, 1925, 16.
45 *Vancouver Daily Province*, May 23, 1925.
46 *Vancouver Daily Province*, June 8, 1927, 22.
47 *Victoria Daily Times*, June 9, 1927, 15; Karen May and George Lewis, "The Deaths of Roald Amundsen and the Crew of the Latham 47," *Polar Record* 51, no. 1 (January 2015): 1–15.
48 *Victoria Daily Times*, May 30, 1927, 8.
49 *Victoria Daily Times*, October 3, 1927, 8.
50 Hallett Abend, *My Life in China, 1926–1941* (New York: Harcourt, Brace, 1941), 59–60, 107.
51 Abend, *My Life in China*, 121.
52 *Vancouver Daily Province*, April 26, 1930, 5.
53 *Victoria Daily Times*, September 28, 1931, 4.

9: Collapsing Thirties

1 *Vancouver Sun*, December 20, 1934, 20.
2 "Personality Traits and Life of Phylis Williams," a profile written by Bill Attwell and shared with the author, January 11, 2022. The material is also based on several interviews the author had with Bill Attwell between November 21, 2021, and January 12, 2022.
3 *Vancouver Daily Province*, August 12, 1932, 4.
4 *Vancouver Sun*, October 31, 1933, 2.
5 *Vancouver Sun*, May 31, 1935, 17.
6 Vancouver-Orient Tour Brochure and Itinerary from organizer and host Annie G. Stewart for tour commencing July 14, 1934, on the *Empress of Asia*, ending August 31, 1934. Nelson Oliver Collection. The document also notes the $375 fare and the tour's inclusions. See www.empressofasia.com.
7 *Vancouver Sun*, May 15, 1934, 8.
8 *Vancouver Sun*, May 23, 1933, 5.
9 *Victoria Daily Times*, November 25, 1922, 22.
10 *Vancouver Daily Province*, July 25, 1930, 11; *Victoria Daily Times*, October 24, 1931, 17.
11 *Vancouver Sun*, May 7, 1934, 1.
12 Turner, *The Pacific Empresses*, 141.

13 Turner, *The Pacific Empresses*, 146.
14 *Vancouver Daily Province*, October 16, 1934, 1.
15 *Vancouver Sun*, December 27, 1934, 4.
16 *Vancouver Daily Province*, September 23, 1929, 24.
17 Harbour and Shipping, LAC F-27-1-5, vol. 16: nos. 1–12 (January–December 1933); Lamb, *BCHQ*, 115.
18 CPFF, General Publicity Department, Canadian Pacific Foundation Library, 1937, 101.
19 *Vancouver Sun*, December 16, 1931, 12.
20 See www.empressofasia.com.
21 CPFF, 207.
22 *Vancouver Sun*, November 7, 1933, 3.
23 *Vancouver Sun*, May 21, 1934, 16.
24 Turner, *The Pacific Empresses*, 271.
25 *Victoria Daily Times*, July 12, 1930, 20.
26 See adp.library.ucsb.edu/index.php/matrix/detail/2000145280/W98698-Sonata_in_D_minor.
27 *Victoria Daily Times*, February 9, 1931, 20.
28 *Asia* Logbook, April 28 to October 13, 1931, 15.
29 *Vancouver Sun*, May 11, 1934, 3.
30 British Columbia Division of Vital Statistics, Death Registration 048670; *Vancouver Sun*, September 18, 1933, 7.
31 Vancouver-Orient Tour on *Empress of Asia*, ending August 31, 1934, Passenger List, September 23, 1933.
32 *Victoria Daily Times*, July 6, 1931, 1.
33 *Vancouver Sun*, May 23, 1933, 3.
34 *Vancouver Daily Province*, March 14, 1932, 5; Steerage Assignment, *Asia* Logbook, Crew List 1931.
35 *Asia* Logbook, April 4 to September 11, 1933, 15.
36 *Asia* Logbook, 17.
37 *Vancouver Sun*, July 14, 1939, 6.
38 George Schuthe, "Canadian Shipping in the British Columbia Coastal Trade" (master's thesis, University of British Columbia, April 1950), 112.
39 *Victoria Daily Times*, January 16, 1933, 1.
40 Ingenium, CP Steamship fonds, Box 130, September 7, 1935, Master-at-Arms Report.
41 *Asia* Logbook, August 13, 1935, to February 15, 1936, 13–14.

42 *Asia* Logbook, February 16 to July 27, 1936, 11.
43 *Victoria Daily Times* Weather Forecast, January 8, 1938, 1.
44 *Victoria Daily Times* Weather Forecast, January 8, 1938, 1.
45 *Victoria Daily Times*, January 12, 1938, 16.
46 Roderick Stewart, *The Mind of Norman Bethune* (Markham, ON: Fitzhenry & Whiteside, 2002), 96.
47 Abend, *My Life in China 1926–1941*, 227–29.

10: Sea upon Sky

1 CVA-AM335-S1, *Empress of Asia*, November 1925 agreement or articles and account of crew. This agreement covers the crew for the period of November 1925 to May 17, 1926. Of the 71 "European crew," 67 signed on in Vancouver and four were signed on in Hong Kong, having been transferred from other Pacific Empress ships. In addition, roughly 450 signed on under Hong Kong Articles.
2 *Victoria Daily Times*, November 11, 1924, 8.
3 Royal BC Museum and Archives, CBC Collection, T1339-0001, Captain Lionel Douglas recording — Imbert Orchard, 1962.
4 Royal BC Museum and Archives, CBC Collection, T1339-0001. The *Silberhorn* disappeared during a South Pacific voyage in 1907.
5 Royal BC Museum and Archives, CBC Collection, T1339-0001.
6 Lamb, *Empress to the Orient*, 147.
7 VAC, Major Campbell Mellis Douglas, Victoria Cross Citation.
8 Leona Kearns Letter to Mother, October 6, 1925.
9 Barbara Gregorich, "American Eye: Stranded," *North American Review* 283, nos. 3–4, May–August 1998, 1.
10 See faroutliers.blogspot.com/2004/08/philadelphia-bobbies-barnstorm-japan.html.
11 CVA-AM335-S1, Logbook for the Canadian Foreign Sea-Going Ship *Empress of Asia*, Vancouver–Hong Kong–Vancouver Voyage, November 12, 1925, to January 29, 1926, 18.
12 CVA-AM335-S1.
13 *Victoria Daily Times*, January 29, 1926, 14.
14 *Victoria Daily Times*, January 29, 1926, 14.
15 *Daily Colonist*, January 30, 1929, 13.
16 Chronometers reflect "real time" as calculated by trigonometric measurement.
17 See worldhistory.us/chinese-history/shanghai-in-the-1930s.php.

18 CVA-AM335-S1, Logbook for Canadian Foreign Sea-Going Ship *Empress of Asia*, Vancouver–Hong Kong–Vancouver voyage, November 12, 1925, to January 29, 1926, 15.

19 *Daily Colonist*, January 30, 1926, 13.

20 *Daily Colonist*, January 30, 1926, 13.

21 Order of Naval Court of Inquiry, Shanghai, January 14–16, 1926, issued by Board of Trade London, April 27, 1926.

22 CVA-AM335-S1, Logbook for Canadian Foreign Sea-Going Ship *Empress of Asia*, July 1, 1919, to December 16, 1919, 17.

23 Henrietta C. Kershaw, "The Storm," July 20, 1978, Palo Alto, California. Unpublished diary provided by Henrietta's daughter, Kathleen Kershaw.

24 *Vancouver Sun*, February 13, 1928, 1.

25 *Daily Colonist*, February 2, 1921, 15.

26 Kathleen Gillison Diary, 1938. Hereafter Gillison Diary.

27 Frances Clemmow Interview, April 18, 2023, hereafter Clemmow Interview; Gillison Diary.

28 *Vancouver Daily Province*, September 13, 1937, 3.

29 Turner, *The Pacific Empresses*, 190.

30 John Alexander died on August 13, 1937, 10 days after he was admitted to hospital in Yokohama. His ashes were committed to the deep at 34.53.5 degrees north, 144.13.5 degrees east, during the *Asia*'s return voyage to Vancouver from Yokohama. Source for geographical coordinates: Turner, *The Pacific Empresses*, 190.

31 *Vancouver Daily Province*, September 13, 1937, 3.

32 *Vancouver Daily Province*, September 11, 1937, 1.

33 Canadian Pacific Steamships Ltd., Commander's General Voyage Report, July–September 1937.

34 *Vancouver Daily Province*, September 8, 1937, 1–2.

35 *Vancouver Daily Province*, September 8, 1937, 1.

36 Ester Benjamin Shifren, correspondence with author, May 3, 2021; Ester Benjamin Shifren, *Hiding in a Cave of Trunks: A Prominent Jewish Family's Century in Shanghai and Internment in a WW II POW Camp* (Ester Benjamin Shifren, 2012).

37 *Vancouver Daily Province*, September 11, 1937, 1.

38 *Vancouver Daily Province*, September 13, 1937, 3.

39 *Sunday Sun* (Vancouver), September 11, 1937, 1.

11: Ship to Shore

1 Low Shau Wah (1898–1994) biographical sketch by granddaughter, Patricia Wong, August 25, 2022; author interviews 2022–23.

2 *Vancouver Dailey Province*, February 1, 1921, 30.

3 Picard Interview, 2022.

4 Yuen-Fong Woon, "Between South China and British Columbia: Life Trajectories of Chinese Women," *BC Studies*, nos. 156–57 (Winter/Spring 2007/08).

5 Government of Canada 1921 census, 542–43.

6 See bankofcanada.ca/rates/related/inflation-calculator.

7 Report of the Royal Commission on Chinese and Japanese Immigration, Order of Parliament, Ottawa, Session, 1902, 299.

8 Arlene Chan, "Chinese Immigration Act," *The Canadian Encyclopedia*, March 7, 2017, thecanadianencyclopedia.ca/en/article/chinese-immigration-act; Anthony B. Chan, "Chinese Canadians," *The Canadian Encyclopedia*, May 22, 2019, thecanadianencyclopedia.ca/en/article/chinese-canadians.

9 Province of British Columbia, Certificate of Marriage, 70684, Licence No. 71213.

10 Harry Con and Edgar Wickberg, *From China to Canada* (Toronto: McClelland & Stewart, 1982), 215–16.

11 Interviews with Albert Hugh's sons Dennis, Lorne, and Richard, and granddaughter Shana Hugh, January 17, 2023. Hereafter Hugh Family Interviews.

12 Hugh Family Interviews.

13 Shana Hugh Email, November 3, 2023.

14 Hugh Family Interviews.

15 Raymond Chong, "Gim Suey Chong — His Life from Hoyping to Gum Saan," *Gum Saan Journal* (2009): 19–53; additional material, Raymond Chong Interviews with author, August 2022, hereafter Chong Interviews.

16 Chong, "Gim Suey Chong"; Chong Interviews.

17 *Vancouver Daily Province*, April 15, 1932, 24.

18 Harold James, *Krupp: A History of the Legendary German Firm* (Princeton, NJ: Princeton University Press, 2012), 2.

19 *Victoria Daily Colonist*, April 12, 1932, 12.

20 *Victoria Daily Times*, April 11, 1932, 18.

21 *Victoria Daily Times*, April 11, 1932.

22 U.S. Department of Labor Immigration Service, File 2500/9540, U.S. Immigration Office, East Boston, Massachusetts, April 21–22, 1932.

23 *Sunday Sun* (Vancouver), September 23, 1923, 32.

24 Official Report of Captain S. Robinson, Royal Navy Reserve, Commander of the Canadian Pacific SS *Empress of Australia*, 1. Hereafter Robinson Report.
25 Robinson Report, 1.
26 Robinson Report, 1.
27 *Victoria Daily Times*, September 22, 1923, 1, 8.
28 Robinson Report, 2.
29 Robinson Report, 3.
30 Robinson Report, 5.
31 Robinson Report, 10–11.
32 *Vancouver Sun*, September 6, 1923, 1.
33 *Vancouver Sun*, September 6, 1923, 13.
34 *Vancouver Sun*, September 6, 1923, 17.
35 *Vancouver Sun*, September 6, 1923, 23.
36 Turner, *The Pacific Empresses*, 139.
37 *Vancouver Daily Province*, December 5, 1927, 20; Bunny (Al) Hailey, correspondence with author, September 5, 2022.
38 *Victoria Daily Times*, June 30, 1925, 1.
39 Gladys Erickson and Gordon R. Erickson, *Freeway West* (Falun, AB: Falun Historical Society, 1974), 101–2.
40 *Victoria Daily Colonist*, July 25, 1925, 7.
41 Hugh Family Interviews.
42 Raymond Chong Correspondence.

12: Smugglers and Stowaways

1 Heavy fog reported by Vancouver and Victoria newspapers, October 8–13, 1921.
2 *Vancouver Daily Province*, December 5, 1923, 1.
3 *Victoria Daily Times*, December 5, 1923, 1.
4 *Vancouver Daily Province*, December 18, 1923, 10.
5 *Vancouver Daily Province*, December 18, 1923, 10.
6 *Vancouver Sun*, March 11, 1924, 14; *Vancouver Evening Sun*, May 22, 1924, 9.
7 Nelson Oliver Interview, October 26, 2023.
8 See web.uvic.ca/vv/student/chinatown/opium/p.3.html.
9 *Vancouver Daily Province*, December 30, 1909, 25.
10 *Vancouver Daily Province*, December 30, 1909, 25.
11 Report, Mr. Justice Dennis Murphy, Royal Commissioner Appointed to Investigate Alleged Chinese Frauds and Opium Smuggling on the Pacific Coast, 1910–11, May 2, 1911, Vancouver, Library of Parliament, Ottawa, 47.
12 *Vancouver Sun*, September 11, 1913, 3.

13 Report of the Department of Customs, Tables of Imports, Exports and Navigation of the Dominion of Canada for Fiscal Year Ended March 31, 1914, 78, and Department of Customs, Tables and Statements of Imports and Exports of the Dominion of Canada for Fiscal Year Ended March 31, 1918, 152.
14 A Century of International Drug Control, United Nations Office on Drugs and Crime, 2008, 3–7.
15 Historical World Population Data, William Robert Johnston, modified February 2015, johnstonsarchive.net.
16 A Century of International Drug Control, UNODC, 2008, 3.
17 *Times* of London, Saturday, April 30, 1927, 11, Simon Fraser University Digital Archive 1785–1985.
18 *Times* of London, Saturday, April 30, 1927, 11.
19 *Times* of London, Saturday, April 30, 1927, 11.
20 *Vancouver Daily Province*, November 18, 1924, 24.
21 *Daily Colonist*, January 24, 1925, 17.
22 *Vancouver Daily Province*, March 25, 1926, 1.
23 Vancouver-Orient Tour on *Empress of Asia*, Notable Passengers, March 22, 1926.
24 Geoffrey Serle, *Australian Dictionary of Biography*, vol. 10, 1986.
25 *Vancouver Daily Province*, October 22, 1926, 25.
26 *Vancouver Daily Province*, July 20, 1926, 1.
27 *Vancouver Sun*, April 1, 1927, 14.
28 *Vancouver Daily Province*, April 15, 1932, 24.
29 *Victoria Daily Times*, August 4, 1933, 20.
30 *Vancouver Sun*, November 15, 1937.
31 *Victoria Daily Times*, June 6, 1924, 8.
32 *Vancouver Sun*, October 27, 1913.
33 *Vancouver Sun*, December 22, 1913.
34 List of Alien Passengers for passengers bound for the United States from Liverpool, August 23, 1918; U.S. Immigration Officer Port of Arrival report, New York, September 1, 1918.
35 *Asia* Logbook, Liverpool–New York, August 24, 1918.
36 *Asia* Logbook, Liverpool–New York, November 5, 1918.
37 *Vancouver Daily Province*, November 3, 1926, 8.
38 *Vancouver Daily Province*, November 10, 1926, 26.
39 *Vancouver Daily Province*, November 11, 1926, 25.
40 See Vancouver-Orient Tour website at www.empressofasia.com.
41 *Asia* Logbook, Vancouver to Hong Kong, March 22, 1921.
42 *Asia* Logbook, Vancouver to Hong Kong, March 5, 1929.

43 *Asia* Logbook, Hong Kong to Vancouver, March 9, 1930.
44 *Vancouver Sun*, August 8, 1939, 18.
45 *Vancouver Daily Province*, July 3, 1920, 32.
46 *Vancouver Daily Province*, July 3, 1920, 32.

13: The "Accidental" Bombing

1 *Victoria Daily Times*, September 24, 1940, 14.
2 Memorandum for Prime Minister, September 27, 1940, from ODS/ET, RG25, External Affairs, Series A-2, vol. 2959, File B-16.
3 RG25, External Affairs, vol. 1874, File 1938 468.
4 RG25, External Affairs, Series A-2, vol. 2959, File B-16.
5 *Vancouver Sun*, September 18, 1940, 9.
6 Herbert Connell, LAC RG 150, Accession 1992-93/166, Box 1911-6.
7 *Sunday Sun* (Vancouver), December 27, 1930, 1.
8 *Vancouver Sun*, September 24, 1940, 1.
9 *Vancouver Daily Province*, September 24, 1940, 2.
10 *Victoria Daily Times*, September 24, 1940, 14.
11 *Victoria Daily Times*, September 24, 1940, 14.
12 *Victoria Daily Times*, September 24, 1940, 14.
13 *Vancouver Sun*, September 24, 1.
14 John Drummond Letter, courtesy Jim Drummond. Hereafter Drummond Letter.
15 *Vancouver Sun*, September 24, 1940, 1.
16 *Vancouver Daily Province*, September 24, 1940, 2.
17 *Vancouver Daily Province*, September 24, 1940, 1–2.
18 Commander's General Voyage Report, August–September 1940.
19 Drummond Letter.
20 Drummond Letter.
21 *Vancouver Daily Province*, September 24, 1940, 1–2.
22 Nelson Oliver Interview with Ernie Higgs, undated.
23 *Victoria Daily Times*, September 24, 1940, 14.
24 John H. Kerr Letter to Harry Schofield, September 17, 1940.
25 *Japan Advertiser*, September 15, 1940.
26 *Japan Advertiser*, September 15, 1940.
27 LAC RG25 External Affairs, Series A-3-b, vol. 2813, File 1079-40, Telegram.
28 LAC RG25, External Affairs, Series A-3-b, vol. 2813, File 1079-40, Bombing of the "Empress of Asia."
29 *Vancouver Sun*, September 19, 1940, 1.
30 LAC RG25, External Affairs, Series A-2, vol. 2959, File: B-16.

31 LAC RG25, External Affairs, Series A-3-b, vol. 2813, File: 1079-40.
32 LAC RG25, External Affairs, Series A-3-b, vol. 2813, File 1079-40, Telegram from Ottawa September 30, 1940.
33 LAC RG25, External Affairs, Series A3b, vol. 2813, File 1079 40, File Part 1. Telegram from Ottawa, September 30 and October 2, 1940.
34 *Victoria Daily Times*, August 7, 1939, 11.
35 *Vancouver Sun*, August 11, 1939, 19.
36 *Vancouver Sun*, August 21, 1939, 5.
37 *Vancouver Sun*, August 28, 1939, 17.
38 Italy entered the Second World War on June 10, 1940.
39 Tozer Email, September 14, 2001.
40 Walsh Email, October 1, 2006.
41 See gltugs.wordpress.com/sulphite.
42 Norma Lumley Interview, June 2021; see also community-focus.com, Local Veterans Recognized by Museum, October 25, 2002.
43 *Vancouver Sun*, February 3, 1947, 4.
44 Dale Mitchell Interview, July 11, 2021.
45 Patricia Slade Interview and Correspondence, January 19, 2022.
46 Frank Davis Interview, May 25, 2021.
47 Frank Davis Interview, May 25, 2021.
48 Oliver Memoir, 81.
49 Frank Davis Interview, May 25, 2021.
50 Frank Davis Interview, May 25, 2021.
51 Atkins Interview, January 29, 2021.

14: Ark of War

1 Oliver Memoir, 82–84; Nelson Oliver Interviews, 2020–24.
2 VAC MNR File AIP202100041.
3 Nelson Oliver Files; VAC MNR, File AIP 202000151.
4 Oliver Memoir, 82.
5 Oliver Memoir, 82.
6 Atkins Interview, January 29, 2021.
7 See thesecondworldwar.org/western-front-1939-1940/the-blitz/liverpool.
8 Oliver Memoir, 84.
9 Kay Hasler Interview, November 22, 2022.
10 Atkins Interview, January 29, 2021.
11 Atkins Interviews, January 29, 2021, and June 18, 2021.
12 Gerry O'Neill Email to Nelson Oliver, February 26, 2007.

13 Louise Ryan, "Aliens, Migrants and Maids: Public Discourses on Irish Immigration to Britain in 1937," *Immigrants & Minorities* 20, no. 3 (2001): 25–42.
14 Mary J. Hickman and Louise Ryan, "The Irish Question: Marginalizations at the Nexus of Sociology of Migration and Ethnic and Racial Studies in Britain," *Ethnic and Racial Studies* 43, no. 16 (2020): 96–114.
15 Patrick James Brown Certificate of Discharge Book.
16 Al Brown Correspondence.
17 Brown Certificate of Discharge Book.
18 Census for 1921; *Liverpool Echo*, August 15, 1939.
19 *A Life at Sea: Tales from Ted Evans BEM*, self-published, Gwen Butler, 34. Hereafter *Life at Sea*.
20 Oliver Perks Blog.
21 *Life at Sea*, 34.
22 Oliver Memoir, 85.
23 Walsh Email, June 22, 2003.
24 Walsh Email, May 1, 2003.
25 Oliver Perks Blog.
26 *Life at Sea*, 35.
27 *Life at Sea*, 36.
28 Atkins Interview, January 29, 2021.
29 *Life at Sea*, 37.
30 See royal.uk.
31 *Life at Sea*, 37; Oliver Memoir, 88.
32 Atkins Interview, January 29, 2021.
33 Oliver Memoir, 88.
34 *New York Times*, September 5, 1941.
35 Oliver Memoir, 90.
36 *New York Times*, September 5, 1941.
37 *New York Times*, September 5, 1941.
38 J. Peter Hvidsten, *Per Hvidsten: An Amazing Journey: Norway to Canada* (J. Peter Hvidsten, 2012).
39 Oliver Memoir, 93.
40 Geoff Tozer Letter, September 14, 1941.
41 Commander Tony German, *The Sea Is at Our Gates: The History of the Canadian Navy* (Toronto: McClelland & Stewart), 108–9.
42 Herbert Roistacher Letters, September 12–15, 1941. Hereafter Roistacher Letters or Letter.
43 Tozer Email, October 4, 2001.

44 Rosalind Brice and accounts from her father, Colin Alt.
45 LAC RG24-C-24-a, File HQS 63-302-249, Microfilm Reel C-5612.
46 *Globe and Mail*, January 28, 1972, 3.
47 Roistacher Letter, September 23, 1941.
48 Roistacher Letter, September 23, 1941.
49 Roistacher Letter, September 28, 1941.
50 Oliver Memoir, 93.

15: Last in Line, First in Trouble

1 Tozer Email, September 14, 2001.
2 Nelson Oliver Interviews, 2020–23.
3 Tozer Interview with Nelson Oliver, 2002.
4 Marcel Denommee Interview, January 5, 2022.
5 *Vancouver Sun*, December 2, 1940, 1.
6 Tozer Email, November 6, 2001.
7 Tozer Email, November 6, 2001.
8 Atkins Interview, January 29, 2021.
9 "Bill Taylor's Memoir: My Unusual Voyage Aboard the Empress of Asia and the Aftermath" by Thomas William (Bill) Taylor Merchant Navy Survivor and former prisoner of war, 7. Hereafter Taylor Memoir. See billtaylorpow.tripod.com/index.html. Courtesy Hightail Tozer Papers.
10 Hosken Diary, November 23, 1941.
11 Taylor Memoir, 7.
12 Atkins Interview, June 18, 2021.
13 Hosken Diary, November 24, 1941.
14 Tozer Email, May 13, 2002.
15 Gary Haines, "The Fall of Singapore: The Fall of Empire," Royal Air Force Museum, August 13, 2020, rafmuseum.org.uk/blog/author/gary-haines-archivist. Hereafter Haines, "The Fall of Singapore."
16 Hosken Diary, December 19, 1941.
17 Hosken Diary, January 4, 1942.
18 Taylor Memoir, 10.
19 Hosken Diary, January 17–18, 1942.
20 Beverley Thackray Interview, March 19, 2021.
21 Haines, "The Fall of Singapore," 11.
22 John Watts Report, 1. Hereafter Watts Report.
23 See cofepow.org.uk/armed-forces-stories-list/125th-anti-tank-regiment-royal-artillery.

24 Diary of Stanley Catherall, courtesy David Catherall, 2001.
25 Watts Report, 3.
26 Watts Report, 4.
27 Winston Churchill, *The Second World War: The Hinge of Fate* (Boston: Houghton Mifflin, 1950), 92.
28 Email Files, Peter Foster.
29 Email Files, Jan Care.
30 Email Files, Terry Loudwell.
31 Jennifer Padley, 2012 profile on Bert Needham, courtesy of Rob Padley. Hereafter Jennifer Padley, Needham Profile.
32 Jennifer Padley, Needham Profile.
33 Watts Report, 4–5.
34 Oliver Memoir, 96.
35 Tozer Email, April 1, 2002.
36 Hosken Diary, January 26, 1942.
37 Hosken Diary, January 31, 1942. Starting in January 1942, Hosken dated the entries in his diary, now more a journal.
38 Hosken Diary, February 2, 1942.

16: Voyage of Doom

1 IWM, Accession 29238.
2 Hosken Diary, February 4, 1942.
3 Hosken Diary, February 4, 1942.
4 Oliver Memoir, 98.
5 Watts Report, 7.
6 IWM, Accession 25212, Reel 5.
7 Atkins Interviews, January 29, 2021, to June 18, 2021.
8 Hosken Diary, February 4, 1942.
9 Allen Tozer Wartime Diary, February 4, 1942. Hereafter Tozer Wartime Diary.
10 Watts Report, 8.
11 Hosken Diary, February 4, 1942.
12 Oliver Memoir, 98.
13 John Drummond, Wartime Journal, February 4, 1942. Hereafter Drummond Wartime Journal.
14 Watts Report, 10.
15 Captain John Bisset Smith, 1942A. Commander's Report, *Empress of Asia*, Voyage 151/3. Report to Canadian Pacific on Final Voyage, February 1942, 1–4. Hereafter Commander's Report.

16 Report of an interview with Master Captain J.B. Smith, SS *Empress of Asia* in Convoy, www. cofepow.org.uk.
17 Watts Report, 9–10.
18 Captain Smith Interview COFEPOW.
19 Atkins Interview, February 2, 2021.
20 Hosken Diary, February 4, 1942, 33.
21 Watts Report, 11.
22 IWM, Accession 15350, Reel 7.
23 Atkins Interview, June 18, 2021.
24 Commander's Report; Nelson Oliver Interviews, 2020–23.
25 IWM, Accession 25212, Reel 5.
26 Oliver Memoir, 98.
27 Watts Report, 11.
28 Watts Report, 12.
29 Taylor Memoir, 11.
30 Watts Report, 14.
31 Watts Report, 13–14.
32 Oliver Memoir, 98.
33 Oliver Memoir, 99.
34 Watts Report, 14–15.
35 Oliver Memoir, 99; Tozer Diary, February 5, 1942; Watts Report, 15.
36 Oliver Memoir, 99.
37 Commander's Report, 1.
38 Atkins Interview, January 29, 2021.
39 Hosken Diary and Tozer Wartime Journal, February 5, 1942.
40 Oliver Memoir, 100.
41 Watts Report, 16–17.
42 Atkins Interview, January 29, 2021.
43 Watts Report, 17.
44 Oliver Memoir, 100.
45 Hosken Diary, February 5, 1942.
46 *Vancouver Daily Province*, April 24, 1942, 34.
47 Drummond Wartime Journal, February 5, 1942.
48 Oliver Memoir, 100.
49 Commander's Report, 1.
50 Watts Report, 24–25.
51 IWM, Accession 29238, Reel 6.
52 IWM, Accession 29238, Reel 6.

53 Watts Report, 19.
54 Oliver Memoir, 102.
55 Oliver Memoir, 101.
56 Taylor Memoir, 12.
57 Atkins Interview, January 29, 2021.
58 Cecil Crofts Letter to Jack Ewart, April 30, 1966.
59 Tozer Email, December 20, 2001.
60 Memories of Richard Farrell's daughter, Deirdre, courtesy of Fiona Allen.
61 *Vancouver Daily Province*, April 24, 1942, 34.
62 Taylor Memoir, 12.
63 IWM, Accession 29238, Reel 6.
64 IWM, Accession 25212, Reel 5.
65 Profile of Captain Wilfred Harrington in the *Australian Dictionary of Biography* states 1,804 men were rescued.
66 *Victoria Daily Times*, May 13, 1942, 18.
67 Hosken Diary, February 5, 1942.
68 Oliver Memoir, 102.
69 Crofts Letter to Ewart, April 30, 1966.
70 Commander's Report, 2.
71 Chief Officer Donald Smith, February 5, 1942, SS *Empress of Asia* Voyage 155/3, Chief Officer's Report to CPR.
72 Oliver Memoir, 102; Nelson Oliver, *Empress of Asia* Casualty Files.
73 *Vancouver Sun*, May 20, 1942, 22.
74 NA-WO361411.

17: Escape from Singapore

1 Oliver Memoir, 103.
2 Drummond Wartime Journal, February 5, 1942, 7.
3 Atkins Interviews, January 29 to February 2, 2021.
4 Oliver Memoir, 103.
5 Commander's Report, 3.
6 LAC RG117, vol. 472.
7 Crofts Letter to Ewart, April 30, 1966.
8 Oliver Memoir, 104.
9 Oliver Memoir, 104.
10 Oliver Memoir, 105.
11 Oliver and Tozer state the *City of Canterbury* and *Félix Roussèl* departed Singapore on February 6, 1942. The movement cards for both ships also give this date.

12 Drummond Wartime Journal, February 8, 1942.
13 Tozer Wartime Journal, February 8, 1942.
14 Oliver Memoir, 105.
15 Hosken Diary, February 9, 1942.
16 Letter to CPR Directors from Medical Department, Government of Straits Settlements, September 11, 1942.
17 Commander's Report, 3.
18 Hosken Diary, February 9, 1942.
19 Oliver Memoir, 106.
20 Oliver Memoir, 106.
21 Tozer Letter to VAC Minister George Baker, July 25, 2000.
22 Hosken Diary, February 10, 1942.
23 Interview Notes by Dr. Wallace Chung from interview with Donald Smith, March 22, 1963.
24 Tozer Wartime Journal, February 11, 1942.
25 Hosken Diary, February 11, 1942.
26 Oliver Memoir, 108.
27 Norman A. Mackenzie Email, March 13, 2021.
28 Hosken Diary, February 11, 1942.
29 Oliver Memoir, 109.
30 IWM Accession 32324, Reel 1.
31 IWM Accession 32324, Reel 1.
32 Oliver Memoir, 109.
33 Drummond Wartime Journal, February 13, 1942.
34 Hosken Diary, February 13, 1942.
35 Tozer Email, November 8, 2001.
36 Oliver Memoir, 111; Drummond Wartime Journal, February 13, 1942, 12.
37 Hosken Diary, February 13, 1942.
38 Oliver Memoir, 111.
39 Drummond Wartime Journal, February 14, 1942, 12.
40 Hosken Diary, February 14, 1942.
41 Tozer Wartime Journal, February 13–14, 1942.
42 *Victoria Daily Times*, May 13, 1942, 18.
43 Tozer Wartime Journal, February 14, 1942.
44 *Victoria Daily Times*, May 13, 1942, 18.
45 Tozer Wartime Journal, February 14, 1942.
46 Tozer Email, April 24, 2002.
47 Tozer Email, October 10, 2001.

48 *Victoria Daily Times*, May 13, 1942, 18.
49 IWM Accession 33518, Reel 2.
50 Watts Report, 25.

18: Homeward

1 Oliver Memoir, 113.
2 Hosken Diary, February 16, 1942.
3 Hosken Diary, February 17, 1942.
4 Atkins Interview, January 29, 2021.
5 Oliver Memoir, 114.
6 WO392/26.
7 IWM Accession 32324, Reel 2.
8 Taylor Memoir, 17–19.
9 IWM Accession 32324, Reel 1.
10 Peter Foster Correspondence, December 28, 2022, to January 2, 2023.
11 Rod Padley Correspondence, July 18–23, 2023.
12 IMW Accession 33518, Reel 2.
13 *Dumfries and Galloway Standard*, June 10, 1942, 5.
14 Tozer Email, January 17, 2002.
15 Drummond Wartime Journal, April 2, 1942.
16 Hosken Diary, February 21, 1942.
17 Oliver Memoir, 117.
18 Oliver Memoir, 117.
19 Atkins Interview, January 29, 2021.
20 Drummond Wartime Journal, April 24, 1942, 78.
21 See warfarehistorynetwork.com.
22 Drummond Wartime Journal, May 6, 1942.
23 Hosken Diary, April 3, 1942.
24 Hosken Diary, April 18, 1942.
25 Tozer Email, September 10, 2001.
26 *Vancouver Daily Province*, April 18, 1942, 13.
27 Nelson Oliver Email, January 15, 2003.
28 Richard Farrell Letter, November 1, 1942.
29 Beverley Thackray Interview, March 19, 1921.
30 Dave Thackray Interview, March 16, 2021.
31 Bree Letter courtesy of Kay Hasler.
32 IWM Accession 33518, Reels 1–3.
33 IWM Accession 33518, Reels 2–3.

34 LAC RG117, vol. 472.
35 Lamb, *BCHQ*, 61.
36 Lamb, *BCHQ*, 63.
37 Nelson Oliver Interview, October 1, 2023.

Epilogue: Signing Off

1 Atkins Interview, February 2, 2021.
2 Crofts Letter to Ewart, April 30, 1966.
3 BC Death Registration, 62-09-011259.
4 BC Death Registration, 007988.
5 *Chilliwack Progress*, April 26, 2002.
6 BC Death Registration, 43-031-23; Dave Thackray Email, April 18, 2023.
7 BC Death Registration, 98-28020.
8 *The Dolphin*, May 1979, 9.
9 Robert Gibson Interview, March 20, 2023.
10 *Scottish Daily Express*, September 6, 1963, 7.
11 BC Death Registration, 71-09-009115.
12 BC Death Registration, 65-09-012660.
13 Bill Attwell Correspondence and Interview, January 2022.

SELECTED BIBLIOGRAPHY

Books

Abend, Hallett. *My Life in China 1926–1941*. New York: Hallett Abend, 1943.

Belchem, John. *Irish, Catholic and Scouse: The History of the Liverpool Irish, 1800–1940*. Liverpool, England: Liverpool University Press, 2007.

Black, Dan. *Harry Livingstone's Forgotten Men: Canadians and the Chinese Labour Corps in the First World War*. Toronto: James Lorimer, 2019.

Canadian Pacific Facts and Figures. Compiled and edited by General Publicity Department. Montreal: Canadian Pacific Foundation Library, 1937.

Churchill, Winston S. *The Second World War: The Hinge of Fate*. Boston: Houghton Mifflin, 1950.

The Fairfield Shipbuilding and Engineering Works: History of the Company: Review of Its Productions and Descriptions of the Works. Govan, Scotland: Fairfield Shipbuilding and Engineering, 1909.

Fraser, John. *The Chinese: Portrait of a People*. Glasgow: William Collins & Sons, 1981.

German, Tony. *The Sea Is at Our Gates: The History of the Canadian Navy*. Toronto: McClelland & Stewart, 1990.

Halford, Robert G. *The Unknown Navy: Canada's World War II Merchant Navy*. St. Catharines, ON: Vanwell Publishing, 1995.

Keegan, John. *The First World War*. Toronto: Key Porter, 1998.

Kemp, Peter, ed. *Oxford Companion to Ships and the Sea*. New York: Oxford University Press, 1976.

Lamb, W. Kaye. *Empress to the Orient*. Vancouver: Vancouver Maritime Museum, 1991.

____. *History of the Canadian Pacific Railway*. New York: Macmillan Publishing, 1977.

Livingston, Joel Thomas. *A History of Jasper County, Missouri, and Its People*. Vol. 2. Chicago: Lewis Publishing, 1912.

Machum, Lieutenant-Colonel George. *Canada's V.C.'s: The Story of Canadians Who Have Been Awarded the Victoria Cross*. Toronto: McClelland & Stewart, 1956.

Macpherson, Ken, and John Burgess. *The Ships of Canada's Naval Forces 1910–1981*. Toronto: Collins, 1981.

Massey, Anne. *Designing Liners: A History of Interior Design Afloat*. 2nd ed. New York: Routledge, Taylor & Francis Group, 2021.

McKee, Fraser. *The Canadian Naval Chronicle, 1935–1945*. St. Catharines, ON: Vanwell Publishing, 1998.

____. *Sink All the Shipping There: Canada's Wartime Merchant Ship and Fishing Schooner Sinkings*. St. Catharines, ON: Vanwell Publishing, 2004.

Moss, Michael S. *Oxford Dictionary of National Biography*. Toronto: Oxford University Press, 2004.

Schull, Joseph. *Far Distant Ships: An Official Account of Canadian Naval Operations in World War II*. Toronto: Stoddart Publishing, 1987.

Shifren, Ester Benjamin. *Hiding in a Cave of Trunks: A Prominent Jewish Family's Century in Shanghai and Internment in a WW II POW Camp*. Ester Benjamin Shifren, 2012.

Stewart, Roderick. *The Mind of Norman Bethune*. Markham, ON: Fitzhenry & Whiteside, 2002.

Thompson, Peter. *The Battle for Singapore: The True Story of the Greatest Catastrophe of World War II*. London: Piakus Books, 2005.

Tolstoy, Ilya. *Reminiscences of Tolstoy*. New York: Century Company, 1914.

Turner, Robert D. *The Pacific Empresses: An Illustrated History of Canadian Pacific Railway's Empress Liners on the Pacific Ocean*. Victoria, BC: Sono Nis Press, 1981.

Magazines, Newspapers, Journal Articles, Essays

Belanger, Damien-Claude. "Marianopolis College, Biographies of Prominent Quebec and Canadian Historical Figures." Montreal: McGill University (2004).

Blackburn, Kevin. "Commemorating and Commodifying the Prisoner of War Experience in South-East Asia: The Creation of Changi Prison Museum." *Journal of the Australian War Memorial* 33 (January 2000).

Chamberlain, Tony. "'Stokers — The Lowest of the Low?' A Social History of the Royal Navy Stokers 1850–1950." University of Exeter (March 2013).

Connelly, Mark L. "The British Campaign in Aden, 1914–1918." *Journal of First World War Studies* 1, no. 3 (2005).

Cousineau, Aimé, and F.G. Legg. "Hydrocyanic Acid and Other Toxic Gases in Commercial Fumigation." *American Journal of Public Health* 25, no. 3 (1935). ajph.aphapublications.org/doi/pdf/10.2105/AJPH.25.3.277.

Davies, John. "Irish Narratives: Liverpool in the 1930s." *The Historic Society of Lancashire & Cheshire* 154 (2005). www.hslc.org.uk/wp-content/uploads/2017/11/154-3-Davies.pdf.

Gregorich, Barbara. "American Eye: Stranded." *North American Review* 283, nos. 3–4 (May–August 1998).

Haines, Gary. "The Fall of Singapore: The Fall of Empire." Royal Air Force Museum (August 13, 2020). rafmuseum.org.uk/blog/fall-of-singapore.

Hétu, Jean. "Pérodeau, Narcisse." *Dictionary of Canadian Biography*. Vol. 16. University of Toronto/Université Laval (2019). biographi.ca/en/bio/perodeau_narcisse_16E.html.

Hickman, Mary J., and Louise Ryan. "The Irish Question: Marginalizations at the Nexus of Sociology of Migration and Ethnic and Racial Studies in Britain." *Ethnic and Racial Studies* 43 (2020).

Johnston, Derek Lukin. "Memories of the Empress." *Vancouver Historical Society Newsletter* 43, no. 9 (June 2004).

Lamb, W. Kaye. "Empress Odyssey: A History of the Canadian Pacific Service to the Orient, 1913–1945." *British Columbia Historical Quarterly* 12, no. 1.

Mauriello, Joseph A. "Japan and the Second World War: The Aftermath of Imperialism" (1999). lehigh.edu/~rfw1/courses/1999/spring/ir163/Papers/pdf/jamm.pdf.

Milling, Marla Hardee. "Cornelia Vanderbilt's Brushes with Death." *Blue Ridge County Magazine* (July 1, 2017).

Packham, Marian A. "Macallum, Archibald Byron." *Dictionary of Canadian Biography*. Vol. 16. University of Toronto/Université Laval, 2003. biographi.ca/en/bio/macallum_archibald_byron_16E.html.

Ryan, Louise. "Aliens, Migrants and Maids: Public Discourses on Irish Immigration to Britain in 1937." *Immigrants & Minorities* 20, no. 3 (November 2001).

Schuthe, George. "Canadian Shipping in the British Columbia Coastal Trade." Master's thesis, University of British Columbia, April 1950. open.library.ubc.ca/media/stream/pdf/831/1.0106747/1.

Smith, Kathryn. "Frank Lloyd Wright and the Imperial Hotel: A Postscript." *The Art Bulletin* 68, no. 2 (June 1985).

Vancouver Daily Province, *Vancouver Sun*, *Victoria Daily Colonist*, *Victoria Daily Times* Archives, 1912–42.

Woon, Yuen-Fong. "Between South China and British Columbia: Life Trajectories of Chinese Women." *BC Studies* 156, no. 7 (Winter/Spring 2007–08).

Personal Journals and Unpublished Memoirs

Drummond, John. Wartime Journal. Courtesy of Bill and Jim Drummond.

Gillison, Kathleen. Diary (1938). Courtesy of Frances Clemmow.

Hosken, Geoffrey. Wartime Diary (1941–42). Courtesy of Brenda Hosken, Nanaimo, BC, and Margot Wallace, North Saanich, BC.

James, Herbert. Journal (1914–15). Courtesy of Nelson Oliver Collection.

Oliver, Walter. Unpublished Memoir. Courtesy of Nelson Oliver, Port Moody, BC.

Robinson, Samuel. Commander Samuel Robinson Diaries. City of Vancouver Archives. AM335-S3, boxes 539-B-05, flds 22-28; 557-A-01, fld 11; 557-C-01, fld. 3.

Russell, William F. William F. Russell Family Archives, Adventures in Asia, 1921 Diary. Courtesy of Sarah Russell Spray.

Tozer, Geoff. Wartime Diary. Hightail Tozer Papers. Courtesy of Allen Tozer.

Yost, Charles W. Unpublished Memoirs (1980). Seeley G. Mudd Manuscript Library, Princeton University, NJ.

IMAGE CREDITS

ACME Newspapers 9/4/41, Newspaper Enterprise Association, Nelson Oliver Collection, *272, 273, 274*

Allen, Fiona/Farrell, Deirdre, *353*

Atkins, Maurice, *251*

Atkinson, Evelyn, *291*

Attwell, Bill, *156*

Australian War Memorial, 016263, *320*

Australian War Memorial, P11611.053.002, *52*

Australian War Memorial, PO1604.001, *321*

Barrett, Cathie, *364*

Butler, Gwen, *264*

Chong, Raymond, *201*

Davis, Kathleen, *247*

Drummond, Bill, *28*

Drummond, Bill/Drummond, Jim, *166*

Drummond, Bill/Drummond, Jim/Nelson Oliver Collection, *114*

Drummond, Bill/Drummond, Jim/Rialto Studio, Manila, Philippines, *152*

Ewart, Jill, *104*

Foster, Peter, *295*

Gibson, Robert, *56*
Gimbel, Richard/Nelson Oliver Collection, *144, 145, 146*
Hasler, Kay, *258, 355*
Higgs, Joan, *106, 362*
Hugh, Shana, *199*
Hutton, Norm, *352*
Hvidsten, J. Peter, *268*
James Crookall fonds AM640-S1: CVA 260-994, *165*
Jennifer Smith genealogical research, Duncan, British Columbia, *338*
Julie Witmer Custom Map Design, *22, 41, 59, 72, 84, 254, 280*
Konzen, Ron, *32*
Le Patourel, Doug, *259*
Linn, Paula, *122, 269*
May, Michael, *270, 271*
McElroy, Doug/Nelson Oliver Collection, *336*
McKinnon, Steve, *305*
Mejan Graphic Design, © Dan Black, 2024, *312*
Mole, Edgar Horace/Gaskell, Denys/Gaskell, Iain, *42, 50*
Nelson Oliver Collection, *4, 73, 74, 88, 125, 147, 173*
Roistacher, Ken, *275*
Rush, Sherri, *178, 179*
Russell Family Photo Archives, *140*
Slade, Patricia, *248*
Stuart Thomson, CVA 99-758, *89*
Stuart Thomson fonds AM1535: CVA 99-2438, *172*
Stuart Thomson fonds AM1535: CVA 99-2639, *157*
Stuart Thomson fonds AM1535: CVA 99-2640, *161*
Thackray, Dave, *117, 118*
Tozer, Allen/High Tail Tozer Papers, *102*
Vancouver Maritime Museum, PA005.009, *66*
Vancouver Maritime Museum, PA005.010, *62*
Vancouver Maritime Museum Collection, Captain Samuel Robinson Records, AM335-S3 CVA, *12*
Wallace, Margo, *109, 284, 363*

Wallace B. Chung and Madeline H. Chung RBSC-ARC-1679-CC-PH-09370-26-004, *159*

Wallace B. Chung and Madeline H. Chung RBSC-ARC-1679-CC-PH-41-1-CC-PH-06054, *83*

Wallace B. Chung and Madeline H. Chung RBSC-ARC-1679-CC-PH-88-1-CC-PH-09386, *17*

Wallace B. Chung and Madeline H. Chung RBSC-ARC-1679-CC-PH-88-1-CC-PH-09390, *16*

Wallace B. Chung and Madeline H. Chung RBSC-ARC-1679-CC-PH-90-1-CC-PH-0943, *19*

Wallace B. Chung and Madeline H. Chung RBSC-ARC-1679-CC-PH-90-1-CC-PH-09453, *15*

Wallace B. Chung and Madeline H. Chung RBSC-ARC-1679-CC-PH-90-1-CC-PH-09497, *24*

Wallace B. Chung and Madeline H. Chung RBSC-ARC-1679-CC-PH-90-1-CC-PH-09530, *154*

Watts, Helen, *304*

Wong, Patricia, *196*

Yost, Felicity, *75*

INDEX

ABOUT THE AUTHOR

Photo by JEMMAN Photography

Dan Black has written and edited hundreds of articles on Canada's military, past and present. He is the former editor of Canada's *Legion Magazine* and is the author or co-author of three previous books, including *Harry Livingstone's Forgotten Men: Canadians and the Chinese Labour Corps in the First World War*, which received excellent reviews in Canada and abroad. He resides near Ottawa.